Oilseed Cake for Nematode Management

Plant-parasitic nematodes (PPNs) devastate many crop plants, causing billions of dollars in agricultural losses each year. Effective management methods to combat PPNs are synthetic nematicides, but most are non-specific and notoriously toxic and threaten the soil ecosystem, groundwater and human health. The plant by-products, such as oilseed cakes, are sources of bioactive compounds with nematicidal potential. Oilseed cakes are an excellent organic fertilizer, and their bioactive compounds are now gaining importance as they are safe for the environment.

This book provides the most comprehensive and up-to-date review of research on the use of oilseed cakes against PPNs. The complete knowledge of better uses of oilseed cakes for nematode management is necessary for developing effective nematode control options to reduce yield loss.

Key features:

- Describes plant by-products such as oilseed cakes and their potential applications
- Explores bioactive compounds from oilseed cakes for agricultural biofertilization and nematicidal activity
- Discusses nematode management in vegetable, fruit and legume crops
- Covers the use of oilseed cakes and management of the associated challenges

This volume is designed and edited to serve as an invaluable resource text for readers associated with plant nematology, plant pathology, plant protection and agricultural science, including researchers, teachers, advanced undergraduates and graduate students and even agricultural extension agents and farmers.

Oilseed Cake for Nematode Management

Edited by
Faheem Ahmad and Rakesh Pandey

CRC Press
Taylor & Francis Group
Boca Raton London New York

CRC Press is an imprint of the
Taylor & Francis Group, an **informa** business

First edition published 2024
by CRC Press
6000 Broken Sound Parkway NW, Suite 300, Boca Raton, FL 33487–2742

and by CRC Press
4 Park Square, Milton Park, Abingdon, Oxon, OX14 4RN

CRC Press is an imprint of Taylor & Francis Group, LLC

ISBN: 9781032333465 (hbk)
ISBN: 9781032333472 (pbk)
ISBN: 9781003319252 (ebk)

DOI: 10.1201/9781003319252

Typeset in Times
by Apex CoVantage, LLC

Contents

Foreword

Plant-parasitic nematodes (PPNs) cause damage to a wide range of economically important crops worldwide. Chemical nematicides have been used successfully to limit the negative impact of PPNs in agriculture, but most nematicides are being restricted because of their non-specific notoriously toxic effects, which pose a threat to the soil ecosystem, groundwater and human health. Thus, there is a need to adopt environmentally friendly strategies to control PPNs that are safe for the soil ecosystem and humans.

Cultural crop protection methods, including applying plant by-products (organic soil amendments), contribute to suppressing soil-borne pathogens. Farmers use organic amendments, including oilseed cake, in management practices to control the abundance of PPNs. Oilseed cake is a plant by-product of the oil industry obtained after oil removal from the seeds. As oilseed cakes are eco-friendly, they are used worldwide due to profound adverse effects on nematode species. This book is directed toward the application of oilseed cake as a management approach to reduce or eliminate the nematode population to a non-injurious level. It provides a resource for studying the impact of various oilseed cakes against nematode pests.

The commitment and the efforts put forth by the authors, A/Prof. Dr Faheem Ahmad, Department of Botany, Aligarh Muslim University, Aligarh, and Prof. Dr Rakesh Pandey, CSIR-Central Institute of Medicinal and Aromatic Plants (CIMAP), Lucknow, in designing this book, are highly appreciated and welcomed. I am sure this book will be a valuable resource for readers.

Prof. Dr. Pravin Chandra Trivedi

Former Vice-Chancellor
Jai Narain Vyas University, Jodhpur (Rajsthan)
DDU Gorakhpur University, Gorakhpur (UP)
Dr. Rml Avadh University, Faizabad (UP)
Maharaja Ganga Sigh University, Bikaner (Rajasthan)
Maharshi Dayanand Saraswati University, Ajmer (Rajasthan)

Preface

This book aims to provide updated and scientifically refined information about using oilseed cakes as plant by-products for nematode management and useful resources for readers to harness. Nematologists have significantly contributed to the development of management strategies to combat plant-parasitic nematodes worldwide using oilseed cakes. Oilseed cakes are sources of bioactive compounds with nematicidal potential. Plant-derived compounds are now gaining importance, as they are safe for the environment. Oilseed cakes containing secondary metabolites are a potential source of biopesticide. Oilseed cakes are also excellent organic fertilizers, and their bioactive compounds protect crops against plant-parasitic nematode infestation and enrich the soil with micro- and macronutrients.

Perhaps the timing of this book publication is appropriate, with the current interest in the potential role of oilseed cakes in pest control to achieve sustainability in agriculture that would be ecologically sound and resource conserving without compromising the environmental cost. The need to focus on the production, chemical compounds and antinematode properties of various oilseed cakes for sustainable agriculture is given in the chapters. Each chapter has attempted to summarize the earlier information and integrate recent findings, including application strategies to control the nematodes in different crops, into this background. This volume comprehensively captures the full breadth of basic and applied information on the use of oilseed cakes for nematode management. Subject experts from different countries contributed authoritative chapters offering a comprehensive account of research findings on nematode management using oilseed cakes in crops. The book begins with Chapter 1, focusing on "Neem Oilseed Cake: A Multipurpose Product for Agricultural Biofertilization and Nematicidal Activity." Subsequent chapters illustrate nematicidal attributes of *Cucusmis*, mustard, cotton, castor, mahua, sesame, flaxseed, sunflower and mahagoni oilseed cakes and their application in nematode management, including associated challenges. The literature on oilseed cakes and their application is a global necessity because of the alarming nematode problem on agri-crops. The focus is also on setting a research agenda that will provide a road map to utilize oilseed cakes to their full potential for managing nematodes. We hope this book's contributory chapters provide new insights, thought-provoking ideas and an exchange of scientific solutions for scientists, graduate students and scientifically trained technicians.

We are indebted to the contributors who made the book possible. We acknowledge the help of research scholar Mr. Zaffar Zahoor for assistance with reference checking. Finally, we acknowledge our publisher, CRC Press, Taylor & Francis Group, LCC, and in particular Renu Upadhyay (commissioning editor) for agreeing to publish the book and Jyotsna Jangra (editorial assistant) for her assistance in this endeavour.

Faheem Ahmad
Rakesh Pandey

Editor Biographies

Faheem Ahmad is a senior assistant professor of plant pathology and nematology at the Department of Botany of Aligarh Muslim University in India. Ahmad studied plant nematology at Aligarh Muslim University, where he obtained his doctoral degree in 2009. He also worked as a postdoctoral research associate at Ehime University, Japan. Before that, he was a postdoctoral fellow at North-West University-Mafikeng Campus (South Africa) and National Sun Yat-Sen University, Kaohsiung (Taiwan). At present, Ahmad is involved with both teaching and research.

He is also an associate editor of the *Journal of Plant Diseases and Protection* (scientific journal of the Deutsche Phytomedizinische Gesellschaft–the German Society of Plant Protection and Plant Health), Springer. Ahmad and his team's research focus is on nematode management and plant-nematode/microbe interaction. Ahmad has published more than 43 peer-reviewed research papers. After joining Aligarh Muslim University as faculty in 2017, he was awarded the University Grants Commission (UGC) Start-up Grant for a two-year research project in 2018.

Rakesh Pandey is professor emeritus AcSIR & CSIR-Emeritus Scientist, Crop Protection and Production Division, CSIR-Central Institute of Medicinal and Aromatic Plants, Lucknow. His research contributions have received wide recognition at both the national and international levels in the field of crop protection. He has authored more than 163 research papers; 35 book chapters; 18 review articles; 4 US-granted patents; and several books, bulletins and popular articles. He has developed eight disease-resistant varieties of medicinal and aromatic plants. He has 29 rhizospheric microbe accession numbers from the National Center for Biotechnology Information (NCBI). He is a fellow of the National Academy of Agricultural Sciences, New Delhi. He is the recipient of several national and international awards, among which are DAAD Fellow (1990–1993) and Senior DAAD Fellow (1999) at the University of Bonn, Germany; visiting scientist (2003–2005), University of Pittsburgh, USA; Golden Peacock Award, 2008; Prof. V.P. Bhide Memorial Award, 2008; Council of Scientific & Industrial Research (CSIR) Technology Award, 1999 and 2015; Outstanding Scientist Award, 2010; Dr. S. L. Mishra Medal, 2011; Young Achiever Award, 2014; Prof. R.K. Srivastava Memorial Oration Award, 2015; Prof. H. S. Srivastava Memorial Award, 2015; Prof. J.F. Dastur Memorial Award, 2017; Prof. H.M. Shah Memorial Award, 2017; Dr. Manmohan Attavar Gold Medal in Floriculture, 2017; Plant Pathology Leadership Award, 2017; Dr. M.R. Siddiqi Memorial Award, 2018; and Prof. Uma Kant Sinha Memorial Award, 2021, among others. He is the president of the Indian Phytopathological Society (2022–23), Indian Agricultural Research Institute (IARI), New Delhi. Dr. Pandey has guided 13 PhD students and several MSc/MTech students. He has been awarded several projects from national funding agencies. Dr. Pandey is an Expert and Task Force member of the Department of Science & Technology (DST), Department of Biotechnology (DBT), Council of Scientific & Industrial Research (CSIR) and Indian Council of Agricultural Research (ICAR) for crop protection–related projects and selections.

Contributors

Amirhossein Abedini
Department of Environmental Health, Food Safety Division, School of Public Health, Tehran University of Medical Sciences, Tehran, Iran

Faheem Ahmad
Department of Botany, Aligarh Muslim University, Aligarh, India

Muneer Ahmad
School of Chemical Engineering and Physical Sciences, Lovely Professional University, Phagwara, India

Dania Ahmed
Department of Botany, Aligarh Muslim University, Aligarh, India

Mohd. Ali
Nematology Lab, Department of Zoology, Guru Nanak Dev University, Amritsar, Punjab, India

Adel Mirza Alizadeh
Department of Food Safety and Hygiene, School of Public Health, Zanjan University of Medical Sciences, Zanjan, Iran

Renu Bhardwaj
Department of Botanical and Environmental Sciences, Guru Nanak Dev University, Amritsar, India

Tamanna Bhardwaj
Department of Botanical and Environmental Sciences, Guru Nanak Dev University, Amritsar, India

B.S. Chandrawat
Department of Nematology, SKN College of Agriculture, Sri Karan Narendra Agriculture University, Jobner, India

Kavita Choudhary
Department of Nematology, SKN College of Agriculture, Sri Karan Narendra Agriculture University, Jobner, India

Kamini Devi
Department of Botanical and Environmental Sciences, Guru Nanak Dev University, Amritsar, India

Raju Dhayal
Department of Nematology, SKN College of Agriculture, Sri Karan Narendra Agriculture University, Jobner, India

Shalini Dhiman
Department of Botanical and Environmental Sciences, Guru Nanak Dev University, Amritsar, India

Saba Fatima
Department of Botany, Aligarh Muslim University, Aligarh, India

Venkadesh G
Division of Nematology, Indian Agricultural Research Institute, New Delhi, India

Rishil Gupta
Department of Botany, Aligarh Muslim University, Aligarh, India

Hemraj Gurjar
Department of Nematology, SKN College of Agriculture, Sri Karan Narendra Agriculture University, Jobner, India

Hedayat Hosseini
Department of Food Sciences & Technology, National Nutrition & Food Technology Research Institute, Faculty of Nutrition Sciences & Food Technology, Shahid Beheshti University of Medical Sciences, Tehran, Iran

Mohd Ikram
Department of Botany, Aligarh Muslim University, Aligarh, India

Dhriti Kapoor
Department of Botany, School of Bioengineering and Biosciences, Lovely Professional University, Phagwara, Punjab, India

Arshad Khan
Department of Botany, Aligarh Muslim University, Aligarh, India

Faryad Khan
Department of Botany, Aligarh Muslim University, Aligarh, India.

Kanika Khanna
Department of Botanical and Environmental Sciences, Guru Nanak Dev University, Amritsar, India

Jaspreet Kour
Department of Botanical and Environmental Sciences, Guru Nanak Dev University, Amritsar, India

Sandeep Kour
Nematology Lab, Department of Zoology, Guru Nanak Dev University, Amritsar, Punjab, India

Deepak Kumar
Nematology Lab, Department of Zoology, Guru Nanak Dev University, Amritsar, Punjab, India

Salah-Eddine Laasli
Nematology Laboratory, Biotechnology Research Unit, National Institute of Agricultural Research (INRA), CRRA-Rabat, Morocco

Rachid Lahlali
Phytopathology Unit, Department of Plant Protection, Ecole Nationale d'Agriculture de Meknes, Meknes, Morocco

Arianna Latini
Italian National Agency for New Technologies, Energy and Sustainable Economic Development. ENEA Casaccia Research Center, Rome, Italy

Ikram Legrifi
Phytopathology Unit, Department of Plant Protection, Ecole Nationale d'Agriculture de Meknes, Meknes, Morocco

Niloufar Mahmoudi
Department of Agrobiotechnology, Peoples' Friendship University of Russia (RUDN University), Moscow, Russia

Susanna Mariani
Italian National Agency for New Technologies, Energy and Sustainable Economic Development. ENEA Casaccia Research Center. Rome, Italy

P. W. Mashela
Department of Plant Production, Soil Science and Agricultural Engineering, Green Biotechnologies Research Centre of Excellence, University of Limpopo, Sovenga, Republic of South Africa

Fouad Mokrini
Nematology Laboratory, Biotechnology Research Unit, National Institute of Agricultural Research (INRA), CRRA-Rabat, Morocco

Hera Nadeem
Department of Botany, Aligarh Muslim University, Aligarh, India

Yousef. Naserzadeh
Department of Agrobiotechnology, Peoples' Friendship University of Russia (RUDN University), Moscow, Russia

Marcello Nicoletti
Department of Environmental Biology, Sapienza University of Rome, Rome, Italy

Puja Ohri
Nematology Lab, Department of Zoology, Guru Nanak Dev University, Amritsar, Punjab, India

Parkirti
Nematology Lab, Department of Zoology, Guru Nanak Dev University, Amritsar, Punjab, India

Parthiban M
Division of Entomology, Sher-e-Kashmir University of Agricultural Sciences and Technology, Kashmir, India

Neha Pathak
Department of Botany, Aligarh Muslim University, Aligarh, India

K. M. Pofu
Department of Plant Production, Soil Science and Agricultural Engineering, Green Biotechnologies Research Centre of Excellence, University of Limpopo, Sovenga, Republic of South Africa

S. Prabhu
Department of Nematology, Tamil Nadu Agricultural University, Coimbatore, India

M. Shanmuga Priya
Agricultural College and Research Institute, Tamil Nadu Agricultural University, Eachangkottai, Thanjavur, India

Ananya Prova
Department of Plant Pathology, EXIM Bank Agricultural University Bangladesh (EBAUB), Chapainawabganj, Bangladesh

Karthik R
Department of Entomology, Assam Agricultural University, Jorhat, India

Sarthak Ranjan
Department of Nematology, SKN College of Agriculture, Sri Karan Narendra Agriculture University, Jobner, India

Parisa Sadighara
Department of Environmental Health, Food Safety Division, School of Public Health, Tehran University of Medical Sciences, Tehran, Iran

Sheila Shahab
Botany Section, Women's College, Aligarh Muslim University, Aligarh, India

Mohammad Shariq
Department of Botany, Aligarh Muslim University, Aligarh, India

Indu Sharma
Department of Botany, Sant Baba Bhag Singh University, Jalandhar, India

Nandni Sharma
Nematology Lab, Department of Zoology, Guru Nanak Dev University, Amritsar, Punjab, India

Neerja Sharma
Department of Botanical and Environmental Sciences, Guru Nanak Dev University, Amritsar, India

Priyanka Sharma
MIT School of Bioengineering Science and Research, MIT-ADT LoniKalbhor Pune, Maharashtra, India

Roohi Sharma
Nematology Lab, Department of Zoology, Guru Nanak Dev University, Amritsar, Punjab, India

R. Sharmila
Department of Nematology, Tamil Nadu Agricultural University, Coimbatore, India

Mansoor Ahmad Siddiqui
Department of Botany, Aligarh Muslim University, Aligarh, India

Amrit Pal Singh
Department of Pharmaceutical Sciences, Guru Nanak Dev University, Amritsar, India

Arun Dev Singh
Department of Botanical and Environmental Sciences, Guru Nanak Dev University, Amritsar, India

Md. Saeed Sultan
Department of Plant Pathology, Bangabandhu Sheikh Mujibur Rahman Agricultural University (BSMRAU), Gazipur, Bangladesh

Raman Tikoria
Nematology Lab, Department of Zoology, Guru Nanak Dev University, Amritsar, Punjab, India

Roberta Tulli
I Consigli dell'Esperto S.r.l. Civitavecchia, Rome, Italy

Tunisha Verma
Department of Botany, School of Bioengineering and Biosciences, Lovely Professional University, Phagwara, Punjab, India

1 Neem Oilseed Cake

A Multipurpose Product for Agricultural Biofertilization and Nematicidal Activity

Arianna Latini, Susanna Mariani, Roberta Tulli and Marcello Nicoletti

CONTENTS

1.1 INTRODUCTION

It is quite difficult to accept for us, but there is more life under the ground than above it. In particular, vegetation strictly depends on what is going on in the adjacent hidden part, wherein many creatures obtain protection, food, collaboration or competition. Therefore, to understand the wellness of a plant, it is necessary to look first down than up. In this way, it has long been considered a need for fertilization, meaning a supply of the elements, like nitrogen, phosphorus and others, whose concentration in the soil is considered insufficient. This concept of fertilization is now revised. The ground must be considered a special habitat, full of interactions among various agents in dynamic equilibria. A biofertilizer should act not only directly on the plant's needs but positively interact in favour of the wellness of the cultivated plant. A biofertilizer should already be part of the organic component, be biodegradable, be sustainable and be able to interact with the living components of the ground as part of the organic chain. In this book chapter, the utilization of neem cake as a biofertilizer is discussed and its use in selected examples presented, with particular focus on its anti-nematode activity.

1.2 NEEM AND NEEM OILSEED CAKE

Azadirachta indica A. Juss (sin. *Melia azadirachta*, family Meliaceae) is an evergreen tree, generally known as neem but also named nimba, margosa, Indian neem or Indian lilac. The last name is

DOI: 10.1201/9781003319252-1

used to distinguish it from the similar species *Melia azedaracht* L., named Melia or Persian lilac. Neem usually grows up to 15–20 m, but under favourable conditions, it can reach a height up to 20–35 m. It is characterized by an irregularly rounded crown, with the leaf canopy making it a useful shade tree. In subtropical regions, it can be widely present, found along roads and avenues in towns and villages. Its cultivation is favourable because it is fast growing. It is mainly present in the Indian subcontinent, like in Tamil Nadu, where it is very commonly used both for shade and as an ornamental plant. However, neem's presence is rapidly increasing worldwide, due to its unique capacity to adapt to hot and dry climates. It is one of the very few trees able to survive in arid zones, and therefore, it is commonly planted and may be encountered in arid and semi-arid areas, where it is one of the very few shade-giving trees able to adapt in drought-prone areas with sub-arid to sub-humid conditions. However, even though it can tolerate high temperatures, it cannot survive low temperatures below 4 °C; thus, its cultivation and diffusion in temperate climates are very difficult. Currently, its presence is expanding rapidly in the world by massive cultivation in sub-tropical regions of America (Caribbean, Cuba, Central and Southern America), Asia (Nepal, Pakistan, Bangladesh, Sri Lanka, Myanmar, Thailand, Malaysia, Indonesia and Iran, China, Turkey, Indonesia) and Africa (Cameroon; Nicoletti 2020). Today, in Western countries, neem is particularly exploited for the insecticidal properties of the seed oil, whereas in Eastern countries, several parts of the tree are useful in many ways, depending on the local tradition, including all over the Indian subcontinent as a fundamental plant of Ayurveda medicine (Kumar and Navaratnam 2013; Del Serrone et al. 2013).

"The marvellous tree, the tree of XXI century, the divine tree, India's tree of life, Nature's drugstore, Panacea for all diseases, a tree for solving global problems": these are only some of the exceptional terms used to describe the importance and the value of the neem tree. For its potential, neem is considered one of the most important plants for humankind's future. In 1989, the World Health Organization/United Nations Environment Programme (WHO/UNEP) considered neem one of the most promising trees of the 21st century for sustainable food production, putting forth evidence of its enormous potential in several fields and applications (Nicoletti 2020).

Cold-pressed neem oil is the most important product, obtained by seed pressure. The seeds may contain up to 50% oil by weight. Cold-pressed neem oil is highly appreciated not only in medicine but also in making shampoos, toothpaste, soaps, cosmetics, mosquito repellent, creams, lotions and pet products, such as pet shampoo and other treatments. Cold-pressed neem oil is also effective as an epidermal treatment against different skin diseases, like eczema, psoriasis and skin allergies.

Insecticidal activity is largely documented and reported in about 100 published papers. Neem oil–derived products showed high efficacy in the control of not only different human pests but also animal pests as the vector of mosquito-borne disease (Nicoletti 2020 and references therein). Many formulations showed evidence of anti-feedancy, fecundity suppression, ovicidal and larvicidal activity (Nicoletti et al. 2012), growth regulation and repellence against almost 600 different species of insects, even when applied at low dosages. On the contrary, useful and beneficial insects are not affected, and this selectivity is an important aspect of neem activity. The wide use of chemical insecticides, although very effective, as in the dichlorodiphenyltrichloroethane (DDT) case, can no longer be considered a practicable solution for two main reasons: (a) we cannot survive if we kill all the insects, and (b) the massive utilization of insecticides gave rise to the resistance phenomenon, meaning that the greater the chemical use, the more ineffective they will be, in addition to environmental damage.

Another important aspect concerns the accumulation in the soil. Neem's active ingredients, like azadirachtins, are degraded by light, with photolysis half-lives of 48 min to 3.98 days in thin films under ultraviolet (UV) light and 2.47 days on leaf surfaces (Johnson et al. 1996). In field trials with olives, azadirachtin residues may have a half-life of 0.8 days (Carboni et al. 2002). These periods of permanence in the soil result in adaption for insecticide activity, but multiple expensive treatments are required. In this context, implementing nanotechnologies could represent a solution to this problem. Several research studies have evidenced the possible use of green nanotechnology to obtain nanoparticles, including neem oil and neem cake, still presenting the biological properties that are able to protect the active constituents from degradation (Murugan et al. 2014; Campos et al. 2016; Anjali et al. 2012; Nazeer et al. 2019).

The main product proceeding from the neem tree is the oil obtained by squeezing the kernels or by extraction by solvents. The name internationally used is seed neem oil, or simply neem oil or margosa oil. Several methods can be utilized to avoid the resulting pungent acrid smell without affecting the main chemical composition.

Insecticide properties of neem oil and related products have been tested and reported by several public institutions, like the U.S. Environmental Protection Agency (U.S. EPA), supported by significant scientific literature. The properties include larvicidal, acaricidal and nematicidal effects, as well as repellent activity. The EPA also verified the environmental safety of neem products and recommended their use in agricultural treatments. The EPA has documented that there are no toxicological or environmental concerns proceeding from the utilization of cold-pressed neem oil in farming or in practical indoor applications.

The insecticidal effects are also joined by relevant antimicrobial activity, giving rise to several products largely used for treating pets and livestock (Foxi and Delrio 2013). In the case of biocidal treatments, neem antimicrobial activities are relevant in consideration of the high possibility of infection and the heavy consequences on animal health (Del Serrone et al. 2013). As a matter of fact, the activity of seeds, leaves and neem extracts has been reported for several pathogens (Badam et al. 1999; Asif 2012; Del Serrone and Nicoletti 2013; Mariani and Nicoletti 2013; Nicoletti et al. 2014).

Neem cake is the by-product of the neem oil cold extraction process, obtained by pressing the neem seeds of *A. indica*, but still contains active constituents and therefore is still present in the neem's biological properties as evidenced by several studies (Nicoletti et al. 2010, 2012; Benelli et al. 2014a, 2014b). Neem cake is a brown-coloured powder, with a bitter taste and a garlic/sulphur smell. Generally, neem cake is applied as organic fertilizer in agriculture and provides major soil benefits. It is de-oiled, representing the residue obtained from pure neem seeds that have been crushed to extract the oil; nonetheless, neem cake can be oily depending on the remaining neem oil after extraction. The cold-pressed extraction method for producing (1) neem oil and (2) de-oiled neem cake as a by-product is generally carried out at the level of small local industrial plants.

In a neem tree, there are about 4,000 clean seeds per kilogram. For the production of neem oil by cold expression, the whole kernel containing the seeds is used. Therefore, the generally used term *seed neem oil* is not correct, considering that after fixed oil production, the expressed kernel becomes the main constituent of neem cake.

The utilization of neem cake is underestimated, but the research has shown that neem cake is important as a biofertilizer for several reasons: (a) the low cost, as a by-product of neem oil; (b) the possibility of great production considering the expanding cultivation of neem; (c) easy and cheap local production, avoiding the need for special industrial facilities; (d) being a powder, it can be easily added to soil without particular protection; and (e) integration into the soil is rapid, and its constituents can act to improve soil habitat. In this regard, the research results evidenced a clear selectivity in favour of beneficial living components.

1.3 THE CHEMICAL COMPOSITION OF NEEM OILSEED CAKE

Phytochemical studies revealed that the chemistry of neem is very complex, and indeed, it has not yet been elucidated in several respects. Hundreds of molecules have been isolated and their structures characterized from different neem tissues/organs, in particular from seeds, which provide the highest potential application and market value. Even though diverse seed compositions have been described, the seeds contain mainly a brownish-yellow oil (approximately 45%) made from numerous fatty acids, that is, oleic (50–60%), palmitic (13–15%), stearic (14–19%), linoleic (8–16%) and arachidic (1–3%). This oil is characterized by an unpleasant strong alliaceous odour and an acrid taste due to the presence of sulphureous constituents. Oilseed extraction can be performed under different thermal and pressure conditions, and the yield is strongly dependent on the final utilized method. In addition, quantitative relevant differences in composition were detected based on the geographic location and the season.

In the seeds, more than 300 phytocompounds have been reported, including several substances so far isolated only from this plant. However, the biological activities are mainly related to specific constituents, generally named azadirachtins, belonging to the nortriterpenoids, accounting for more than one-third of identified constituents. Nortriterpenoids are triterpenoids lacking a methylene group; they are chemotaxomically well located in a few related families of Rosidae Angiosperm Dicotyledons, that is, Rutaceae, Simarubaceae, Cucurbitaceae and Meliaceae. In contrast with the most common steroidal model, the partial loss of the lateral chain in these molecules determines a complicated structural rearrangement, giving rise to different polycyclic molecular skeletons showing oxygenated and partially acylated functional groups (Nicoletti 2020). In neem, two chemical groups of nortriterpenoids have been classified: limonoids (C26) and quassinoids (C20 and C19). While limonoids present only a partial loss of the lateral chain, quassinoids lack it completely. Limonoids detected in neem encompass three main skeleton types: the azadirachtins, the nimbins and a type similar to the azadirachtins containing the dihydrofurane ring. Biological activity and decomposition rate change in accordance with these structural types. Different formulations of neem oil may show different azadirachtin amounts (in the range from 1,000 up to 4,000 mg/Kg) based on the kind of product used.

As evidenced by high-performance thin-layer chromatography (HPTLC) analysis (Toniolo et al. 2013), the composition of neem oil and neem cake may be very diverse, according to the raw material of origin, production and conservation. The HPTLC chromatographic profile reveals the type of raw material used by the fingerprint profile. Indeed, neem oils on the market are very different from each other; therefore, this factor is also crucial for the activity and the properties of use.

In the samples of neem cake so far analysed, the main differences in comparison with the oil are the prevalent presence of nimbin against azadirachtin A and B, and the oil's persistence is very variable (Figure 1.1).

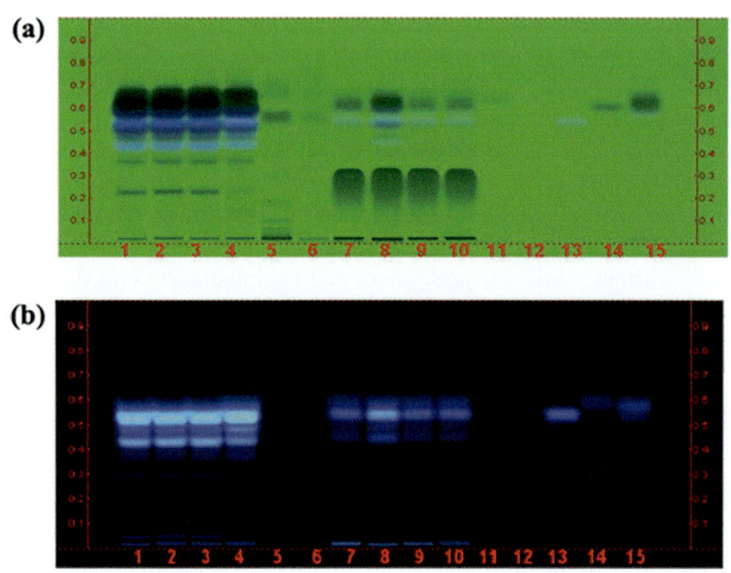

FIGURE 1.1 HPTLC analysis of neem products. (a) Imagine under 254 nm UV; (b) under 366 nm UV.

1.4 NEEM OILSEED CAKE AS BIOFERTILIZER

In the underground habitat, organisms of various types compete for natural resources in different ways depending on the season. Even in the subsoil, the environmental conditions are fundamental, exactly as in the air; for example, the seeds of many plants germinate only at certain temperatures, and cyclic dominances of microorganisms are recorded. Once the process of vegetative development has begun, competition for mineral resources becomes fundamental since these are not sufficient to support the development of all possible plant forms, as the number of seeds is much higher than what the environment can support. Something similar also happens at the top, with the production of buds by multi-annual species. Overproduction of progeny is an often-adopted mechanism of species to overcome possible adverse conditions, even unexpected ones, which can lead to the loss of a large part of the development potential but always taking into account environmental resources. To improve this situation, humans traditionally intervene in two ways. With fertilization, we try to avoid the insufficient availability of certain elements, especially N and P, which the soil is generally lacking. For trees, we intervene with pruning. The treatment of the soil, obviously coupled with the elimination of weeds by means of inorganic fertilization, although it guarantees initial successes, has proved very harmful for cultivation. Basically, the use of this type of fertilizer on a large scale and in massive doses, for optimizing the production per hectare, may cause a severe intoxication in the plants, in terms of matter and energy. In these conditions, in the long term, fertilizer overdoses and their effects in the plants may turn from positive to negative, and the continuous increase of fertilizer dosage to achieve higher and higher production levels has become unsustainable and unnatural.

These are limited and superficial cultivation attitudes, spoiled by the fact that we tend to consider what we see, neglecting the hidden part. A plant is a molecular mechanism of conversion of matter and energy. The matter is located below and energy above. In most cases, it is, therefore, a question of projecting what is found in the soil, transforming it into organic matter capable of trapping light and converting it into binding energy. It is a temporary situation because when the leaves fall or the plant dies, matter and energy are returned to the ground to be recycled.

An alternative treatment to inorganic fertilization of the biofertilization type must mainly consider various factors that intervene in the underground habitat, avoiding limiting oneself to sectoral and partial intervention. Regarding the living part of the soil, (a) the matter contained in the soil is in limited quantities and not always available; (b) consequently, there is strong competition between living organisms, some that live permanently in the subsoil and others destined to develop above; (c) in the multitude of inhabitants of the underground habitat, exactly as it happens on the surface, two strategies prevail, the individual one of struggle and selection to the detriment of other individuals, that is, on the one hand, pathogens, infectious agents and parasites, and, on the other, cooperation, symbiosis and concerted development in superorganism situations.

Regarding the non-living part, (a) supplying inorganic material containing limiting elements is certainly simpler, less expensive and more immediate but goes against the logic of the recycling of organic matter mentioned earlier; (b) the additional input, which we call fertilization, acts not only on the cultivated plant but ends up affecting the entire habitat, altering the normal equilibrium and development of the underground habitat; and (c) these variations, considering the habitat as an interconnected dynamic complex system, generate a sequence of events that affect every creature of the subsoil and are therefore very important for the development of the cultivated plant. Consequently, a biofertilizer must meet the following fundamental requirements, which refer to the natural recycling of energy and matter: (a) be organic in nature while containing limiting elements but in a bioavailable form; (b) be bio-sustainable and compatible with the environmental situation; and (c) play a selective role, ensuring the survival and development of positive, collaborative and/or symbiont organisms and limiting, controlling or eliminating harmful ones. To this, we must add a series of other requirements: (a) low cost, which is obtained by using recycled organic material, exactly as occurs in nature; (b) easy production and use in great quantities; and (c) biodegradability, in order to avoid accumulation in the soil, contrary to what happens with traditional fertilizers.

The withdrawal of polluting agrochemicals from the market makes neem cake exploitation even more interesting, given the large availability of neem cake products on the world market. Neem cake, a by-product obtained from the production of neem oil, is proposed here as a multipurpose product, in accordance with the previous definition, used as a low-cost biofertilizer, insecticide and nematicide in agriculture and zootechnics. Agricultural application of neem cake presents several benefits: It is an excellent biological soil amendment (BSA), a nutrient-rich organic product that can replenish soil organic matter (Latini et al. 2021); and it is natural and consequently biodegradable; it may be used mixed to other organic fertilizers; it inhibits nitrogen nitrification and at the same time increases the availability of nitrogen for plants; it ameliorates the physical soil texture, aeration and soil water-holding capacity, playing a main role in root development.

As neem cakes are sustainable, it is very cost efficient in the long term. During Expo 2015, a world expo hosted by Milan (Italy), focusing on "Feeding the Planet, Energy for Life", the project "Neemagrimed" was awarded the "Best Sustainable Development Practice" (BSDP) in organic agriculture (Latini et al. 2018).

1.5 CASE STUDY: EXAMPLE OF UTILIZATION OF NEEM OILSEED CAKE AS BIOFERTILIZER

In Italy, neem cake is provided in the RADISANA organic fertilizer product (www.iconsiglidellesperto.it/en/products/orchard-and-garden-line/neem/item/94-radisana.html, last access on July 8, 2022). This is formulated by the company I Consigli dell'Esperto S.R.L., operating in the field of fertilizers for the care of all house plants, orchards and gardens, located in Civitavecchia (Rome, Italy). Made entirely from the neem tree, it contains a high percentage of organic nitrogen that improves the soil's physical structure. It has a marked repellent action toward soil insects and favourably influences the absorption of nutrients, particularly nitrogen and iron (Latini et al. 2021). RADISANA gradually releases its organic nitrogen to the plants and has the main property of slowing down the nitrification of nitrogen (Mohanty et al. 2008), depending on the soil, neem cake quality and the modality of application (Marcolini et al. 2016). Thus, the other nitrogen forms present in the soil, such as ammonia and ureic nitrogen, are made available and integrated progressively, sustaining regular plant development and, at the same time, avoiding the accumulation of nitrates in the agricultural final product. It is worthy of note to claim here that in Europe there are maximum permitted limits of nitrates (NO_3 mg/kg) allowed in horticultural products, as established by the European Directive on nitrates (https://ec.europa.eu/environment/water/water-nitrates/index_en.html, last access on July 8, 2022). The gradual nitrogen supply also improves iron adsorption by plants. A multi-residual analysis has been conducted according to specifications for the determination of pesticide residues on foodstuffs EN 15662:2009–QuEChERS-method (Kemmerich et al. 2015), confirming the absolute absence of pesticides. Furthermore, the presence of the main soil-polluting chemical elements has been investigated in different RADISANA samples, resulting in very low concentrations (Table 1.1) and following within limits established by the N° 152 Italian National Legislative Decree of 3 of April 2006 related to soil amendments (https://leap.unep.org/countries/it/national-legislation/legislative-decree-3-april-2006-n-152-environmental-regulations, last access on July 8, 2022).

For all its properties, neem cake is suggested for use in organic farming. Unpublished results of the application of RADISANA in tunnel greenhouse trials, performed in Pontecagnano Faiano (Salerno, Italy) by the I Consigli dell'Esperto farm, have been obtained about arugula for gamma IV fresh-cut salads intended for English organic markets. Arugula (*Eruca sativa* subsp. *sativa* [Miller] Tell., family Grassicaceae), also known as rocket, is an edible annual herbaceous plant, used as a leaf edible vegetable, which is appreciated for not only its fresh, bitter and peppery flavour but also its richness in vitamins A and C and minerals. Present in the wild but also cultivated for the salad market, it is known to exhibit a short biological cycle, with differences in production and quality according to the environmental conditions.

TABLE 1.1
Content of Main Soil-Polluting Chemical Elements Expressed as Means ± Standard Deviations in RADISANA Samples and Respective Maximum Amount Limits Established by Italian Law

Soil-Polluting Chemical Element	Mean ± St. Dev. in RADISANA (in mg/Kg)	Maximum Limit (in mg/Kg) Allowed by Italian Legislative Decree N° 152 for Soil Amendments
Total copper (Cu)	15.1 ± 0.8	230
Total zinc (Zn)	49.3 ± 3.7	500
Total lead (Pb)	0.260 ± 0.012	140
Hexavalent chromium (Cr)	<0.2 ± 0.0	0.5
Total cadmium (Cd)	<1.0 ± 0.0	1.5
Total nichel (Ni)	3.1 ± 0.6	100
Total mercury (Hg)	<0.01 ± 0.0	1.5

Source: An accredited laboratory carried out the analyses; mean values are proceeded by three independent RADISANA® samples.

FIGURE 1.2 (a) Arugula trial under greenhouse in Pontecagnano Faiano (Salerno, Italy); (b) sector planted without soil application of RADISANA; (c) sector planted with application of RADISANA. Pictures were taken on January 2014, 4 weeks after germination.

RADISANA was applied before sowing in a field sector at a concentration of 1 ton/ha, premixed with the soil at depth of 10–15 cm. Frequent irrigation was applied immediately after seeding and during the first week, and then gradually it was reduced, according to common farming procedures under tunnel greenhouse conditions in a Mediterranean climate. Arugula seeds sown in the field sector treated with RADISANA germinated 15 days earlier than seeds sown in a field sector without adding the biofertilizer. Furthermore, the green colour of the leaves that resulted was visibly more intense, as proof that neem cake enhanced iron absorption by plants (Figure 1.2).

Interestingly, RADISANA has also shown beneficial activity against gall-forming nematodes (see Section 1.7).

1.6 THE NEMATICIDAL ACTIVITY OF NEEM OILSEED CAKE

Plants can exudate active molecules endowed with inhibitory effects against several soil pathogens, including nematodes (Desmedt et al. 2020). Neem cake soil amendment is known for ameliorating the damping-off severity of plants and reducing concentrations of plant-parasitic nematodes (PPNs) and soil-borne pathogens of plants (Abbasi et al. 2005). Phytoparasitic nematodes are a very abundant group of plant pathogens, reaching more than 4,100 species, responsible worldwide for an estimated agricultural economic loss of US$125 billion per year (Mesa-Valle et al. 2020). Neem cake has been assessed to be able to provide effective control of PPNs, as reported from different experimentations. Such nematicidal action of neem cake, conferred by the presence of specific bioactive substances, particularly the azadirachtin active compounds, shows high potential, with results that are particularly intriguing. More in depth, neem cake has been reported to significantly reduce PPNs such as *Pratylenchus goodeyi* (Sher and Allen 1953) and *Meloidogyne* spp. in banana plants (Musabyiman and Saxena 1999). Under greenhouse conditions, the application of neem cake in pot soil at 1% (mass/mass) resulted in a reduction up to 90% of the number of lesions due to *Pratylenchus penetrans* and the root-knot *Meloidoygene hapla*. At the same concentration but under field conditions, neem cake limited up to 23% of the lesions in corn roots and up to 70% of the lesions in soil roots (Abbasi et al. 2005). Nazir et al. (2006) showed that application of crude neem cake formulations post-invasion by the root-knot *Meloidogyne javanica* reduced female and egg masses in tomatoes and that neem nematicidal metabolites were absorbed by the roots, having a negative effect on nematode development and fertility. The same authors also attempted to quantify the persistence of various neem formulations in soil for up to 4 months, concluding that repeated neem cake applications are necessary to reduce nematode populations effectively and avoid economic yield-related losses. Neem seed extracts were effective restraints for *Heterodera glycines* Ichinohe, a cyst nematode that damages soybean, considered one of the main soybean pests: A reduction of up to 90% of the number of *H. glycine* females and eggs on soybeans under greenhouse conditions was reported (Silva et al. 2008). The root lesion nematode *Pratylenchus delattrei* can be managed in the tropical flowering plant *Crossandra undulaefolia* L. by using oil cakes, and a synergistic effect was observed coupling neem cake with carbofuran and castor cake (Jothi et al. 2004). Neem cake could be considered a natural pesticide; indeed, it has been shown to be eco-friendly, safe to non-target organisms and an optimal product for sustainable pest management and environmental conservation (Saxena 2015).

1.7 EFFECTS OF NEEM OILSEED CAKE ON GALL-FORMING NEMATODES

PPNs represent a very abundant group of nematodes, showing at the same time a large variety of interactions with their host. Gall-forming PPNs known as "root-knot nematodes" (RKNs) represent a major problem for many crops of primary economic importance (vine, fruit, horticultural, chard, tomato, tobacco, potato, flower growing), especially in sandy soils, typical in coastal areas, such as those of Mediterranean, with hot climates and short winters (Archidona-Yuste et al. 2018). RNKs are obligate endoparasites. For example, two important species attacking vines belong to the *Meloidogyne* genus: *Meloidogyne incognita* and *Meloidogyne javanica*. These nematodes are conserved as active or quiescent larvae or as quiescent eggs for up to 3 years, and in the presence of attackable roots, the eggs hatch and the emerging larvae immediately penetrate the roots (https://fitogest.imagelinenetwork.com/it/malattie-piante/malattie-parassiti/altri-organismi-nocivi/nematodi/nematodi-galligeni/1026, last access on July 8, 2022). Concerning the harmful effects on plants, the larvae penetrate the roots, damaging the tissues; the plant reacts with abnormal growth of the tissues (due to hypertrophy, *i.e.* due to the larger size of the cells) with the formation of root galls. These modifications alter the root functionality, especially in the case of smaller roots, compromising their absorption capacity. Finally, this can cause the death of young plants and a huge decrease in the yield when the infection occurs in mature plants. Among the conventional strategies employed in farming for their management, soil application of chemical nematicides and biofumigation were the most common (Pulavarty et al. 2021). Currently, different agronomic interventions and other control measures can be considered to reduce the significant damage of these nematodes: solarization, summer soil tillage, cultivation and green manure of biocidal plants (Brassicaceae), using panels or pellets of brassica seeds (e.g., *Brassica carinata*), crop rotations with insensitive species (wheat, corn, etc.), using nematode-free material (certified material), using resistant or tolerant rootstocks, avoiding regrowth, avoiding water stagnation and destroying the residues of infected vegetation. As reported in different studies, neem cake has been shown to be highly effective against *Meloidogyne incognita* (Akhtar and Mahmood 1995; Rao et al. 1997) and *Meloidogyne javanica* (Ashraf and Khan 2010; Javed et al. 2008) infestations.

RADISANA has been also tested for its efficacy in contrasting gall-forming nematodes, which are highly present in the field trial location at Maccarese, Fiumicino (Rome, Italy). The soil had a loose sandy texture, with a high sand content (92%), a pH almost equal to 8 and a low total nitrogen content (0.4 g/kg). The neem cake product was applied in a field sector at a concentration of 1.5 tons per hectare, in the vegetable pre-transplanting phase, in the soil at a depth of 10–15 cm (Figure 1.3a-b). Watermelon, specifically Minirossa variety provided by Lamboseeds (www.lamboseeds.com/en/sementi/watermelon/minirossa/, last accessed on July 8, 2022), very adaptive to the local Mediterranean climate and with an average yield of 650 quintals per hectare, was grown in parallel with and without RADISANA treatment, respectively, in two different separated field sectors. Irrigation was supplied frequently and stopped during fruit maturation. Plants were transplanted in 2014 on June 20; a first harvest was performed on August 13 and a second one on September 4. As a result, at harvesting time, watermelon roots grown in soil with neem cake were found to be free from attack from nematodes, as demonstrated by the lack of root galls (knots; Figure 1.3c–d). A similar result was obtained in carrots (data not shown).

FIGURE 1.3 Distribution of neem cake in the field for watermelon cultivation. Neem cake is distributed into the soil at the pre-transplant phase at a depth of 15 cm (a, b). Damaged root with galls caused by root-knot nematodes of a watermelon grown without RADISANA neem cake (c); healthy root of a watermelon grown with RADISANA neem cake (d). Picture kindly provided by Caramadre organic farm (https://venditadiretta0.wixsite.com/biocaramadre, last access on July 8, 2022).

1.8 CASE STUDY: MORINGA BIOFERTILIZATION WITH NEEM OILSEED CAKE, A COUPLED SYSTEM FOR IMPROVING AGRI-FOOD VALUE CHAINS OF BOTH NEEM AND MORINGA

Moringa oleifera Lam, indigenous in Northern India and found worldwide in several tropical and subtropical areas, is considered a "magic tree" with miraculous nutritional potential of its fruit, known as *drumstick* (Tshabalala et al. 2019; Trigo et al. 2021; Patil et al. 2022). In this section, we report a proposal for an interesting study that the authors and other colleagues would like to address in the future in developing countries, with a focus on the employment of neem cake for fertilization and pest defence in moringa, with further optimization of the two agro-industrial processes (Figure 1.4). The rationale of this idea is that, in several undeveloped tropical and subtropical countries, the neem tree is widely spread, while its processing chain is not, and there is a need for agro-industrial plant equipment for neem oil extraction with neem cake recovery. However, in these countries, moringa is a common food in the local tradition and even considered a superfood; its production/processing chain is well established.

Moringa plantation is kept almost organic, but it would need fertilization to boost its productivity; hence, it would benefit from amendment and pest management with neem cake. Thus, in synthesis, this study would foresee (a) the development of a process of a circular economy for neem, with the recovery of neem cake to be used for boosting the moringa production, and (b) the implementation of novel local small businesses exploiting a single plant for the processing of both neem and moringa fruits and seeds, which could be possible thanks to their ripening complementarity (neem oil extraction from seeds is generally performed during the wet periods according to

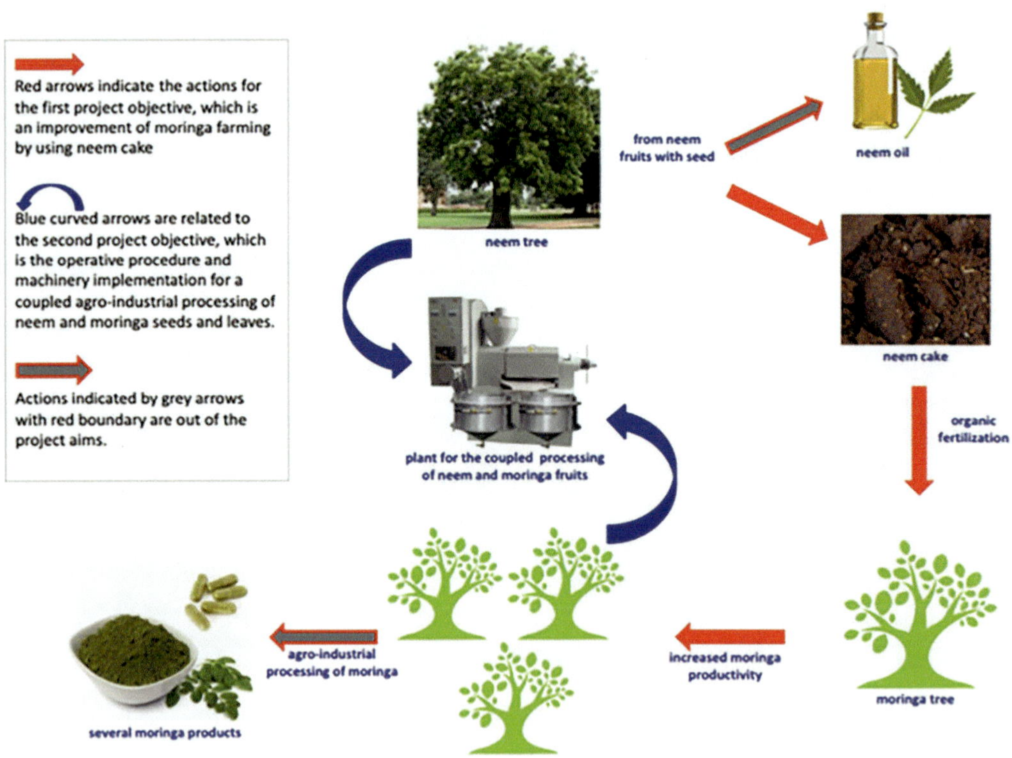

FIGURE 1.4 Graphical abstract of the study proposal on a coupled sustainable system for neem and moringa trees for the enhancement of their industrial processing value chains.

neem harvesting, while the processing of moringa occurs during the drought season). The optimization of energy use and the modulation of the operative processes (particularly the extraction) would allow the obtainment of high-quality products from the two chains of neem and moringa trees (Figure 1.4).

The exploitation of neem and its valuable properties in the food chain for improving human health, increasing the yield of agricultural produce and "feeding" the planet may represent an effective strategy in response to the urgency of realizing novel sustainable agricultural systems, taking into account that the soil application of several polluting chemicals can no longer be allowed. Moreover, the use of neem cake as biofertilizer for the plantation of another crop worthy of note, as moringa is, will boost the production and processing chain of moringa. Being a drought-tolerant crop, for moringa we can foresee a tendency to increase its cultivation sites, thus enhancing yield and the further related agro-industrial transformation. A modern neem industry in developing countries would make available to farmers a new rich soil amendment, easily available on the market given its large diffusion and cheapness, for attaining a sustainable fertilization of soils and crop plantations without improper and excessive use of chemicals. Not only developing countries but also developed ones would gain from exploiting neem organic waste. Indeed, the withdrawal from the market of highly toxic pesticides is also forcing non-organic farmers to modify for agricultural and livestock extremely polluting practices, pointing to the rediscovery of ancient and more sustainable techniques. In both cases, this fertilization would also be functional for pest and nematode control.

Drylands cover a big part of African, Caribbean and Pacific areas, and most soils are characterized by drought and low nutrient content, strongly limiting agricultural production, which is also debilitated by insects and nematode pathogens. The couple neem cake–moringa could positively address these issues, representing an opportunity for reversing agricultural land degradation by the adoption of eco-friendly agricultural practices for sustainable agriculture and providing several benefits for third-world countries. Indeed, the increased use of neem cake is expected to improve soil fertility and replenish soil organic matter and crop yield; as a consequence, the reduction in the use of chemical fertilizers would also result in fewer greenhouse gas emissions.

1.9 THE NEMATICIDAL ACTIVITY OF MORINGA

Moringa oleifera Lam. has potential as an anticancer, antidiabetes and antimicrobial agent (Patil et al. 2022). Furthermore, numerous studies suggest that moringa can effectively control *M. incognita* and *M. javanica* infections (Murslain et al. 2014; El-Ansary and Al-Saman 2018; Ladi et al. 2019). Páez-León et al. (2022) demonstrated that the ethyl acetate extracted from *M. oleifera* Lam. leaves represents a useful solution against agricultural nematodes *Haemonchus contortus* and *Nacobbus aberrans*. Moringa extracts have been shown to be a good and low-cost method for the control of nematodes, being environmentally friendly and safe both to farmers and consumers. This opens the way for researchers to test different combinations of neem and moringa products for obtaining a biofertilizer with nematicidal action. Added value may also be directed against a broader range of nematode pests.

1.10 CONCLUSION

The importance of neem and its derived products will be helpful in the agriculture industry, medicine and the environment, but its impact must be finalized by research (Campos et al. 2016; Uchegbu et al. 2011). Neem oilseed cake is an optical biofertilizer used in organic farming, regenerative agriculture and agroecology. In addition, it could represent a win–win strategy against nematodes, much better than sprinkling chemicals into the soil. In agriculture, it prevents crop damage due to nematodes. Considering that a single product could not be a solution worldwide for all agricultural environments and locations, research on sustainable and environmentally safe management practices and products still needs to be encouraged and performed. Moringa cake represents another product with an extraordinary capacity against nematodes. Here, a possible study aiming at increasing the sustainability of neem and moringa trees and enhancing their circular economies was proposed. Among the novel products that should be given more attention, there are microbial cocktails and fermentation-based bionematicides, commonly referred to as microbial biofertilizers. Besides the nematicidal effect, they exhibit the added value of enhancing plant nutrient uptake by colonizing the rhizosphere, thus facilitating nutrient access to plant root hairs.

REFERENCES

Abbasi, P.A., Riga, E., Conn, K.L., Lazarovits, G. 2005. Effect of neem cake soil amendment on reduction of damping-off severity and population densities of plant-parasitic nematodes and soilborne plant pathogens. *Canadian Journal of Plant Pathology* 27:38–45. https://doi.org/10.1080/07060660509507191

Akhtar, M., Mahmood, I. 1995. Evaluation of a neem-based product against root-knot nematode *Meloidogyne incognita*. *Tests of Agrochemicals and Cultivars* 126:6–7.

Anjali, C.H., Sharma, Y., Mukherjee, A., Chandrasekaran, N. 2012. Neem oil (*Azadirachta indica*) nanoemulsion-a potent larvicidal agent against *Culex quinquefasciatus*. *Pest Management Science* 68:158–163. https://doi.org/10.1002/ps.2233

Archidona-Yuste, A., Cantalapiedra-Navarrete, C., Liébanas, G., Rapoport, H.F., Castillo, P., Palomares-Rius, J.E. 2018. Diversity of root-knot nematodes of the genus *Meloidogyne* Göeldi, 1892 (Nematoda: Meloidogynidae) associated with olive plants and environmental cues regarding their distribution in southern Spain. *PLoS One* 13:e0198236. https://doi.org/10.1371/journal.pone.0198236

Ashraf, M.S., Khan, T.A. 2010. Integrated approach for the management of *Meloidogyne javanica* on eggplant using oil cakes and biocontrol agents. *Archives of Phytopathology and Plant Protection* 43:609–614. https://doi.org/10.1080/03235400801972434

Asif, M. 2012. Antimicrobial potential of *Azadirachta indica* against pathogenic bacteria and fungi. *Journal of Pharmacognosy and Phytochemistry* 1:78–83.

Badam, L., Joshi, S.P., Bedekar, S.S. 1999. In vitro antiviral activity of neem (*Azadirachta indica*, A. Juss) leaf extract against group B Coxsackie viruses. *The Journal of Communicable Diseases* 31:79–90.

Benelli, G., Bedini, S., Cosci, F., Toniolo, C., Conti, B., Nicoletti, M. 2014a. Larvicidal and ovideterrent properties of neem oil and fractions against the filariasis vector *Aedes albopictus* (Diptera: Culicidae): A bioactivity survey across production sites. *Parasitology Research* 114:227–236. https://doi.org/10.1007/s00436-014-4183-3

Benelli, G., Conti, B., Garreffa, R., Nicoletti, M. 2014b. Shedding light on bioactivity of botanical by-products: Neem cake compounds deter oviposition of the arbovirus vector *Aedes albopictus* (Diptera: Culicidae) in the field. *Parasitology Research* 113:933–940. https://doi.org/10.1007/s00436-013-3725-4

Campos, E.V.R., de Oliveira, J.L., Pascoli, M., de Lima, R., Fraceto, L.F. 2016. Neem oil and crop protection: From now to the future. *Frontiers in Plant Science* 7:1494. https://doi.org/10.3389/fpls.2016.01494

Carboni, P., Cabras, M., Angioni, A., Russo, M., Cabras, P. 2002. Persistence of azadirachtin residues on olives after field treatment. *Journal of Agricultural and Food Chemistry* 50:3491–3494. https://doi.org/10.1021/jf020076+

Del Serrone, P., Guendalini, E., Mariani, S., Toniolo, C., Nicoletti, M. 2013. Neem, una promessa del XXI secolo. *Phytojournal* 9:9–12.

Del Serrone, P., Nicoletti, M. 2013. Antimicrobial activity of a neem cake extract in a broth model meat system. *International Journal of Environmental Research and Public Health* 10:3282–3295. https://doi.org/10.3390/ijerph10083282

Desmedt, W., Mangelinckx, S., Tina, K., Vanholme, B. 2020. A phytochemical perspective on plant defense against nematodes. *Frontiers in Plant Science* 11:602079. https://doi.org/10.3389/fpls.2020.602079

El-Ansary, M.S.M., Al-Saman, M.A. 2018. Appraisal of *Moringa oleifera* crude proteins for the control of root-knot nematode, *Meloidogyne incognita* in banana. *Rendiconti Lincei. Scienze Fisiche e Naturali* 29:631–637. https://doi.org/10.1007/s12210-018-0692-9

Foxi, C., Delrio, G. 2013. Efficacy of a neem cake for the control of Culicoides biting midges larvae. *Pharmacology OnLine* 9:21–24.

Javed, N., Gowen, S.R., El-Hassan, S.A., Inam-ul-Haq, M., Shahina, F., Pembroke, B. 2008. Efficacy of neem (*Azadirachta indica*) formulations on biology of root-knot nematodes (*Meloidogyne javanica*) on tomato. *Crop Protection* 27:36–43. https://doi.org/10.1016/j.cropro.2007.04.006

Johnson, S., Morgan, E.D., Peiris, C.N. 1996. Development of the major triterpenoids and oil in the fruit and seeds of neem (*Azadirachta indica*). *Annals of Botany* 78:383–388. https://doi.org/10.1006/anbo.1996.0133

Jothi, G., Sundara Babu, R., Ramakrishnan, S., Rajendran, G. 2004. Management of root lesion nematode, *Pratylenchus delattrei* in crossandra using oil cakes. *Bioresource Technology* 93:257–259. https://doi.org/10.1016/j.biortech.2003.11.002

Kemmerich, M., Rizzetti, T.M., Martins, M.L., Prestes, O.D., Adaime, M.B., Zanella, R. 2015. Optimization by central composite design of a modified QuEChERS method for extraction of pesticide multiresidue in sweet pepper and analysis by ultra-high-performance liquid chromatography–tandem mass spectrometry. *Food Analytical Methods* 8:728–739. https://doi.org/10.1007/s12161-014-9951-2

Kumar, V.S., Navaratnam, V. 2013. Neem (*Azadirachta indica*): prehistory to contemporary medicinal uses to humankind. *Asian Pacific Journal of Tropical Biomedicine* 3:505–514. https://doi.org/10.1016/S2221-1691(13)60105-7

Ladi, B.Y., Muhammad, A.K., Joy, B., Alake, N.M., Sarah, O.J. 2019. Inhibitory effect of neem (*Azadirachta indica*) and moringa (*Moringa oleifera*) leaf extracts on egg hatch of root knot nematode *Meloidogyne incognita*. *World Journal of Advanced Research and Reviews* 1:028–033. https://doi.org/10.30574/wjarr.2019.1.2.0001

Latini, A., Foxi, C., Nicoletti, M., Serafini, M., Toselli, M. 2018. Neemagrimed: Una “Best Practice” in agricoltura biologica. La Rivista di Scienza dell’Alimentazione. FOSAN Fondazione per lo Studio degli Alimenti e della Nutrizione. N° 1, January-April. ISSN: 1128–7969.

Latini, A., Giagnacovo, G., Campiotti, C.C., Bibbiani, C., Mariani, S. 2021. A narrative review of the facts and perspectives on agricultural fertilization in Europe, with a focus on Italy. *Horticulturae* 7:158. https://doi.org/10.3390/horticulturae7060158

Marcolini, G., Toselli, M., Quartieri, M., Gioacchini, P., Baldi, E., Sorrenti, G., Mariani, S. 2016. Nitrogen and carbon mineralisation of different Meliaceae derivatives. *Plant Soil Environment* 62:121–127. https://doi.org/10.17221/637/2015-PSE

Mariani, S., Nicoletti, M. 2013. Antilarval activity of neem cake extract against *Aedes aldopictus*. *Pharmacology OnLine* 111:137–140.

Mesa-Valle, C.M., Garrido-Cardenas, J.A., Cebrian-Carmona, J., Talavera, M., Manzano-Agugliaro, F. 2020. Global research on plant nematodes. *Agronomy* 10:148. https://doi.org/10.3390/agronomy10081148

Mohanty, S., Patra, A.K., Chhonkar, P.K. 2008. Neem (*Azadirachta indica*) seed kernel powder retards urease and nitrification activities in different soils at contrasting moisture and temperature regimes. *Bioresource Technology* 99:894–899. https://doi.org/10.1016/j.biortech.2007.01.006

Murslain, M., Javed, N., Khan, S.A., Khan, H.A., Huma Abbas, H., Kamran, M. 2014. Combined efficacy of *Moringa oleifera* leaves and a fungus, *Trichoderma harzianum* against *Meloidogyne javanica* on eggplant. *Pakistan Journal of Zoology* 46:827–832.

Murugan, K., Pitchai, G.J., Madhiyazhagan, P., Nataraj, T., Nareshkumar, A., Hwang, J.S., Chandrasekar, R., Nicoletti, M., Amsath, A., Bhagooli, R. 2014. Larvicidal, repellent and smoke toxicity effect of neem products against malarial vector, *Anopheles stephensi*. *International Journal of Pure and Applied Zoology* 2:71–83.

Musabyiman, T., Saxena, R. 1999. Efficacy of neem seed derivatives against nematodes affecting banana. *Phytoparasitica* 27:43–49. https://doi.org/10.1007/BF02980726

Nazeer, A.A., Rajan, H.V., Vijaykumar, S.D., Saravanan, M. 2019. Evaluation of larvicidal and repellent activity of nanocrystal emulsion synthesized from *F. glomerata* and neem oil against mosquitoes. *Journal of Cluster Science* 30:1649–1661. https://doi.org/10.1007/s10876-019-01611-x

Nazir, J., Dowen, S.R., Inam-ul-Haq, M., Abdullah, K., Shahina, F. 2006. Systemic and persistent effect of neem (*Azadirachta indica*) formulations against root-knot nematodes, *Meloidogyne javanica* and their storage life. *Crop Protection* 26:911–916. https://doi.org/10.1016/j.cropro.2006.08.011

Nicoletti, M. 2020. *Insect-borne diseases in the 21st century*. New York: Elsevier, Academic Press. https://doi.org/10.1016/C2018-0-04116-X

Nicoletti, M., Murugan, K., Benelli, G. 2013. Emerging Insect-Borne Diseases of Agricultural, Medical and Veterinary Importance. Insecticides Resistance. InTech. http://dx.doi.org/10.5772/61467

Nicoletti, M., Mariani, S., Maccioni, O., Coccioletti, T., Murugan, K. 2012. Neem cake: Chemical composition and larvicidal activity on Asian tiger mosquito. *Parasitology Research* 111:205–213. https://doi.org/10.1007/s00436-012-2819-8

Nicoletti, M., Murugan, K., Del Serrone, P. 2014. Current mosquito-borne disease emergencies in Italy and climate changes. The neem opportunity. *Trends in Vector Research and Parasitology* 1:2. http://dx.doi.org/10.7243/2054-9881-1-2

Nicoletti, M., Serafini, M., Aliboni, A., D'Andrea, A., Mariani, S. 2010. Toxic effects of neem cake extracts on *Aedes albopictus* larvae. *Parasitology Research* 107:89–84. https://doi.org/10.1007/s00436-010-1838-6

Páez-León, S.Y., Carrillo-Morales, M., Gómez-Rodríguez, O., López-Guillén, G., Castañeda-Ramírez, G.S., Hernández-Núñez, E., Wong-Villarreal, A., Aguilar-Marcelino, L. 2022. Nematicidal activity of leaf extract of *Moringa oleifera* Lam. against *Haemonchus contortus* and *Nacobbus aberrans*. *Journal of Helminthology* 23:e13. https://doi.org/10.1017/S0022149X22000025

Patil, S., Mohite, B.V., Marathe, K.R., Salunkhe, N.S., Marathe, V., Patil, V.S. 2022. Moringa tree, gift of nature: A review on nutritional and industrial potential. *Current Pharmacology Reports* 8:262–280. https://doi.org/10.1007/s40495-022-00288-7

Pulavarty, A., Egan, A., Karpinska, A., Horgan, K., Kakouli-Duarte, T. 2021. Plant parasitic nematodes: A review on their behaviour, host interaction, management approaches and their occurrence in two sites in the republic of Ireland. *Plants* 10:2352. https://doi.org/10.3390/plants10112352

Rao, M.S., Reddy, P.P., Nagesh, M. 1997. Management of root-knot nematode, *Meloidogyne incognita* on tomato by integration of *Trichoderma harzianum* with neem cake. *Journal of Plant Disease and Protection* 104:423–425.

Saxena, R.C. 2015. Neem for sustainable pest management and environmental conservation. *ECHO Asia Notes*. Issue 24. Available at www.echocommunity.org/resources/d7cd259f-eb9e-4a04-bc17-b26bdaf7e42d (last access on July 8, 2020).

Silva, J.C.T., Oliveira, R.D.L., Jham, G.N., Aguiar, N.D.C. 2008. Effect of neem seed extracts on the development of the *Soybean Cysts Nematode*. *Tropical Plant Pathology* 33:171–179. https://doi.org/10.1590/S1982-56762008000300001

Toniolo, C., Nicoletti, M., Murugan, K. 2013. The HPTLC approach to metabolomic determination of neem products composition. *Pharmacology OnLine* 3:123–127.

Trigo, C., Castelló, M.L., Ortolá, M.D., García-Mares, F.J. 2021. *Moringa oleifera*: An unknown crop in developed countries with great potential for industry and adapted to climate change. *Foods* 10:31. https://doi.org/10.3390/foods10010031

Tshabalala, T., Ncube, B., Madala, N.E., Nyakudya, T.T., Moyo, H.P., Sibanda, M., Ndhlala, A.R. 2019. Scribbling the cat: A case of the "miracle" plant, *moringa oleifera*. *Plants* 15:510. https://doi.org/10.3390/plants8110510

Uchegbu, M., Okoli, I., Esonu, B., Iloeje, M. 2011. The growing importance of neem (*Azadirachta indica* A. Juss) in agriculture, industry, medicine and environment: a review. *Research Journal of Medicinal Plant* 5:230–245.

2 *Cucumis* Oilseed Cake
Nematicidal Attributes and Management of Associated Challenges

P. W. Mashela and K. M. Pofu

CONTENTS

DOI: 10.1201/9781003319252-2

2.1 INTRODUCTION: BACKGROUND AND DRIVING FORCES

Direct association of fumigant synthetic chemical nematicides to climate change through ozone breakdown and various other biocidal aspects and persistence meant the products were proved to be environmentally unfriendly, resulting in the ultimate withdrawal of the products from agrochemical markets (Mashela et al. 2017a). The substitutes, non-fumigant synthetic chemical nematicides, had their own share of associated challenges, including but not excluding being nematostatic, due to their basipetal and acropetal movements that were accompanied by high chemical residues in produce, with occasional incidents of unprecedented consumer fatalities (Mendes et al. 2005). The withdrawal of synthetic chemical nematicides from agrochemical markets resulted in unprecedented research and development of various alternatives in attempts to manage nematode population densities, particularly the root-knot (*Meloidogyne* species) nematodes, which were previously not viewed as pests of economic importance in most cropping systems (Mashela et al. 2017a). Among the alternatives was the use of nematode resistance due to its simplicity (Pofu 2012), whereas knowledge about the host status for most commercial crops was limited. Subsequent studies demonstrated that plant species such as potato (*Solanum tuberosum* L.) did not have any genotypes with resistance to *Meloidogyne* species, with yield reduction due to nematodes being as high as 50% to complete crop failure (Pofu et al. 2012). Additionally, crops with resistant genotypes also suffered losses due to the existence of biological races in *Meloidogyne* species, which are morphologically similar but could be separated by lengthy differential host plant tests or costly molecular techniques (Pofu 2012).

In addition to nematode resistance, the conventional use of organic amendments in managing nematode population densities was resorted to (Mashela 2002). However, the advantages of this strategy were outweighed by the disadvantages, which, to list a few, included huge quantities, availability challenges, high transport costs, long waiting periods to avoid a negative period and inconsistent results in nematode suppression (Stirling 2014). In various attempts to ameliorate the disadvantages associated with the listed challenges, *Cucumis* oilseed cakes were among the botanical products that were instituted as an alternative strategy to conventional organic amendments for managing nematode population densities (Mashela 2002). The objective of this study was therefore to review the nematicidal attributes of *Cucumis* oilseed cakes and how their associated challenges were empirically resolved.

2.2 BIOSYNTHESIS OF CUCURBITACINS IN PLANTS

Cucurbitacins are secondary metabolites that are biosynthesized within the Cucurbitaceae family and at least 11 other plant families (Kaushik et al. 2015). These unique secondary metabolites are biosynthesized as highly oxygenated triterpenoids built on a tetracyclic cucurbitane nucleus skeleton, chemically the 19-(10→*β*-abeo-10α-lanost-5-ene base structure (Chen et al. 2005, 2014). In all cucurbitacin-producing families, the primary cucurbitacins are biosynthesized exclusively as cucurbitacin B and E, each with an acetyl function at C-25 (Gry et al. 2006). At least 20 other cucurbitacins are produced from cucurbitacin B and E through chemical reactions that include hydrogenation by cucurbitacin Δ^{23-} reductase, deacetylation by cucurbitacin acetylesterases, hydroxylation, dehydrogenation and isomerization (Schabort and Teijema 1968; Gry et al. 2006; Zhou et al. 2016). Almost always, cucurbitacins differ from one another primarily due to four different chemical reactions, namely (a) hydroxylation at C-2, C-3, C-19 or C-24, (b) double chemical bonds between C-1 and C-2 or between C-23 and C-24, (c) acetylation of C-25 hydroxyl groups and (d) occurrence of ketone function at C-3 (Chen et al. 2005, 2014). These chemical reactions confer distinct pharmacological attributes to cucurbitacins in cucurbitacin–animal cell, cucurbitacin–plant cell or cucurbitacin–nematode cell relations (Chen et al. 2005, 2014; Mphahlele et al. 2012; Abbas et al. 2013; Mashela et al. 2017a; Mirr et al. 2019). Although limited work had been done on cucurbitacin–insect (Zhao et al. 2021) and cucurbitacin–fungus relations, most groundbreaking research had been conducted on cucurbitacin–human (Kaushik et al. 2015; Lee et al. 2010; Mirr et al. 2019; Mphahlele et al. 2012; Van Wyk and Wink 2014) and cucurbitacin–nematode relations (Dube 2016; Mashela et al. 2017a), with the latter being the focus of the current review.

2.3 PREPARATION OF *CUCUMIS* OILSEED CAKE

In the cucurbitacin–nematode relation, dried fruits of two wild *Cucumis* species, wild cucumber (*Cucumis myriorcapus* Naude.) and wild watermelon (*Cucumis africanus* L.) are used. Both plant species are indigenous to tropical South Africa (Kristkova et al. 2003). In *C. myriocarpus*, the active ingredient cucurbitacin A ($C_{32}H_{46}O_9$) accumulates in fruit and roots, whereas in *C. africanus*, cucurbitacin B ($C_{32}H_{48}O8$) accumulates in all organs of the plant (Shadung 2016). In each plant species, fruits contain numerous seeds with limited fruit carpus and therefore are ideal for making *Cucumis* oilseed cake. The fruits of the two *Cucumis* species are illustrated in Figure 2.1.

FIGURE 2.1 *Cucumis africanus* (A) and *Cucumis myriocarpus* (B) fruits.

TABLE 2.1
Raw Materials for Preparation of *Cucumis* Oilseed Cakes per 20-L Plastic Container

Material	Nemarioc-AL	Nemafric-BL
Ground *Cucumis* fruits	80 g	40 g
Chlorine-free water	16 L	16 L
Sugar	100 g	100 g
Molasses	100 mL	100 mL
Effective microorganisms	100 mL	100 mL

In both *Cucumis* species, harvested mature fruits are cut into pieces and preserved through oven-drying. Generally, medicinal plant produce is dried at 40°C for 72 h in order to preserve the secondary metabolite active ingredients (Müller and Heindl 2006). However, when fresh fruit pieces from test *Cucumis* species were dried at 40°C, hyphae of *Penicilium simplicissimum* proliferated, resulting in the rotting of the materials (Mphahlele et al. 2012). The identified fungus is resident to fruits in the Cucurbitaceae family, where fruits are naturally in direct contact with soil during fruit development. The optimum drying temperature for cut fruit pieces of *Cucumis* species was, as illustrated later, a compromise at 52°C for 72 h (Shadung et al. 2016). After drying fruit pieces, materials (seeds + carpus) are ground in a Wiley mill (Thomas Scientific, Swedesboro, NJ, USA) to pass through a 1-mm sieve and then stored in either closed or open containers (Shadung and Mashela 2016). Effective microorganisms (EMs) used comprise unidentified yeast bacteria, lactic acid bacteria, photosynthetic bacteria and actinomycete bacteria, with each playing a distinct role in ensuring that the final product was of high quality, without the EMs having any degradable effect on cucurbitacin A and B (Mashela et al. 2020). Raw materials used in the fermentation process of the two products are summarized in Table 2.1 in order to enhance clarity on the ingredients of *Cucumis* oilseed cake in liquid formulation.

A mini-fermentation plant, as illustrated in Figure 2.2, comprising air-sealed 400-L container systems, each with an outlet dangled in a 2-L bottle half-filled with water to allow for the escape

FIGURE 2.2 Mini-fermentation plant for *Cucumis* oilseed cake.

of toxic gaseous end-products that occur during the fermentation process, was established for research purposes. In liquid formulation (L), *Cucumis* oilseed cake is available in 5- and 20-L plastic containers as Nemarioc-AL and Nemafric-BL, with A and B being the active ingredients cucurbitacin A and B, respectively, and with L representing the liquid formulation. Similarly, in granular formulation (G), the products are available as Nemarioc-AG and Nemafric-BG (Mashela et al. 2017a).

2.4 APPLICATION METHODS OF *CUCUMIS* OILSEED CAKE

Three application methods, empirically determined, are available for the application of *Cucumis* oilseed cake: ground leaching technology (GLT), botinemagation technology (BOT) and priming-and-drying (PAD) technology, each with advantages and disadvantages.

2.4.1 GLT

The maiden application method in *Cucumis* oilseed cake was the GLT, where 2 g of material were spread in a circular furrow around the stem seedlings at transplanting and covered with soil (Mashela 2002). This is user-friendly and can be used in smallholder farming systems, with the major drawback being its labor cost. Active ingredients in GLT are released through irrigation water, without evidence of microbial degradation, at least when *Cucumis* oilseed cake was used (Mashela et al. 2017a). Generally, the application interval for managing population densities of r-strategist nematodes, such as *Meloidogyne* species and the citrus nematode (*Tylenchulus semipenetrans* Cobb), using the test products is 56 days. However, after 56 days, at 120 days after applying *Cucumis* oilseed cake, the material increased population densities of *T. semipenetrans* on citrus seedlings under greenhouse conditions (Maile et al. 2013).

2.4.2 BOT

BOT involves the application of fermented *Cucumis* oilseed cake through irrigation water (Mashela et al. 2017a). Although the cost related to the irrigation and fermentation infrastructure is often cited as a drawback, BOT also has much shorter application intervals, with at most about half of the life cycle of the nematode being managed. Generally, r-strategist nematodes are small in size, have high reproductive rates and are quite sensitive to environmental disturbances (Mulder et al. 2005).

2.4.3 PAD Technology

PAD technology involves subjecting seeds with hypogenous germination to solutions produced through the fermentation of *Cucumis* oilseed cake (Mashela et al. 2022). Seeds with hypogenous germination leave the seed residue as a carrier of cucurbitacins belowground, where the radicle is positioned just below the seed remnant. The seed residue gradually releases cucurbitacins in the root zone, which would affect second-stage juveniles (J2) that search for penetration sites. The drawback of this technology is that seeds with hypogenous germination are limited in scope, with those for major agronomic crops, namely, maize (*Zea mays* L.), pea (*Pisum sativum* L.) and certain beans being under this category.

2.5 EFFICACY OF *CUCUMIS* OILSEED CAKE ON NEMATODE SUPPRESSION

The product from fresh fruits of *C. myriocarpus* reduced population densities of *M. incognita* in roots by 46–99% and in soil by 53–96% (Pelinganga and Mashela 2012; Pelinganga et al. 2011). Nemarioc-AL reduced nematode numbers in roots by 78–99% and in soil by 70–90% (Pelinganga et al. 2013). Products from dried fruits have had efficacy rates from as high as 90–100% suppression of nematode population densities (Mashela et al. 2017a), especially on r-strategist nematodes such as *Meloidogyne* species (Mulder et al. 2005).

2.5.1 Suppression of Root Galls

Products from *Cucumis* oilseed cake significantly reduce galls in plant roots infected with *Meloidogyne* species (Mashela et al. 2017a), suggesting either the reduction of penetration of roots by J2 or failure of J2 to reach the feeding site and successfully establishing the feeding site. In *Meloidogyne* species, it had been established that the absence of galls does not imply that J2 did not establish the feeding site, develop and reproduce. In certain commercial *Cucumis* species (Fassuliotis 1970) and maize (Windham and Williams 1988), the successful infection by *Meloidogyne* species did not result in the formation of root galls.

2.5.2 Suppression of Nematode Juveniles in Soil

The eggs of *Meloidogyne* species are deposited into a protective structure, the egg mass, which can be seen on roots. Due to the nature of the life stages of nematodes, the first-stage juveniles (J1s) molt in the egg to the J2, resulting in a J2 hatch instead of an egg hatch. After hatching, the J2 search for penetration sites on roots, guided by various abiotic and biotic factors (Hida et al. 2015; Nyasha 2021; Rasmann et al. 2012; Reynolds et al. 2010; Wuyts et al. 2006). In most instances, the penetration location is on the elongation zone of roots. In the event J2 cannot locate the penetration site as soon as possible, juveniles are able to survive in a soil solution using their body lipid resources for at least 12 weeks (Mashela et al. 2017b). The strategy of exiting the egg masses into the soil solution to search for penetration sites and the ability to spend extended periods in soil solution enhance the possibility of J2 exposure to the test products. Generally, when suitable factors are not detected while J2 are still in the egg, J2 hatching might be delayed, which could also arrest developmental activities in eggs and J1s, thus serving as a survival strategy (Doncaster and Shepherd 1967; McSorley 2003; Zheng et al. 1991). *Cucumis* oilseed cake products were shown on various crops to reduce J2 in soil from as high as 70–90% in Nemarioc-AL and from 65–100% Nemafric-BL (Mashela et al. 2017a).

2.5.3 Suppression of Nematode Juveniles in Root

After penetration at the elongation zone of rootlets, J2 of *Meloidogyne* species move through the symplast and apoplastic spaces of root cells to the tip of the rootlet, then enter the vascular bundle and move upward to the opposite of the elongation zone, where the feeding sites are established (Mashela et al. 2017b). Since cucurbitacins are able to move into apoplastic spaces of roots, this, along with J2 movement in such spaces, enhances J2 exposure to active ingredients of the products. *Cucumis* oilseed cake products were, on various crops, able to reduce J2 in roots from 80% to 90% in Nemarioc-AL and from 87% to 100% in Nemafric-BL (Mashela et al. 2017a).

2.6 COMPARING EFFICACY OF PURE CUCURBITACINS TO *CUCUMIS* OILSEED CAKE

Using J2 hatch, J2 immobility and J2 mortality as indicators, products of *Cucumis* oilseed had significantly higher efficacies than those of purified cucurbitacins (Dube 2016). The latter agreed with observations in products of other oilseed cakes in comparison with their purified active ingredients (Dube 2016). Generally, in cucurbitacin-containing oilseed cakes, the existence of different chemical reactions during the biosynthesis pathway demonstrates the potential existence of various precursors, which, over time, continue with the biosynthesis of the test active ingredients, thus resulting in significant increases in the concentration of cucurbitacins (Shadung 2016).

2.7 COMPARING EFFICACY OF *CUCUMIS* OILSEED CAKE TO SYNTHETIC CHEMICAL NEMATICIDES

Generally, both watermelon (*Citrullus lanatus* (Thunb Matsum & Nakai var. lanatus)) and potato (*Solanum tuberosum* L.) rank high in commercial crops that do not have genotypes with resistance to *Meloidogyne* species (Seshweni 2016; Thies and Levi 2007). The efficacy of *Cucumis* oilseed cake products and velum on soil and root stages of guava root-knot (*Meloidogyne enterolobii*) on watermelons cv. 'Congo' (Pofu and Mashela 2021) and potatoes (Seshweni 2016), along with that of the test products with fenamiphos and aldicarb on tomatoes against *M. incognita* (Mashela et al. 2008) were significantly similar. Previously, saponins ($C_{58}H_{94}O_{27}$), a triterpenoid glycoside from alfalfa (*Medico sativa L.*), significantly reduced *M. incognita* population densities more than did aldicarb (D'Addabbo et al. 2010).

2.8 BIOACTIVITIES OF *CUCUMIS* OILSEED CAKE ON NEMATODES

Bioactivities of *Cucumis* oilseed cake on nematodes were extensively investigated using both extracts and their related pure active ingredients with the focus being on J2 hatch, J2 mobility, J2 mortality and J2 infectivity.

2.8.1 Juvenile Hatch

Juvenile hatch in nematodes involves increased J2 movements and continuous pressing of stylets against the egg shells, resulting in breaking up the entities in the process (Bohlmann 2015; Curtis 2008; Perry and Moens 2011). Even though J2 hatch is a physical process, in most plant nematodes, it is stimulated by environment-friendly external cues (Mashela et al. 2015). The stimulation is made possible by the existence of a number of chemoreceptors, which mostly cover the frontal and cervical regions (Bird and Bird 1991; Matsuura et al. 2007). Negative quadratic relations of J2 hatch versus plant extracts had been observed (Akyazi 2014; Javed et al. 2008; Kalaiselvam and Devaraj 2011). Similarly, *in vitro* inhibition of J2 hatch in *Cucumis* oilseed cake was observed for *Meloidogyne* species (Dube 2016) and *T. semipenetrans* (Muedi et al. 2005). Generally, in studies in which the concentration straddles all three phases of density-dependent (DDG) patterns, a gradual inhibition of J2 hatch occurs, but after the minimum concentration, significant increases in J2 hatch occur. The actual bioactivities related to the inhibition of J2 hatch include interference with stylet development, disruption of the lipid parts of cell membranes by lipophilic extracts, interference with cytokinesis without affecting karyokinesis resulting in multinucleated cells and possible induction of cell cycle arrest (Lee et al. 2010). Although J2 hatch is mainly a physical process, characterized by increased J2 movement, with J2 continuously pressing stylets against egg shells, with resultant shell cracking and the emergence of J2, any factor that interferes with J2 behavioral activities would invariably affect hatching (Doncaster and Shepherd 1967).

At low concentrations, aldicarb was shown to stimulate J2 hatch through increased bioactivities of juveniles inside the egg shells (Hough and Thomason 1975). Similarly, certain botanical chemical compounds could either have stimulative, neutral or inhibitive effects on J2 hatch of the burrowing nematode (*Radopholus similis* Cobb; Wuyts et al. 2006), which confirmed the established DDG patterns. Additional inhibitive effects on J2 hatch for *Meloidogyne* species were observed in plant extracts of wild wormwood (*Artemisia vulgaris* L.; Costa et al. 2003), garlic (*Allium sativum* L.), fennel (*Foeniculum vulgare* Mill; Ibrahim et al. 2006), *Tagetes* species (Kalaiselvam and Devaraj 2011), cedar (*Melia azedarach* L.), elderberry (*Sambucus nigra* L.; Akyazi 2014), neem (*Azadirachta indica* A. Juss; Javed et al. 2008) and chrysanthemum (*Chrysathemum coronarium* L.). However, stimulated J2 hatch of *M. incognita* (Ibrahim et al. 2006), along with neutral effects on J2 hatch of stem and bulb nematode (*Ditylenchus dipsaci* Kuhn) post-exposure to extracts of rock fleabane (*Inula viscose* L.), were observed (Oka et al. 2001). All the cited examples support the principles of DDG patterns, which had been quantified using a curve-fitting algorithm response data (CARD) computer-based model, characterized by quadratic curves with distinct phases, namely stimulative, neutral and inhibitive phases (Liu et al. 2003). Generally, in nematode oilseed cake relations, DDG patterns had been observed in a limited studies at post-exposure of nematode eggs to increasing concentration of botanical products, primarly due to limitations imposed by limited concentration ranges used in the studies (Abdul 2013; Azhagumurugan and Rajan 2014; Costa et al. 2003; Ibrahim et al. 2006).

In most J2 hatch studies, a lack of standardization of concentration ranges, or the use of untreated control and one positive treatment, has since resulted in inconsistent observations in nematode suppression, with incidents of stimulated, neutral or inhibition effects on nematode population densities. For instance, Ibrahim et al. (2006) observed that extracts of *C. coronarium* and French marigold (*Tagetes patula* L.), which was later shown to inhibit J2 hatch (Kalaiselvam and Devaraj 2011), had no effect on J2 hatch of *M. incognita*, whereas Pérez et al. (2003) observed similar contradictory effects on J2 hatch while using extracts of *C. coronarium*. Maile et al. (2013), when investigating the effects of *Cucumis* oilseed cake on population densities of *T. semipenetrans*, observed that when experiments were terminated at 120 days instead of the usual 56 days for the granular products, relative to the untreated control, the products increased nematode population densities. The listed contradictions amplified the rallying point which suggested that plant-based cake products are, due to inconsistent results, not fit for commercialization. In another study, Mashela et al. (2017c) demonstrated that post-application of Nemarioc-AL and aldicarb in citrus seedlings, nematode population densities of *T. semipenetrans* soon after application nematodes increased slightly to a short but visible peak and then gradually decreased to the minimum population density thereafter, with the quadratic curves showing an inclination to increase, thereby suggesting the existence of cyclic population growth, which subscribes to biological behavioral patterns that necessitate repeated application, along with timed-harvesting of experiments. However, in certain crops such repeated applications are not essential since the influence of nematode damage on yield is directly proportional to initial nematode population densities at sowing or transplanting (Seinhorst 1965).

2.8.2 Juvenile Mortality

In vitro mortalities of J2 in *Cucumis* oilseed confirmed J2 mortalities induced by the extracts of angel's trumpet (*Datura stramonium* L.), neem (*Azadirachta indica* L. (Nelaballe and Mukkara 2013), various *Moringa* species (Claudius-Cole et al. 2010), *Artemisia vulgaris* (Costa et al. 2003) and *Allium sativum* (Ibrahim et al. 2006). The mortality of *M. incognita* J2 versus *Cucumis* oilseeds exhibited DDG patterns, an observation less common in other studies due to limited applied concentration ranges. Usually, under limited applied concentration ranges, results depicted either stimulation through positive linear models (Azhagumurugan and Rajan 2014; Pavaraj et al. 2012), which should be viewed as a decrease in population densities of the test nematodes. In contrast, incidents where no significant effects on J2 mortality were observed (Ardakani et al. 2013) are analogous to the neutral phase of DDG patterns, where entities are saturated with the active ingredient. Alternatively, as illustrated by biological indices of the CARD model, when treatment effects are not significant ($P > 0.05$), it could be suggested that the concentration used was, for the entity, below the stimulation threshold dose (D_m) (Liu et al. 2003; Mashela et al. 2017a). Below D_m, the test product does not induce any response.

Dube (2016) used a CARD model to explain the relationship between J2 mortality versus increasing concentration of *Cucumis* oilseed cake and noted that the sensitivity values of J2 mortality for Nemarioc-AL and Nemafric-BL products were both high and comparable. In the study, the minimum lethal concentration (MLC) values for J2 mortality of *M. incognita* induced by *Cucumis* oilseed products were 1.12% and 0.67% for Nemarioc-AL and Nemafric-BL, respectively (Dube 2016). In comparison, MLC values of clove (*Syzygium aromaticum* L.) on *M. incognita* and *R. similis* mortality were at 1% (Mustika and Slamet 1994). In contrast, that of neem on J2 mortality of *M.*

incognita was at 5% (Taye et al. 2013), whereas MLC of 5% dry neem extract to *M. incognita* J2 was 10% at 3 h post-exposure (Agbenin et al. 2005). Low MLC values and low sensitivity values of J2 bioactivities to *Cucumis* oilseed cake provide evidence of the high potency of Nemarioc-AL and Nemafric-BL when compared to most other botanical products. Additionally, MLC values provide a universal measure that can be used in comparing the toxicity of a range of crude extracts across a variety of trials. However, the frequent use of unstandardized units such as parts per million, percentage or mass, in accordance with DGG principles, as shown in Figure 2.3, renders the findings incomparable. The MLC values of *Cucumis* oilseed cake on J2 mortality, when compared with those of most other botanical products (Agbenin et al. 2005; Taye et al. 2013), were mostly low. For example, neem leaf extracts at 20% and 30% could, respectively, cause 50% and 90% J2 mortality of *M. incognita*, but these are at the phytotoxic range. In contrast, extracts of castor bean (*Ricinus communis* L.) and lemongrass (*Cymbopogon citratrus* Stapf) could not induce 100% J2 mortality even when used at 100% (Mukesh and Sobita 2013). The lethal concentration (LC_{100}) of African marigold (*Tagetes erecta* L.) at 10% (Kalaiselvam and Devaraj 2011) was comparable to those of products from *Cucumis* oilseed cake (Dube 2016). White cedar (*Melia azedarach* L.) and black elderberry (*Sambucus nigra* L.) had fairly low LC_{100} values against *M. incognita* J2 at 5 and 2.5%, respectively, when compared with those of products from *Cucumis* oilseed cake (Akyazi 2014). The CARD model-generated sensitivity values demonstrated that J2 mortality was highly sensitive to both products of *Cucumis* oilseed when compared to a number of other plant extracts.

2.8.3 Juvenile Mobility

The effects of plant extracts on nematode mobility have received less attention when compared with either J2 hatch or J2 mortality. Oka et al. (2000) during their investigation of 27 essential oils on J2 mobility observed that 12 could inhibit 80% J2 mobility of *M. javanica*. In contrast, when J2 of *M. incognita* were exposed to increasing concentrations of crude extracts of two neem species, *A. sativum* and *A. indica*, J2 mobility versus each neem product exhibited positive quadratic relations (Agbenin et al. 2005) and by extension the DDG patterns. In certain extracts, mobility not only exhibited positive DDG patterns, but J2 mobility responses were nematode species–specific (Wuyts et al. 2006). Generally, DDG patterns with clear stimulation of J2 mobility at low concentrations are common when extract concentrations have multiple levels (Abdul 2013; Azhagumurugan and Rajan 2014), which should at least straddle the three DDG phases, as conceptualized in Figure 2.2. In J2 mobility of the free-living *Caenorhabditis elegans* and the plant soybean cyst nematode (*Heterodera glycine* Ichinohe) versus a series of increasing concentrations of geldanamycin ($C_{29}H_{40}N_2O_9$) exhibited significant positive quadratic relations (Skantar et al. 2005). Geldanamycin is a 1,4-benzoquinone ansamycin antitumor antibiotic that inhibits the function of Hsp90 by binding to the unusual ADP/ATP-binding pockets of proteins. In contrast, J2 mobility of *M. javanica* versus limited levels of neem extract concentrations exhibited negative linear relations (Javed et al. 2007), which is further clarified using the conceptualized dosage model in Figure 2.2.

In nematology, the phenomenon where nematode movement follows or avoids an increasing gradient of a specific chemical cue is technically termed chemotaxis (Hida et al. 2015; Rasmann et al. 2012; Prot 1980; Reynolds et al. 2010). Movement towards the chemical cue is positive chemotaxis, with such chemicals described as having chemoattractant attributes. In contrast, moving away from the chemical cues is negative chemotaxis, with such chemicals having chemorepellent attributes. In soil spaces, nematode J2 are usually exposed to both liquid and volatilized chemicals, which could have either chemoattractant or chemorepellent attributes. Nematode bodies are well equipped for such environments due to the extensive chemoreceptors, primarily on the cephalic and cervical regions (Bird and Bird 1991; Hida et al. 2015; Matsuura et al. 2007; Rasmann et al. 2012). The nematode response to chemoattractant and chemorepellent chemicals plays a critical role in the behavioral capabilities of nematodes, with the body structures being such that the nematode can easily adjust to certain environmental conditions as a survival strategy (Mashela et al. 2020a). In the soil, nematodes have to contend with a wide range of chemicals since plants continuously release copious quantities of chemicals through exudation, leaching, volatilization and microbial degradation. As indicated, such chemicals induce a range of responses on J2 mobility (Mashela et al. 2015), with chemorepellents conferring pre-nematode resistance in certain plants (Nyasha 2021; Pofu and Mashela 2021). Wuyts et al. (2006), when investigating J2 mobility of various nematode species versus pure active ingredients from the rain tree (*Philenoptera violacea* Klotzsch) in the Fabaceae family, showed that chemotaxis effects were nematode species specific. Among the test chemicals, 26% had chemorepellent attributes to *R. similis*, and 2.6% had chemoattractant qualities to the same nematode species, whereas 45% had no significant effects (Wuyts et al. 2006). In most cases, botanicals with chemorepellent attributes induce disorientation of nematode, thereby displaying delays in host finding and root penetration, which is analogous to the effects of non-fumigant systemic nematicides (Noling 2019). Nyasha (2021) demonstrated that plants with chemorepellent nematode resistance were not suitable for crop rotations that were intended to suppress nematode population densities for subsequent nematode-susceptible crops such as watermelons and potatoes. In contrast, most plants with chemoattractant attributes are either susceptible or have post-infectional nematode resistance. In addition to causing behavioral adjustment in J2, most chemoattractant chemicals have the ability to induce paralysis in J2 and invariably death (Hida et al. 2015; Rasmann et al. 2012; Reynolds et al. 2010). In contrast to J2 hatch and J2 mortality, where bioactivities of each versus increasing concentration of *Cucumis* oilseed cake exhibited negative quadratic relations, J2 mobility versus increasing concentration of *Cucumis* oilseed cake exhibited positive quadratic relations (Dube 2016).

2.9 MECHANISM OF ACTION ON NEMATODES IN *CUCUMIS* OILSEED CAKE

Fumigant synthetic nematicides induced paralysis that invariably led to death. In contrast, non-fumigant systemic nematicides such as organophosphates, carbamates and oxycarbamates have had nematostatic effects on nematodes, characterized by notable reversible effects (Noling 2019). Generally, the bioactivity effects of *Cucumis* oilseed cake products had been irreversible (Dube 2016), suggesting that the products had nematicidal properties as opposed to nematostatic properties. In fumigant nematicides the induced irreversible effects had mainly been thought to be through paralysis, which had been confirmed in higher animals but never in nematodes. Paralysis would have been well suited for nematode bodies due to the existence of an unique network of sensory systems (Bird and Bird 1991). As illustrated in subsequent sections, our work suggests that *Cucumis* oilseed cake could be having a totally different *modus operandi* on the irreversibility effects on nematodes.

2.9.1 Attraction of Cucurbitacins by Nematode Cuticles

Nematodes are described as being a "tube-in-a-tube", with the outermost tube being the multi-layered cuticle separated from the internal tube by a hydrostatic liquid (Bird and Bird 1991). The cuticle comprises a single-layered epicuticle, which consists of lipids. The constituents of the subepicuticular layers include soluble and insoluble proteins stacked into distinct layers (Bird and Bird 1991). Cucurbitacins have lipophilic properties and are, conversely, hydrophobic (Van Wyk and Wink 2014). First, cucurbitacins are fished from aqueous solutions by lipids on the epicuticles, which provides a sufficient explanation for why nematodes are highly sensitive to *Cucumis* oilseed cakes (Dube and Mashela 2016, 2017, 2018, 2019; Mashela et al. 2022). Second, once cucurbitacins have been removed from soil solution by epicuticles, due to their hydrophobic nature, cucurbitacins are attracted to proteins in the subepicuticular layers, with proteins undergoing isoprenylation and farneslation reactions (Mashela et al. 2022). By definition, isoprenylation is the addition of hydrophobic molecules such as cucurbitacins through the prenyl groups (3-methylbut-2-en-1-yl) to proteins, whereas the addition of lipids to proteins is lipidation (Marshall 1993; Casey and Seabra 1996; Novelli and D'Apice 2012). The prenyl groups are important for protein-to-protein binding through specialized prenyl-binding domains (Novelli and D'Apice 2012). Actually, prenylation involves the transfer of either a farneslation or a geranylgeranyl moiety to C-terminal cysteine(s) of the target protein, using any of the three enzymes, farnesyl transferase, Caax protease or geranylgeranyl transferase I (Novelli and D'Apice 2012). The biosynthesis pathway of cucurbitacins in the MEP/DORXP pathway (Chen et al. 2005) includes farnesyl as one of the enzymes required in the biosynthesis pathway (Mashela et al. 2022). The reviewed mechanism provides a glimpse to the observation where the bioactivities of *Cucumis* oilseed cakes on nematodes are irreversible (Dube 2016).

2.9.2 Evidence of Isoprenylation and Farneslation Induced by *Cucumis* oilseed Cakes in Nematodes

In a recent study, Mashela et al. (2022) demonstrated that after 72 h- exposure of *M. incognita* J2 to a geometric series (0, 2, 4, 8, 18, 32 and 64%) of Nemafric-BL, total protein and *Cucumis* oilseed cake exhibited negative quadratic relations. Using the $x = -b_1/2b_2$ relation from the quadratic relation, $Y = b_2x^2 + b_1x + c$, the minimum total protein was accrued at 4.9% Nemafric-BL (transformed data), after which there were clear increases in the total protein. The gradual decrease (breakdown of proteins) and then increase (biosynthesis) in total protein depicted the existence of the isoprenylation and farneslation of proteins, respectively (Casey and Seabra 1996; Marshall 1993; Novelli and D'Apice 2012). The observed isoprenylation and farneslation supported the irreversibility of certain bioactivities of J2 post-exposure to *Cucumis* oilseed cake or their active ingredients (Dube 2016), which translated to death that did not involve paralysis but the fragmentation of the cuticle.

2.10 AVOIDING PHYTOTOXICITY OF *CUCUMIS* OILSEED CAKE

In most botanical products intended for use in managing nematode population densities, the major drawback that precludes potential products from entering the agrochemical markets is their high levels of phytotoxicity (Mashela et al. 2015). *Cucumis* oilseed cake was shown to be highly phytotoxic to various monocotyledonous and dicotyledonous crops (Mafeo and Mashela 2010; Mafeo et al. 2011a, 2011b), which ruled out their potential registration for use in nematode management, due to the existence of zero tolerance to phytotoxicity in agricultural inputs (Mashela et al. 2015). However, our insistence that all agricultural inputs when wrongly applied were phytotoxic resulted in developing a series of empirically based solutions, which were intended to solve the challenges of phytotoxicity.

2.10.1 Application Concentration

The application concentration is basically equivalent to half the stimulation phase of CARD-generated quadratic curves (Figure 2.3), which we technically code-named the mean concentration stimulation point (MCSP). The MCSP is that concentration of *Cucumis* oilseed cake that would stimulate plant growth while suppressing nematode population densities (Mashela et al. 2015, 2017a). The MCSP is derived by first subjecting the test plant to increasing concentrations of test *Cucumis* oilseed cake, applied weekly for 8 weeks, which is equivalent to three life cycles of *Meloidogyne* species (Mashela et al. 2017a). The MCSP is a concept that was developed for avoiding phytotoxicity (Mashela et al. 2015) using the biological indices from the CARD algorithm model (Liu et al. 2003). All CARD-derived curves were skewed to the right (Liu et al. 2003; Mafeo 2012), which suggested that the data did not have a normal distribution. Thus, inference for the population mean could not be made from the means derived from such data. Consequently, the x-axis data were transformed in order to normalize the interval between the concentrations, thus ensuring the fulfillment of the inference requirement (Causton 1977). The geometric series of the phytonematicide concentrations was further transformed using $\log_2 2^x$, where $\log_2 2 = 1$, with the x-axis resulting in 0%, 1%, 2%, 3%, 4%, 5% and 6% phytonematicide (Tseke and Mashela 2018).

The seven generated biological indices included (1) threshold stimulation (D_m), (2) saturation point (R_h), (3) 0% inhibition (D_0), (4) 50% inhibition (D_{50}), (5) 100% inhibition (D_{100}), (6) degree of sensitivity (k) and (7) coefficient of determination (R^2; Liu et al. 2003). Opposite the biological index D_{50} on the stimulation phase was supposed to be the mid-stimulation point; however, the CARD model did not provide that dose. After analyzing CARD-generated curves from various experiments, Mashela et al. (2015) derived MCSP relation using biological indices on the stimulation phase as shown in Figure 2.3:

$$\mathbf{MCSP = D_m + (R_h/2)}$$

The relationship was validated on a wide range of crops (Maake 2017; Masenya 2022; Mashela et al. 2017a, 2021; Sebothoma 2018; Selomo 2019; Sithole et al. 2016) and demonstrated that at MCSP, the phytonematicide would not induce phytotoxicity on the protected plant but would consistently suppress nematode population densities (Mashela et al. 2015). Generally, most allelochemicals affect biological systems through DDG patterns (Liu et al. 2003), which comprise three phases, stimulation, neutral and inhibition, depending on the level of concentration and the degree of sensitivity of the test plant organs (Liu et al. 2003; Mashela et al. 2015).

The successful use of *Cucumis* oilseed cake depended on the establishment of MCSP, which was crop specific (Mashela et al. 2017a). Importantly, in the empirical establishment of MCSP, in addition to conceptualizing and using a geometric series of concentrations that allowed the inclusion of the three DDG phases, various plant organs were collected in order to base MCSP on the whole plant rather than one organ. The generated MCSP values were interpreted alongside sensitivity (k) values for separate plant variables and the overall sensitivity of the plant (Σk) to the *Cucumis* oilseed cake (Liu et al. 2003; Mashela et al. 2015). Basically, Σk is a function of the *Cucumis* oilseed cake concentration, application rate, plant organs used and plant species (Mashela et al. 2015). The closer the (Σk) of the plant is to zero, the higher the sensitivity of the plant to the test *Cucumis* oilseed cake, while the opposite is true (Liu et al. 2003).

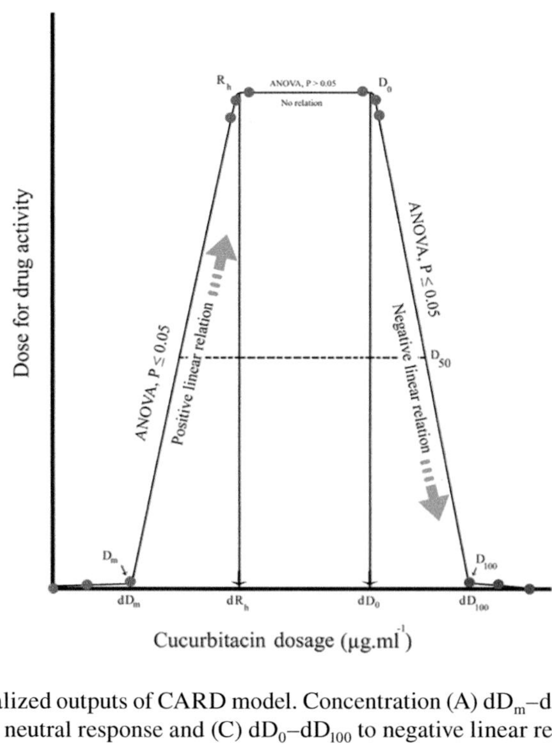

FIGURE 2.3 Conceptualized outputs of CARD model. Concentration (A) dD_m–dR_h results in positive linear response, (B) dR_h–dD_0 to neutral response and (C) dD_0–dD_{100} to negative linear response.

Research findings on MCSP and Σk values of *Cucumis* oilseed cake to several plant species have already been investigated. Sithole et al. (2016) developed MCSP for Nemarioc-AL and Nemafric-BL for *Pelargonium sidoides* as 6.2% and 2.9%, respectively, along with Σk values of 3 units for both products. Such values were also investigated for sweet potatoes (*Ipomoea batatas* L.; Selomo 2019), citrus seedling rootstocks (Mathabatha et al. 2017), green beans (*Phaseolus vulgaris* L. cv. "Tahoe"), and squash (*Cucurbita pepo* L.) were 11.9% and 2.8%, with the Σk values of 0 and 3 units, respectively (Lebea 2017). Cancer bush (*Sutherlandia frutescens* (L. R.) Br.; Masenya 2022), Swiss chard (*Beta vulgaris* var. Cicla; Mashela et al. 2021) and many other crops all demonstrated that MCSP values were plant specific but with the major focus being on the concentration that would significantly reduce population densities of the test nematode species without inducing phytotoxicity to the plant protected against nematode infection and damage. The empirically established MCSP had since been referred to as the first law of phytonematicides (Masenya 2022).

2.10.2 Application Interval

The application interval represents an interval adjacent to the successive applications of the phytonematicide that would not be phytotoxic but would successfully suppress nematode population densities (Mashela et al. 2015). The MCSP for the test crop, along with the life span of the test nematode, is used in the empirical determination of the application interval (Mashela et al. 2015) using the relationship:

$$\text{Application interval} = \text{Weeks-per-month} \times \text{Nematode life cycle}.$$

This application interval relation is dubbed the second law of phytonematicides (Masenya 2022). We provided an illustration of the law using *M. incognita* with a life cycle of 30 days. The relationship for *M. incognita* could first be expressed in days as 0, 7.5, 15, 22.5 and 30 days, which is equivalent to 0, 1, 2, 3 and 4 weeks, which constitutes the application interval for the previously derived MCSP value. The best would be in weeks since the equal distances would enhance normality and thereby comply with the concept of inference. After the experiment had been conducted, with weekly application of MCSP, plant variables are collected as in the determination of MCSP, subjected to lines of the best fit, with the maximization of the derived positive quadratic relations using the relation $x = -b_1/2b_2$ for different plant variables (Mashela et al. 2017a). The derived maximization average ($\bar{x}$) in weeks is then converted to application interval in days. Using MCSP 2.18% for Nemafric-BL on sweet potato cv. "Bophelo" for managing population densities of *M. incognita*, the application interval was 19 days (Selomo 2019), with that for cancer bush being 29 days and 17 days for 3% for Nemarioc-AL and Nemafric-B, respectively (Masenya 2022). In contrast using MCSP of 3% for Nemarioc-AL and Nemafric-BL on tomatoes, the application intervals for *M. incognita* were 20 and 16 days, respectively (Pelinganga et al. 2013). In *T. semipenetrans*, with a life cycle of 8 weeks, the application intervals of *Cucumis* oilseed cake were relatively higher (Mathabatha et al. 2017). The intended purpose of the law is to ensure that the life cycle of the nematode being managed is interrupted on several occasions during the crop cycle, as further illustrated through the application frequency.

2.10.3 Application Frequency

The application interval allows the computation of the application frequency (T_f), which is defined as the proportion of the crop cycle to the application interval [T_f = crop cycle (days)/application interval (days)], a unit-less relation (Mashela et al. 2017a) is viewed as the third law of phytonematicides (Masenya 2022). The application frequency provides the information required for planning purposes, and it is also used in the establishment of the dosage model.

2.10.4 Dosage Model and its Role

The dosage model (DM) is a product of MCSP (%) and the unit-less application frequency (T_f), summarized as (Mashela et al. 2017a):

$$DM\ (\%) = C\ (\%) \times T_f.$$

This DM constitutes the fourth law of phytonematicides and predicts the total amount of the active ingredient that would have been placed in a given environment by the end of the crop cycle (Mashela et al. 2015). Generally, active ingredients of most pesticides are removed from the environment through bacteria, fungi and plants in a process referred to as bioremediation (Norris 1993). Bioaccumulation of cucurbitacins in plants does not occur since there are hardly any cucurbitacin residues in the fruit and leaves of various crops (Dube 2016; Maake 2017; Shadung 2016). Additionally, bacteria and fungi hardly produce enzymes that degrade cucurbitacins (Mashela et al. 2022). However, it was shown that another group of animals, the ecdyzoans, which include nematodes, were highly effective in the bioremediation of cucurbitacins due to the unique properties that confer these group of chemical compounds the capabilities of isoprenylation (protein breakdown) and farneslation (protein biosynthesis). Furthermore, Mashela et al. (2021) argued that due to the existence of a large number of ecdosomes in the soil, their role in bioremediation could explain the relatively shorter application intervals of cucurbitacin phytonematicides when compared with fumigant nematicides, which were renowned for their persistence in the soil, at times lasting for as long as 12 years (Van Gundy and McKenry 1977).

2.11 MANAGING CHALLENGES ASSOCIATED WITH *CUCUMIS* OILSEED CAKE

The use of phytonematicides, in addition to the primary challenges of phytotoxicity which were resolved using the dosage model, has had other associated challenges due to limited knowledge about phytochemical–nematode relations. In this study, we focused primarily on managing challenges associated with *Cucumis* oilseed cake.

2.11.1 Inconsistent Results in Nematode Suppression

Various workers in botanical pesticides suggested that the products were invariably associated with inconsistent results in nematode suppression and that they could, therefore, not fulfill the important principle of "fit for purpose" as a requirement for registration of *Cucumis* oilseed products (Mashela et al. 2015). The inconsistent results were in accordance with the basic principle of DDG patterns, which is illustrated in Figure 2.3, which we have observed in various studies and succinctly reviewed (Mashela et al. 2017a). Briefly, any observation is concentration dependent, which, in turn, is phase dependent in relation to the three phases: stimulation, neutral and inhibition. Additionally, as we illustrated in Figure 2.3, should, for the test bioactivity, the concentration be lower than that of D_m or should that bioactivity be subjected to the concentration within doses of its neutral phase, namely R_h and D_0, the effects might not be statistically different. However, should the concentration be within doses of the stimulation phase ($D_m - {}_{Rh}$), the bioactivity might be depicted through a positive linear relationship. In contrast, if the concentration is within the doses of the inhibition phase ($D_0 - D_{100}$), the bioactivity might be depicted as a negative linear relationship.

2.11.2 Managing Skewness of CARD-Generated Quadratic Curves

Historically, quadratic curves as outputs of the CARD model were invariably skewed to the right (Alghasham 2013; An et al. 1993; De Liu and Graham 2005; Calabrese and Baldwin 2003; De Liu and Graham 2005; Duangmano et al. 2012; Liu et al. 2003; Mafeo and Mashela 2010; Mafeo et al. 2011a, 2011b; Pelinganga and Mashela 2012; Pelinganga et al. 2012). The developers of the model alluded to the skewness as its main shortcoming. Also, in our repeated assessment of the model using both secondary and primary data (Mafeo and Mashela 2010; Mafeo et al. 2011a, 2011b; Pelinganga and Mashela 2012; Pelinganga et al. 2012), the skewness was as prominent as it was in the original trials (Liu et al. 2003). One of the basic requirements in procedures with "scientific merit" is that the sample mean ($\bar{x}$) should be used to make inferences about the population mean (μ) (Hosmand 1998). The μ is a central point of a population with normal distribution. In order to achieve a normal distribution for data that generate $\bar{x}$ as a proxy of μ, the sample size (n) should be at least 30 ($n \geq 30$), with measurements collected from randomly selected experimental units (Hosmand 1998). Should the assumptions of the analysis of variance still not be met, measurements could be transformed in order to meet the minimum requirements for normal distribution (Hosmand 1998), thereby allowing an inference to be made about μ from $\bar{x}$. We have earlier shown that the success of the CARD model depends on whether the increasing concentration of the test phytonematicide straddles all three dose-response phases in Figure 2.3. In reality, straddling all three phases with equidistant concentration intervals would be an impossible feat since excessive treatment numbers would be required. We solved the challenge of excessive treatments by increasing concentrations of phytonematicide using a geometric series that included 0%, 2%, 4%, 8%, 16%, 32% and 64% phytonematicide as illustrated in Figure 2.3. Alternatively, 0%, 3%, 9%, 27%, 81% and 243% phytonematicide would serve the purpose. Notwithstanding including the three phases on x-axis variables, in a series of tests with treatments straddling the three dose-response phases, the generated quadratic curves were still skewed to the right in Figure 2.4. A detailed inspection of skewness to the right suggested that the challenge of skewness originated because of unequal concentration intervals between adjacent treatments. In order to have equidistant concentration intervals on the x-axes, we first expressed the geometric series as an exponential series [(2^0, 2^1, 2^2, 2^3,2^4, 2^5 and 2^6)% *Cucumis* oilseed cake]. Second, the generated exponential series was log-transformed using an appropriate base. In the transformation of the exponential series (2^0, 2^1, 2^2, 2^3, 24, 2^5 and 2^6)%, since $\log_2 2^x = x \times (\log_2 2)$ and $\log_2 2 = 1$ (Causton 1977), then the transformed series became 0%, 1%, 2%, 3%, 4%, 5% and 6% phytonematicide. Variables versus transformed series were then subjected to the CARD model to provide quadratic curves with parabolic shapes that emulated those of data with normal distribution, with the models explained by high R^2 values as illustrated in Figure 2.3. Additionally, the output information from the CARD model included the much-improved biological indices, showing vast differences between those from untransformed and transformed datasets.

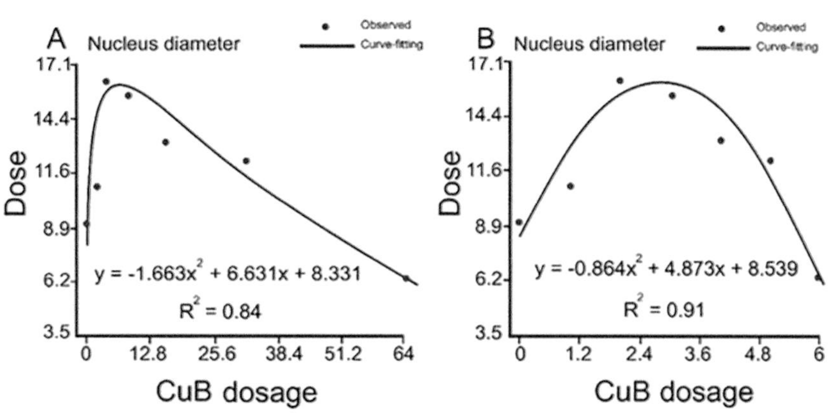

FIGURE 2.4 Untransformed and transformed data: (A) skewness to the right and (B) normal distribution for nucleus diameter over increasing cucurbitacin B concentration.

2.11.3 Variability in Quantity of Cucurbitacins

A large number of factors that have negative effects on the quantity of active ingredients in plants as alternative medicine have been identified but are mainly phytochemical specific. Thus, the focus of the current study was limited to those factors with the potential of affecting the quantity of cucurbitacins in *Cucumis* oilseed cake.

2.11.4 Suitable Organs for Cucumis Oilseed Cake

Secondary metabolites in plant organs, especially those in leaf tissues, fluctuate with seasonal changes (Alves et al. 2007; Mudau et al. 2008). In contrast, secondary metabolites in fruits tend to be more stable (Moco et al. 2007). Although in *C. africanus* cucurbitacin B occurs in all organs of the plant, to avoid seasonal variability, only fruit are used as the sole source in the production of Nemafric-BL, resulting in consistent nematode suppression (Mashela et al. 2011).

2.11.5 Optimum Fruit Harvest Time

Generally, there are five developmental stages during the formation of fruits in *Cucumis* species (Shaik et al. 2011; Shadung 2016). Flowering is followed by fruit formation, which is characterized by rapid cell division of the fertilized ovary. The elongation stage, which proceeds for 2–3 weeks, ensues. Thereafter, the fruit maturation stage occurs, characterized by fruit swelling without a noticeable increase in fruit size. The color of the fruit changes from green to slightly yellow to signify fruit ripening. Finally, senescence sets in, resulting in a softening of fruit carpus and thereafter overall fruit shriveling (Shaik et al. 2011). Consequently, harvesting fruits indiscriminately at different developmental stages might serve as a source of variation in the test active ingredients. Optimization of cucurbitacin concentration versus time from transplanting of 4-week-old seedlings showed that the optimum fruit harvesting times for *C. myriocarpus* and *C. africanus* were at 89 ± 5 and 95 ± 9 days, respectively (Shadung 2016).

2.11.6 Preservation of Fruits for *Cucumis* Oilseed Cake

Different preservation methods had been serving as the initial process in the preparation of botanical drugs in attempts to preserve the active ingredients (Lusia et al. 2015), which are lost through exudation but mainly through microbial. Drying lowers the moisture content and thereby prevents microbial growth and degradation (Oztekin and Martinov 2007). Drying methods can either be thermal (sun- and oven-drying) or nonthermal (shade- and freeze-drying) methods (Chan et al. 2008; Lusia et al. 2015). However, drying can result in unintended consequences by destroying potent chemicals through biochemical degradation and/or the volatilization of target active ingredients (Sharma and Prasad 2003). The unintended consequences could result in the reduction of bioactive compounds, thereby affecting the quality of the products. The reduction of target active ingredients is undesirable because this might have a direct effect on "fit-for-purpose" attributes.

Cucurbitacin A and B are thermo-stable, with boiling points at 731 and 699°C at sea level (760 mmHg), respectively (Krieger 2001). In most medicinal plants, low drying temperatures from 30 to 50°C are being used to protect the integrity of the active ingredients (Müller and Heindl 2006). However, when fruit pieces from *C. myriocarpus* and *C. africanus* were dried within the temperature ranges recommended for medicinal plants, the materials were lost to decay, with a blue, bluish-green or olive-green color, surrounded by white mycelium and a band of water-soaked tissues that characterize *Penicillium simplicissimum* Oudem. In optimization trials of cucurbitacins, relative to 52°C, a temperature of 60°C significantly increased cucurbitacin A by 4%, with reductive effects starting from 70 to 100°C. At 52°C, both cucurbitacin A and B were significantly higher than those at higher temperatures, which were actually not different from one another from 60 to 100°C (Shadung et al. 2016; Shadung 2016). Similarly, relative to 52°C, increasing temperature significantly reduced cucurbitacin B from 28 to 85%. Thus, drying fruits from the two *Cucumis* species at 52°C for 72 h was viewed as a balance between the retention of cucurbitacins from thermal degradation and preserving fruits from post-harvest decay induced by *P. simplicissimum.*

2.11.7 Storage Duration of Inventories

Poor-quality plant material at pre- and post-harvests causes huge losses in botanical products (Isman 1997; Niazi 2009; Orwa et al. 2004). Most such losses are incurred during inventory prior to processing (Kader 2002, 2005). At pre- and post-harvests, both extrinsic and intrinsic factors contribute to the change in the stability of active ingredients, which could inevitably alter the quality and quantity of the active ingredients (Gobbo-Neto and Lopes 2007; Isman 1997).

2.11.8 Shelf Life of *Cucumis* Oilseed Cake

In product development, behavioral reaction orders of the Arrhenius model are generally used to provide shelf-life in food products. Cucurbitacins did not conform to the behavioral reaction orders of the Arrhenius model, and thus, an alternative quadratic model consistent with the behavioral reaction of cucurbitacins was developed under two fixed conditions: a chilled condition (5°C + 95–98% RH) and a fixed tropical condition (38°C at 90% RH) (Mashela et al. 2020b). Sampling for cucurbitacin concentration was done at timeframes compliant with geometric series, where cucurbitacin analysis was performed using high-performance liquid chromatography techniques (Shadung 2016). Under chilled conditions, the quadratic model could not predict the shelf life for Nemarioc-AL, while that for Nemafric-BL was 35 weeks. Cucurbitacin A in Nemarioc-AL is generally unstable and disintegrates into cucumin ($C_{36}H_{46}O_9$) and leptodermin ($C_{36}H_{46}O_8$) (Jeffrey 1980). In contrast, under tropical conditions, Nemarioc-AL and Nemafric-BL had shelf lives of 35 and 825 weeks, respectively (Mashela et al. 2020b).

2.11.9 Fruit Harvest Location for *Cucumis* Oilseed Cake

The fruits of *C. myriocarpus* and *C. africanus* collected from five districts in Limpopo Province, South Africa, with different coordinates, rainfall and parent soil groups, suggested that location did not have significant effects on the concentration of cucurbitacins (Shadung 2016). However, for commercial production, fruits from cultivated *Cucumis* species are preferable.

2.12 MINIMIZING THE IMPACT OF *CUCUMIS* OILSEED CAKE ON THE ENVIRONMENT

The existing application methods, particularly GLT and BOT, place large quantities of cucurbitacins into the soil environment. In a study to assess the residual effects of *Cucumis* oilseed on subsequent plant growth of sweet stem sorghum (*Sorghum bicolor* L.) post-cultivation of cowpea (*Vigna unguiculata* (L.) Walp) where nematode population densities were managed using the test products, the residual effects significantly promoted plant growth and sucrose content (Mashela 2014). As indicated in the current study, phytotoxicity by *Cucumis* oilseed cake to plants protected against nematode damage was successfully avoided through the empirically based dosage model (Mashela et al. 2015).

2.12.1 Dosage Model

The dosage model, which is a product of MCSP, application interval and application frequency, was intended to reduce the quantity of the applied cucurbitacin into soil solutions. The model accommodates avoidance of phytotoxicity, which is crop specific; the nematode life cycle; and the crop life cycle (Mashela et al. 2017a), which suggests that it is a biological model. During recent technological developments, it was shown that PAD technology using *Cucumis* oilseed cake on seeds with hypogenous germination, followed by short-term drying, could innovatively serve as an important application technology, with the least active ingredients being released into the soil environment (Mashela et al. 2022).

2.12.2 Bioremediation of Cucurbitacins

Cucurbitacin is mostly nonpolar and cannot be absorbed by plant roots (Van Wyk and Wink 2014). Although the soil residual effects of *Cucumis* oilseed cake were demonstrated (Mashela 2014; Mashela and Dube 2014), the effects of cucurbitacin in soil environments are short lived, with liquid formulations applied biweekly, whereas granular formulations are applied every 8 weeks (Mashela et al. 2017a), suggesting the existence of bioremediation of cucurbitacins. Recently, Mashela et al. (2022) demonstrated that cucurbitacin B in nematode cuticles induced isoprenylation and farneslation, which were the likely mechanism through which cucurbitacins were bioremediated from soil solutions. In addition to the large nematode population densities in soil solutions, the soil is also replete with another class of cuticle-bearing organisms, the ecdosomes, which could help to clarify the issue of bioremediation of cucurbitacins, which is commendable when contrasted with the environment-unfriendly fumigant synthetic chemical nematicides (Van Gundy and McKenry 1977).

2.12.3 Developing an Environment-Friendly Application Method

Conventional application methods of *Cucumis* oilseed cakes, namely GLT and BOT, due to the need to repeatedly apply the products during the crop cycle, place substantial quantities of cucurbitacins in the soil environment–as confirmed by soil residues that were significant on nematode suppression and growth of the successor crops (Mashela 2014; Mashela and Dube 2014). However, due to the existence of bioremediation of cucurbitacins by ecsodomes which exist in substantial numbers in the soil, as confirmed by the short application intervals of the test products (Mashela et al. 2017a). The latter provides some assurance that the products are less persistent when compared with the fumigant nematicides, which, among other reasons, were withdrawn from the agrochemical markets due to their exceptionally long persistence in the environment (Van Gundy and McKenry 1977).

The PAD technology involves priming seeds with hypogenous germination in MCSP-diluted solutions of *Cucumis* oilseed cakes for a short period and then, after wiping off the excess solution, drying the seeds and then storing at room temperature until use. Tests using such seeds demonstrated that the underground cotyledons and related layers successfully served as carriers of cucurbitacins (Mashela et al. 2022). In addition to improving the growth of pea plant variables, PAD technology was also successful in suppressing nematode population densities of *M. incognita* throughout the crop cycle (Mashela et al. 2022).

2.13 IMPACT OF *CUCUMIS* OILSEED CAKE ON PRODUCE QUALITY

Produce quality in crops is assessed in various ways, which include but are not limited to yield, nutrition-related attributes and mineral malnutrition elements, which are briefly discussed.

2.13.1 Crop Yield

During the inception of *Cucumis* oilseed cake in 2002, it was observed that the product stimulated plant growth, with the resultant high tomato fruit yield (Mashela 2002), which was then viewed as a "fertilizer effect". The observation was confirmed in various studies on tomato plants but with a limited accumulation of nutrient elements until the dosage model with its three phases was well understood (Mashela et al. 2017a). Generally, when used in accordance with the principles of the dosage model, products of *Cucumis* oilseed cake invariably increase crop yield as observed in tomatoes (Dube 2016; Maake 2017; Pelinganga et al. 2012; Shadung 2016), potatoes (Seshweni 2016; Thopola 2019), sweet potato roots (Selomo 2019), watermelons (Nhlane 2017), Swiss chard leaf yield (Mashela et al. 2021) and many other crops (Mashela et al. 2017a).

2.13.2 Physical Produce Quality

Infection by *Meloidogyne* species on root and tuber crops results in visible symptoms such as cracks, galls and various unsightly deformations, which consumers view as quality-related concerns. *Cucumis* oilseed cake, by virtue of reducing nematode population densities in various crops, reduces the incidence of visible symptoms on produce. In addition to visible symptoms, quantitative analysis has since demonstrated that the use of *Cucumis* oilseed cake had significant improvement in internal produce quality that is invisible but important to consumers, which are briefly outlined next.

2.13.3 Internal Produce Quality

The link of diets to human diseases is well established (Mabhaudhi et al. 2016). The internal produce quality is therefore as important as the external produce quality as consumers are more inclined to demand quality produce through their own or various other consciousness structures.

2.13.3.1 Mineral Malnutrition Elements

The link between diets and human diseases is well established (Mabhaudhi et al. 2016). Mineral malnutrition (MMN) elements, which include Ca, Mg, Na, K, P, Cl, Zn and S, are those mineral elements without which, in diets, over an extended period, an animal would suffer from the incidence of malnutrition (Mabhaudhi et al. 2016). Generally, nematode infection results in a situation in which MMN elements are negatively affected. For example, infection by the citrus nematode (*Tylenchulus semipenetrans* Cobb) on citrus reduces most MMN elements in both leaf and fruit tissues but increased the accumulation of Na and Cl ions to toxic levels in leaf tissues while reducing the two in root tissues along with K in both leaf and root tissues. Similar effects on MMN elements were observed when plants were infected by other r nematode strategists such as *Meloidogyne* species (Ramputla 2019). In contrast, the application of *Cucumis* oilseed cake on various crops resulted in increased accumulation of certain MMN elements in leaf tissues in test crops (Mashela et al. 2017a).

2.13.3.2 Cucurbitacin Residues in Produce

Cucurbitacin A and B are slightly polar and nonpolar, respectively, and thus are, respectively, slightly soluble and insoluble in water (Jeffrey 1980; Van Wyk and Wink 2014). Generally, regardless of the size of the molecules, root cells do not allow the passage of nonpolar molecules (Van Wyk and Wink 2014). Currently, there is no evidence that plants play any role in bioremediation of cucurbitacins from soil solution (Mashela et al. 2022). Evidence-based studies suggested that in tomato plants, there was hardly any trace of cucurbitacin A and B as chemical residues (Dube 2016; Maake 2017; Shadung et al. 2017). Similar active ingredients with large molecular structures such as azadirachtin ($C_{35}H_{44}O_{16}$) in neem (*Azadirachta indica* A. Juss) had negligible chemical residues in olives after a field treatment (Caboni et al. 2002).

2.13.3.3 Sucrose Content in Produce

During the investigation of soil allelochemical residue effects on cowpea–sweet stem sorghum rotation, soil residues significantly increased sucrose content of the successor crop (Mashela and Dube 2014). The increased effects on sucrose in sweet stem sorghum due to the application of *Cucumis* oilseed cake were confirmed in various experiments when *Cucumis* oilseed cakes were used to complement nematode resistance in sweet stem sorghum cv. "Ndendani-X1" (Mashela and Dube 2014; Maleka 2020).

2.14 CONCLUSION AND FUTURE PERSPECTIVES

Like the fumigant synthetic chemical nematicides, *Cucumis* oilseed cake had challenges related to phytotoxicity and inconsistent results in suppression of nematode population densities, which were successfully resolved using the CARD algorithm computer model, which laid the foundation of the dosage model. Unlike the fumigant synthetic chemical nematicides that had nematicidal properties on both r and K nematode strategists, products from *Cucumis* oilseed cake have nematicidal effects on r nematode strategists alone, with limited effects on K nematode strategists, which was similar to the effects of systemic chemical nematostic products on r nematode strategists, except that the effects of *Cucumis* oilseed cake were irreversible. Additionally, *Cucumis* oilseed cake hardly had any chemical residues in produce, whereas the challenge was common in crops where nematode population densities were managed using systemic non-fumigant nematicides. Currently, the mechanism through which K nematode strategists are tolerant to *Cucumis* oilseed cake at concentrations that are not phytotoxic is little understood. Yet, most such K nematode strategists are vectors of various nematodes. Future studies would apply *Cucumis* oilseed cake at pre-planting using high phytotoxic concentrations, followed by a waiting period in the management of population densities of K nematode strategists. Furthermore, the mechanisms through which *Cucumis* oilseed cake improves crop produce quality and absorption of certain mineral malnutrition elements, along with the extent of bioremediation and persistence of cucurbitacins from *Cucumis* oilseed cake in different ecdosomes, soil types and bodies of water, would be investigated, along with their potential application in control of zoonematodes.

REFERENCES

Abbas, S., Vincourt, J., Habib, L., Netter, P., Greige-Gerges, H. and J. Magdalou. The cucurbitacins E, D and I: Investigation of their cytotoxicity towards human chondrosarcoma SW 1353 cell line and their biotransformation in man liver. *Toxicology Letters* 216 (2013):189–199.

Abdul, N. C. Effect of four leaf extracts on egg hatching and juvenile mortality of root-knot nematode *Meloidogyne incognita*. *International Journal of Advanced Life Sciences* 6 (2013):68–74.

Agbenin, N. O., Emeche, A. M., Marley, P. S. and A. D. Akpa. Evaluation of nematicidal action of some botanicals on *Meloidogyne incognita in vitro*. *Journal of Agriculture and Rural Development in the Tropics and Subtropics* 1 (2005):29–39.

Akyazi, F. Effect of some plant methanol extracts on egg hatching and juvenile mortality of root-knot nematode *Meloidogyne* incognita. *American Journal of Experimental Agriculture* 4 (2014):1471–1479.

Alghasham, A. A. A. Cucurbitacins–A promising target foe cancer therapy. *International Journal of Health Sciences* 7 (1) (2013):77–89.

Alves, M. N., Sartoratto, A. and J. R. Trigo. Scopolamine in *Brugmansia suaveolens* (Solanaceae): Defense, allocation, costs and induced response. *Journal of Chemical Ecology* 33 (2007):297–309.

An, M., Johnson I. R. and J. V. Lovett. Mathematical modeling of allelopathy: Biological response to allelochemicals and its interpretation. *Journal of Chemical Ecology* 19 (1993):2379–2388.

Ardakani, A. S., Hosyninej, S. A. and A. Pourshirz. Killing effects of *Myrtus communis* L. essential oil on *Meloidogyne incognita*. *International Journal of Agriculture and Crop Science* 5 (2013):806–810.

Azhagumurugan, C. and M. K. Rajan. Effect of leaf extract of nilakumil (*Gmelina asiatica*) against the root-knot nematode (*Meloidogyne incognita*). *Research Journal of Recent Sciences* 3 (2014):264–266.

Bird, A. F. and J. B. Bird. *The structure of nematodes*. New York: Academic Press, 1991.

Bohlmann, H. Introductory chapter on the basic biology of cyst nematodes. *Advances in Botanical Research* 73 (2015):33–59.

Caboni, P. Cabras, M., Angioni, A., Russo, M. and P. Cabras. Persistence of azadirachtin residues on olives after field treatment. *Journal of Agricultural and Food Chemistry* 50 (12) (2002):3491–3494.

Calabrese, E. J. and L. A. Baldwin. Hormesis: The dose-response revolution. *Annual Review of Pharmacology and Toxicology* 43 (2003):175–197.

Casey, P. J. and M. C. Seabra. Protein prenyltransferases. *Journal of Biology and Chemistry* 27 (10) (1996):5289–5292.

Causton, D. R. *Contemporary biology: A biologist's mathematics*. London: Bedford Square, 1977.

Chan, W. W. C., Lim, Y. Y., Wong, S. K., Lim, K. K., Tan, S. P., Tan, F. S. and M. Y. Yong. Effects of different drying methods on the antioxidant properties of leaves and tea of ginger spices. *Food Chemistry* 113 (2008):166–172.

Chen, C., Kuo, T. C., Yang, M., Chien, T., Chu, M., Huang L., Chen, C., Lo, H. Jeng, S. and O. L. Chen. Identification of cucurbitacins and assembly of a draft genome for *Aquilaria agallocha*. *BMC Genomics* 15 (2014)15(578): 1471–2164.

Chen, J. C., Chiu, M. H., Nie, R. L., Cordell, G. A. and S. X. Qiu. Cucurbitacins and cucurbitane glycosides: Structures and biological activities. *Nature Product Reports* 22 (2005):386–399.

Claudius-Cole, A. O., Aminu, A. E. and B. Fawole. Evaluation of plant extracts in the management of root-knot nematode *Meloidogyne incognita* on cowpea (*Vigna unguiculata* (L.) Walp). *Mycopathology* 8 (2010):53–60.

Costa, S. S. R., Santos, M. S. N. A. and M. F. Ryan. Effect of *Artemisia vulgaris* rhizome extracts on hatching, mortality, and plant infectivity of *Meloidogyne megadora*. *Journal of Nematology* 35 (2003):437–442.

Curtis, R. H. C. Plant-nematode interactions: Environmental signals detected by the nematode's chemosensory organs control changes in the surface cuticle and behavior. *Parasite* 15 (2008):310–316.

D'Addabbo T., Carbonara T., Leonettii P., Radicci V., Tava A. and P. Avato. Control of plant parasitic nematodes with active saponins and biomass from *Medicago sativa*. *Phytochemistry Reviews* 10 (2010). doi: 10.1007/s1101-9180-2.

De Liu, L. and E. H. Graham. Implementation of CARD: Curve-fitting allelochemical response data. *Nonlinearity in Biology, Toxicology and Medicine* 3 (2005):235–244.

Doncaster, C. C. and A. M. Shepherd. The behavior of second-stage *Heterodera rostochiensis* larvae leading to their emergence from the egg. *Nematologica* 13 (1967):476–478.

Duangmano, S., Sae-Lim, P., Suksamrarn, A., Domann, F. E. and P Patmasiri. Cucurbitacin B inhibits human breast cancer cell proliferation through disruption of microtubule polymerize and nucleophosmin/B23 translocation. *Complementary and Alternative Medicine* 12 (1) (2012):185. doi: 101186/1472-12-185.

Dube, Z. P. *Nemarioc-AL and Nemafric-BL phytonematicides: Bioactivities in nematodes, crops and soil*. PhD thesis, University of Limpopo, Sovenga, South Africa, 2016.

Dube, Z. P. and P. W. Mashela. Response of *Meloidogyne incognita* egress and overall sensitivity active ingredients of Nemarioc-AL and Nemafric-BL phytonematicides. *Transylvanian Review* 7 (2016):954–960.

Dube, Z. P. and P. W. Mashela. Confirmation of bioactivities of active ingredients of Nemarioc-AL and Nemafric-BL phytonematicides. *Acta Agriculturae Scandinavica Section B-Soil and Plant Science* 67 (6) (2017):571–575.

Dube, Z. P. and P. W. Mashela. Effects of cucurbitacin A on mobility of *Meloidogyne incognita* second-stage juveniles. *Research on Crops* 19 (3) (2018):504–508.

Dube, Z. P., Mashela, P. W. and D. D. Waele. Sensitivity of *Meloidogyne incognita* second-stage juvenile hatch, motility and viability to pure cucurbitacins and cucurbitacin phytonematicides. *South African Journal of Plant and Soil* 36 (1) (2019):29–32.

Fassuliotis, G. Resistance of *Cucumis* spp. to the root-knot nematode, *Meloidogyne incognita acrita*. *Journal of Nematology* 2 (1970):174–177.

Gobbo-Neto, L. and N. P. Lopes. Medicinal plants: Factors of influence on the content of secondary metabolites. *Quimica Nova* 30 (2007):374–381.

Gry, J., Søborg I. and A. H. Christer. *Cucurbitacins in plant food*. Copenhagen: Nordic Council of Ministers, 2006.

Hida, H., Nishiyamad, H., Sawad, S., Higashiyama, T. and H. Arata. Chemotaxis assay of plant-parasitic nematodes on a gel-filled micro-channel device. *Sensors and Actuators B* 221 (2015):1483–1491.

Hosmand, A. R. *Statistical methods for environmental and agricultural sciences*. Boca Raton, FL: CRC Press, 1998.

Hough, A. and I. J. Thomason. Effects of aldicarb on the behavior of *Heterodera schachtii* and *Meloidogyne javanica*. *Journal of Nematology* 7 (1975):221–229.

Ibrahim, S. K., Traboulsi A. F. and S. El-Haj. Effect of essential oils and plant extracts on hatching, migration and mortality of *Meloidogyne incognita*. *Phytopathologia Mediterranea* 45 (2006):238–246.

Isman, M. B. Neem and other botanical insecticides: Barriers to commercialization. *Phytoparasitica* 25 (1997):339–344.

Javed, N., Gowen, S. R., El-Hassan, S. A., Inam-uL-Haq, M., Shahina, F. and B. Pembroke. Efficacy of neem (*Azadirachta indica*) formulations on biology of root-knot nematodes (*Meloidogyne javanica*) on tomato. *Crop Protection* 27 (2008):36–43.

Javed, N., Gowen, S. R., Inam-Ul-Haq, M., Abdullah, K. and F. Shahina. Systemic and persistent effect of neem (*Azadirachta indica*) formulations against root-knot nematodes, *Meloidogyne javanica* and their storage life. *Crop Protection* 26 (2007):911–916.

Jeffrey, C. A review of *Cucurbitaceae*. *Botany Journal of Linnaeus Society* 81 (3) (1980):233–247.

Kader, A. A. Standardization and inspection of fresh fruits and vegetables. In *Postharvest technology of horticultural crops*, edited by A. A. Kader, 122–130. Oakland, CA: University of California ANR Publication, 2002.

Kader, A. A. Increasing food availability by reducing postharvest losses of fresh produce. *Acta Horticulturae* 682 (2005):2169–2175.

Kalaiselvam, I. and A. Devaraj. Effect of root exudates of *Tagetes* sp. On egg hatching behavior of *Meloidogyne incognita*. *International Research Journal of Pharmacy* 2 (2011):93–96.

Kaushik, U., Aeri, V. and S. R. Mir. Cucurbitacins–An insight into medicinal leads from nature. *Pharmacognosy Reviews* 9 (2015):12–18.

Krieger, R. *Handbook of pesticides toxicology*. San Diego, CA: Academic Press, 2001.

Kristkova, E., Lebeda, A., Vinter, V. and O. Blahousek. Genetic resources of genus *Cucumis* and their morphological description. *HortScience* 30 (2003):14–42.

Lebea, M. P. *Mean concentration stimulation point for Nemarioc-AL and Nemafric-BL phytonematicides on Cucurbita pepo cv. 'Canserta'*. MSc Dissertation, University of Limpopo, Sovenga, South Africa, 2017.

Lee, D. H., Lwanski, G. B. and N. H. Thoennissen. Cucurbitacin: Ancient compound shedding new light on cancer treatment. *The Scientific World Journal* 10 (2010):413–418.

Liu, D. L., An, M., Johnson, I. R. and J. V. Lovett. Mathematical modeling of allellopathy. III. A model for curve-fitting allellochemical dose responses. *Non-linearity Biology, Toxicology and Medicine* 1 (2003):37–50.

Lusia, B. M., Hasmadi, M., Zaleha, A. Z. and F. A. B. Mohd. Effect of different drying methods on phytochemicals and antioxidant properties of unfermented and fermented teas from Sabah snake grass (*Clinacanthus nutans* Lind.) leaves. *International Food Research Journal* 22 (2015):661–670.

Maake, M. V. *Potential cucurbitacin chemical residues in tomato fruits post-managing Meloidogyne species treated with Nemarioc-AL and Nemafric-BL phytonematicides*. MSc Dissertation, University of Limpopo, Sovenga, South Africa, 2017.

Mabhaudhi, T., Chirabada, T. and T. Modi. Water-food-nutrition-health nexus: Linking water to improving food nutrition and health in sub-Saharan Africa. *International Journal of Environmental Research and Public Health* 13 (107) (2016). doi: 10.3390/ijerph13010107

Mafeo, T. P. *Responses of economically important crops to crude extracts of Cucumis myriocarpus fruit when used as a pre-emergent bio-nematicide*. PhD thesis, University of Limpopo, Sovenga, South Africa, 2012.

Mafeo, T. P. and P. W. Mashela. Allelopathic inhibition of seedling emergence in dicotyledonous crops by *Cucumis bio-nematicide*. *African Journal of Biotechnology* 9 (2010):8349–8354.

Mafeo, T. P., Mashela, P. W. and M. S. Mphosi. Sensitivity of selected Alliaceae seedlings to crude extracts of *Cucumis myriocarpus* fruits. *African Journal of Agricultural Research* 6 (2011a):158–164.

Mafeo, T. P., Mashela, P. W., Mphosi, M. S. and K. M. Pofu. Modelling responses of maize, millet and sorghum seedlings to crude extracts of *Cucumis myriocarpus* fruit as pre-emergent bio-nematicide. *African Journal of Agricultural Research* 6 (2011b):3678–3684.

Maile, K. D., Mashela, P. W. and P. E. Tseke. Responses of the citrus nematode to a phytonematicide Nemarioc-AG with and without micro-organisms in citrus production. *African Crop Science Conference Proceedings* 11 (2013):333–337.

Maleka, K. G. *Aphid-nematode interaction effects on growth and sucrose production in nematode-resistant sweet stem (Sorghum bicolor) and the sustainable management strategies against both pests*. PhD thesis, University of Limpopo, Sovenga, South Africa, 2020.

Marshall, C. J. Protein prenylation: A mediator of protein-protein interactions. *Science* 259 (5103) (1993):1865–1866.

Masenya, T. A. *Nodulation bacteria, cucurbitacin-containing phytonematicides, dosage model and nutritional water productivity of Sutherlandia frutescens in the context of climate-smart agriculture*. PhD thesis, University of Limpopo, Sovenga, South Africa, 2022.

Mashela, P. W. Ground wild cucumber fruits suppress numbers of *Meloidogyne incognita* on tomato in microplots. *Nematropica* 32 (2002):13–19.

Mashela, P. W. Soil allelochemical residue effects in a tomato-cowpea rotation–Nodulation and productivity of cowpea and nematode suppression. *Acta Agriculturae Scandinavica, Section B–Soil and Plant Science* 64 (2014):372–375.

Mashela, P. W., De Waele, D., Z. Dube, Z., Khosa, M. C., Pofu, K. M., Daneel, M. S., Tefo, G. and D. Fourie. Alternative nematode management strategies. In *Nematology in South Africa: A view from the 21st century*, edited by H. Fourie, V. W. Spaull, R. K. Jones, M. S. Daneel and D. De Waele, 151–181. Cham: Springer Nature, 2017a.

Mashela, P. W. and Z. P. Dube. Soil allelochemical residue effects of Nemafric-BL and Nemarioc-AL phytonematicides on soil health, growth of sweet sorghum and *Meloidogyne* species. *Acta Agriculturae Scandinivica Section B: Soil & Plant Science* 64 (2014):79–84.

Mashela, P. W., Dube, Z. P. and K. M. Pofu. Managing the phytotoxicity and inconsistent nematode suppression in soil amended with phytonematicides. In *Organic amendments and soil suppressiveness in plant disease management, soil biology 46*, edited by M. K. Meghvansi and A. Vorma, 147–173. Cham: Springer International Publishers, 2015.

Mashela, P. W., Ndhlala, A. R., Pofu, M. and Z. Dube. Phytochemicals of nematode-resistant transgenic plants. In *Transgenesis and secondary metabolism* (Reference Series in Phytochemistry), edited by S. Jha, 1–16. Cham: Springer Nature, 2017b.

Mashela, P. W., Pofu, K. M. and M. P. Bopape-Mabapa. Efficacy of seed priming with cucurbitacin phytonematicides against *Meloidogyne incognita* on pea. *Frontier in Microbiology* 10 (13) (2022):863808. doi: 10.3389/fmicb.2022.863808

Mashela, P. W., Pofu, K. M. and J. T. Madaure. Responses of the citrus nematode population densities post-application of Nemarioc-AL phytonematicide and aldicarb synthetic nematicide. *Research on Crops* 18 (3) (2017c):523–527.

Mashela, P. W., Pofu, K. M. and E. Shokoohi. Biological sterilization, detoxification and stimulation of cucurbitacin-containing manure. In *Biology of composts*, edited by M. K. Meghvansi and A. Varma, 31–47. Cham: Springer International, 2020c.

Mashela, P. W., Shimelis, H. A. and F. N. Mudau. Comparison of the efficacy of ground wild cucumber fruits, aldicarb and fenamiphos on suppression of the root-knot nematode in tomato. *Journal of Phytopathology* 156 (2008):264–267.

Mashela, P. W. and E. Shokoohi. Morphometric and total protein responses in *Meloidogyne incognita* second-stage juveniles to Nemafric-BL phytonematicide. *Scientific Report* 11 (2021):1135–1147.

Mashela, P. W., Shokoohi, E. and K. M. Pofu. Morphological adjustment in free-living *Steinernema feltiae* infective juveniles to increasing concentration of Nemafric-BL phytonematicide. *PLoS One* 15 (1) (2020a):e0227448.

Mashela, P. W., Shokoohi, E. and K. M. Pofu. Shelf-life in cucurbitacin phytonematicides: Non-conformity to Arrhenius model. *PLoS One* 15 (2) (2020b):e0227959.

Mashela, T. S., Shadung, K. G., Mafeo, T. P. and P. W. Mashela. Non-phytotoxic concentration of cucurbitacin containing phytonematicides on Swiss chard (*Beta vulgaris* var. Cicla) under greenhouse conditions. *Research on Crops* 22 (2) (2021):327–333.

Mathabatha, R. V., Mashela, P. W. and M. N. Mokgalong. Non-phytotoxic concentration of cucurbitacin-containing phytonematicides and the overall sensitivities to Swingle citrumelo seedling rootstock. *Research on Crops* 18 (3) (2017):518–522.

Matsuura, T., Endo, S., Iwamoto, R., Takahashi, H. and M. Ichinose. Developmental changes in chemotactic response and choice of two attractants, sodium acetate and diacetyl, in the nematode *Caenorhabditis elegans*. *Comparative Biochemistry and Physiology* 147 (2007):920–927.

McSorley, R. Adaptations of nematodes to environmental extremes. *Florida Entomologist* 86 (2003):138–142.

Mendes, C. A. C., Mendes, G. E., Cipullo, J. P. and E. A. Burdmann. Acute intoxication due to ingestion of vegetables contaminated with aldicarb. *Clinical Toxicology (Phila)* 43 (20) (2005):117–118.

Mirr, S. A., Mukherjee, S., Makar, S. and G. Pal. Cucurbitacins a vibrant triterpenoid: A review on its anticancer property. *PharmaTutor* 7 (2) (2019):43–54.

Moco, S., Capanoglu, E., Tikunov, Y., Bino, R. J., Boyacioglu, D., Hall, R. D., Vervoort, J. and R. C. H. De Vos. Tissue specialization at the metabolite level is perceived during the development of tomato fruit. *Journal of Experimental Botany* 58 (2007):4131–4146.

Mphahlele, R., Mashela, P. W. and K. M. Pofu. Post-harvest fruit decay-inducing pathogen in medicinally important *Cucumis* species indigenous to South Africa. *African Journal of Agricultural Research* 6 (26) (2012):3786–3791.

Mudau, F. N., Mogotlane, I. D., Mashela, P. W. and P. Soundy. Seasonal variation of total antioxidant contents of wild bush tea. *Acta Horticulturae* 802 (2008):273–276.

Muedi, H. T. H., Mashela, P. W. and M. C. Mathabe. Bioactivity of wild cucumber fruit extracts on plant-parasitic nematodes. *Proceedings of Nematological Society of Southern Africa* 17 (2005):27.

Mukesh, D. and S. Sobita. Efficacy of certain botanical extracts in the management of *Meloidogyne graminicola* of rice. *International Journal of Agricultural Science and Research* 3 (2013):91–98.

Mulder, C., Schouten, A. J., Hund-Rinke, K. and A. M. Breure. The use of nematodes in ecological soil classification and assessment concepts. *Ecotoxicology and Environmental Safety* 62 (2005):278–289.

Müller, J. and A. Heindl. Drying of medicinal plants. In M*edicinal and aromatic plants*, edited by R. J. Bogers, L. E. Craker and D. Lange, 237–252. Dordrecht: Springer, 2006.

Mustika, I. and A. R. Slamet. *In efficacy of clove products and other botanical plants against nematodes attacking black pepper, conference on results of the experiments in order to utilize botanical pesticides*. Bogor: Research Institute for Spice and Medicinal Crops, 1994.

Nelaballe, V. K. and L. D. Mukkara. A preliminary study on the nematicidal effect of some local flora on *Meloidogyne incognita* Chitwood infesting Mulberry. *International Journal of Chemical, Environmental and Biological Sciences* 1 (2013):475–477.

Nhlane, R. N. Response of '*Citrullus lanatus*' 'Charleston Gray' and 'Congo' cultivars grafted on *Cucumis africanus* and *Cucumis myriocarpus* to *Meloidogyne incognita* race 2. Master of Agricultural Management dissertation, University of Limpopo, Sovenga, Republic of South Africa, 2017.

Niazi, S. K. *Handbook of pharmaceutical manufacturing formulations*. New York: Informa Healthcare, 2009.

Noling, J. W. Movement and toxicity of nematicides in the plant root zone (2019). htts://edis.ifas.ufl.edu (Accessed June 15, 2022).

Norris, R. *Handbook of bioremediation*. Orlando, FL: CRC Press, 1993.

Novelli, G. and M. R. D'Apice. Protein farneslation and disease. *Journal of Inherited Metabolic Disease* 35 (5) (2012):917–926.

Nyasha, N. E. Influence of pre-infectional and post-infectional nematode resistance mechanisms in crop rotation sequences on population densities of *Meloidogyne* species and soil health. PhD thesis, University of Limpopo, Sovenga, Republic of South Africa, 2021.

Oka, Y., Ben-Daniel, B. and Y. Cohen. Nematicidal activity of powder and extracts of *Inula viscose*. *Nematologica* 3 (2001):735–742.

Oka, Y., Nacar, S., Putievsky, E., Ravid, U., Yaniv, Z. and Y. Spiegel. Nematicidal activity of essential oils and their components against the root-knot nematode. *Phytopathology* 90 (2000):710–715.

Orwa, J. A., Keter, L. K., Ouko, S. P., Kibwage, I. O. and G. M. Rukunga. Influence of manufacturing practices on quality of pharmaceutical products manufactured in Kenya. *Journal of East African Medicine* 81 (2004):287–292.

Oztekin, S. and M. Martinov. *Medicinal and aromatic crops: Harvesting, drying and processing*. New York: Haworth Food Agricultural Products Press, 2007.

Pavaraj, M., Bakavathiappan, G. and S. Baskaran. Evaluation of some plant extracts for their nematicidal properties against root-knot nematode, *Meloidogyne incognita*. *Journal of Biopesticides* 5 (2012):106–110.

Pelinganga, O. M. and P. W. Mashela. Mean dosage stimulation range of allelochemicals from crude extracts of *Cucumis africanus* fruit for improved growth of tomato plant and suppressing *Meloidogyne incognita* numbers. *Journal of Agricultural Science* 12 (2012):8–12.

Pelinganga, O. M., Mashela, P. W., Mphosi, M. S., Mafeo, T. P. and Z. P. Dube. Using density-dependent growth patterns of tomato plants to establish application intervals for 3% Nemarioc-A phytonematicide. *African Crop Science Conference Proceedings* 11 (2013):343–347.

Pelinganga, O. M., Nzanza, B., Mamphiswana, N. and P. W. Mashela. Influence of fermented fruit extracts of *Cucumis africanus* and *Cucumis myriocarpus* on nematode numbers and tomato productivity. *Symposium of Nematological Society of Southern Africa* 20 (2011):71.

Pérez, M. P., Navas-Cortés, J. A., Pascual-Villalobos, M. J. and P. Castillo. Nematicidal activity of essential oils and organic amendments from Asteraceae against root-knot nematodes. *Plant Pathology* 52 (2003):395–401.

Perry, R. N. and M. Moens. *Molecular and physiological basis of nematode survival*. Wallingford: CAB International, 2011.

Pofu, K. M. *Potential uses of indigenous Cucumis africanus and Cucumis myriocarpus as root-knot nematode resistant rootstocks in watermelon (Citrullus lanatus) husbandry*. PhD thesis, University of Limpopo, Sovenga, South Africa, 2012.

Pofu, K. M. and P. W. Mashela. Comparative efficacy of cucurbitacin phytonematicides and Velum on growth and fruit quality of watermelon cultivar 'Congo' and suppression of *Meloidogyne enterolobii* under field conditions. *Acta Agriculturae Scandinavica, Section B: Soil & Plant* 72 (1) (2021):105–110.

Pofu, K. M., Mashela, P. W. and H. Shimelis. Host-status and host-sensitivity of wild *Cucumis* species to *Meloidogyne incognita* race 4. *Acta Agriculturae Scandinavica, Section B: Soil & Plant* 62 (4) (2012):329–334.

Prot, J. C. Migration of plant-parasitic nematodes towards plant roots. *Revue de Nematologie* 3 (1980):305–318.

Ramputla, M. J. *Nutritional water productivity of chilli (Capsicum annuum) infected with Meloidogyne incognita.* MSc dissertation, University of Limpopo, Sovenga, South Africa, 2019.

Rasmann, S., Ali, J. G., Helder, J. and W. H. Van Der Putten. Ecology and evolution of soil nematode chemotaxis. *Journal of Chemical Ecology* 38 (2012):615–628.

Reynolds, A. M., Dutta, T. K., Curtis, R. H. C., Powers, S. J., Gaur, H. S. and B. R. Kerry. Chemotaxis can take plant-parasitic nematodes to the source of a chemo-attractant via the shortest possible routes. *Journal of the Royal Society Interface* 8 (2010):568–577.

Schabort, J. C. and H. L. Teijema. The role of cucurbitacin Δ^{23} reductase in the breakdown pathway of toxic bitter principles in *Cucurbita maxima. Phytochemistry* 7 (1968):2107–2110.

Sebothoma, C. *Development of non-phytotoxic concentration and application interval of Nemarioc-AL phytonematicide in management of Meloidogyne javanica on sweet potato cultivar 'Bophelo'.* MSc dissertation, University of Limpopo, Sovenga, Republic of South Africa, 2018.

Seinhorst, J. W. The relationship between nematode density and damage to plants. *Nematologica* 11 (1965):137–154.

Selomo, M. D. *Development of non-phytotoxic concentration and application interval of Nemafric-BL phytonematicide in management of Meloidogyne incognita race 2 on sweet potato cultivar 'Bophelo'.* MSc dissertation, University of Limpopo, Sovenga, Republic of South Africa, 2019.

Seshweni, M. D. *Integrated system for the management of population densities of Meloidogyne javanica in potato production.* Master of Agricultural Management dissertation. University of Limpopo, Sovenga, South Africa, 2016.

Shadung, K. G. *Quality protocols for Nemarioc-AL and Nemafric-BL phytonematicides.* PhD thesis, University of Limpopo, Sovenga, South Africa, 2016.

Shadung, K. G. and P. W. Mashela. Storage dynamics of cucurbitacin A in phyto-inventories of Nemarioc-AL phytonematicide. *Transylvanian Review* 7 (2016):962–967.

Shadung, K. G., Mashela, P. W. and M. S. Mphosi. Response of cucurbitacin B concentration in Nemafric-BL phytonematicide to increasing storage period. *Journal of Stored Products and Postharvest Research* 7 (2016):32–36.

Shadung, K. G., Mashela, P. W., Mphosi, M. S. and V. L. Mulaudzi. Study of chemical residues from Nemafric-AL and Nemafric-BL phytonematicides in tomato fruit. *African Agricultural Research* 68 (2017):1164–1168.

Shaik, R. S., Weston, L. A., Burrows, G. and D. A. Gopurenko. *Comparative phenological and genetic diversity analysis of two invasive weeds, camel melon (Citrullus lanatus (Thunb.) matsum. and nakai var. lanatus) and prickly paddy melon (Cucumis myriocarpus L.) in inland Australia.* The Sebel Cairns, Australia: 23rd Asian-Pacific Weed Science Society Conference, 2011.

Sharma, G. P. and S. Prasad. Drying of garlic *(Allium sativum)* cloves by microwave hot air combination. *Journal of Food Engineering* 50 (2003):99–105.

Sithole, N. T., Pofu, K. M. and P. W. Mashela. Overall sensitivity of *Pelargonium sidoides* and root-knot nematodes to Nemarioc-AL phytonematicide. *Transylvanian Review* 24 (2016):2996–3000.

Skantar, A. M., Agama, K., Meyer. S. L. F., Carta, L. K. and B. T. Vinyard. Effects of geldanamycin on hatching and juvenile motility in *Caenorhabditis elegans* and *Heterodera glycines. Journal of Chemical Ecology* 31 (2005):2481–2491.

Stirling, G. R. *Biological control of plant parasitic nematodes: Soil ecosystem management in sustainable agriculture.* Wallingford: CAB International, 2014.

Taye, W., Sakhuja, P. K. and T. Tefera. Root-knot nematode (*Meloidogyne incognita*) management using botanicals in tomato (*Lycopersicon esculentum*). *Academia Journal of Agricultural Research* 1 (2013):9–16.

Thies, J. A. and A. Levi. Characterization of watermelon (*Citullus lanatus* var. citroides) germplasm for resistance to root-knot nematodes. *HortScience* 42 (7) (2007):1530–1533.

Thopola, T. E. *Influence of priming cultivar 'Mondial 3G'seeds in 3% Nemarioc-AL, 3% Nemafric-BL and 3% Mormofly phytoinsecticides on plant growth and suppression of Meloidogyne javanica.* Master of Agricultural Management dissertation. University of Limpopo, Sovenga, South Africa, 2019.

Tseke, P. E. and P. W. Mashela. Influence of arbuscular mycorrhiza fungi on growth of *Cucumis myriocarpus* indigenous leafy vegetable. *Research on Crops* 19 (2) (2018):289–293.

Van Gundy, S. D. and M. V. McKenry. *Action of nematicides*, edited byJ. G., and E. B. Cowling, 263–283. New York: Academic Press, 1977.

Van Wyk, B. E. and M. Wink. *Medicinal plants of the world.* Pretoria: Briza Publications, 2014.

Windham, G. L. and W. P. Williams. Reproduction of *Meloidogyne javanica* on corn hybrids and inbreds. *Annals of Applied Nematology* 2 (1988):25–28.

Wuyts, N., Swennen, R. and D. De Waele. Effects of plant phenylpropanoid pathway products and selected terpenoids and alkaloids on the behavior of the plant-parasitic nematodes *Radopholus similis*, *Pratylenchus penetrans* and *Meloidogyne incognita*. *Nematology* 8 (2006):89–101.

Zhao, C., Ma, C., Luo, J., Niu, L., Hua, H., Zhang, S. and J. Cui. Potential of cucurbitacin B and epigallocachin gallate as biopesticides against *Aphis gossypii*. *Insects* 12 (1) (2021):32. https://doi.org/10.3390/insects12010032

Zheng, L. and H. Ferris. Four types of dormancy exhibited by eggs of *Heterodera schachtii*. *Revue de Nematologie* 14 (1991):419–426.

Zhou, Y, Ma, Y., Zeng, J., Duan, L., Xue, X., Wang, H., Lin, T., Liu, Z., Zeng, K., Zhong, Y., Zhang, S., Hu, Q., Liu, M., Zhang, H., Reed, J., Moses, T., Liu, X., Huang, P., Qing, Z., Liu, X., Tu, P., Kuang,H., Zhang, Z., Osbourn, A., Ro, D., Shang, T. and S. Huang. Convergence and divergence of cucurbitacin biosynthesis and regulation in Cucurbitaceae. *Nature Plants* 2 (2016):16183–16202.

3 Mustard Oilseed Cake
Chemical Compounds and Nematicidal Potential

Amirhossein Abedini, Adel Mirza Alizadeh, Hedayat Hosseini and Parisa Sadighara

CONTENTS

3.1 INTRODUCTION

The by-products of the food and agricultural industries are among the high-use products, one of which is oilseed cakes (OCs), which contain a variety of nutrients for human and animal nutrition, as well as functional components that bring uniqueness to food quality. Mustard seed cake, flaxseed cake, sunflower seed cake, pomegranate seed cake, sesame seed cake, canola seed cake, almond seed cake, palm kernel cake, peanut cake, olive cake, and pistachio cake are all examples of oil cakes. There are two types of OCs produced by the food industry: edible and non-edible, with worldwide output reaching 317,000,000 tons in 2016 and estimated to be 386,000,000 tons by 2025. To obtain excellent yield and maximum utilization of OC potential, the nutritional value must be increased while anti-nutrients and toxins must be reduced (OECD/FAO 2020; Abedini et al. 2022b).

OCs include an abundance of protein, energy, carbohydrates, and minerals. The oilseed cake has been proved to boost the weight of young calves (Singh et al. 2022). The oilseeds include between 15–50 percent protein and several fatty acids and other nutrients. As a result, next to grains, they represent the second-largest provider of food stores worldwide. Also, the content of desirable antioxidant, fiber, and proteins compounds in oilseed cake makes it a suitable source of nutritious food for cutting down expenses, creating economic uplift, and enhancing the agricultural and food business (Zhao et al. 2012; Abedini et al. 2022b). Uses of OC in the food industry are shown in Figure 3.1 (Ayrilmis et al. 2013; Purnawan et al. 2017).

The cake might be used as a solution for the growth of microorganisms, which can be employed to produce nutrients and even nutritious foods like protein bars (Milani et al. 2017; Bello et al. 2018). According to some research, the high nutritional content of OCs may improve the gelatinization, emulsion, and taste aspects of designed meals. Consequently, it was suggested that the cake might be utilized as a dietary supplement to enhance the quality of meals (Sunil et al. 2015; Gültekin Subaşı et al. 2021; Rani and Badwaik 2021). When treating some disorders, like celiac disease, substituting

DOI: 10.1201/9781003319252-3

FIGURE 3.1 Oilseed cake applications in food sciences.

certain cakes, such as mustard oilseed cake (MSC), for wheat-carrying gliadin or rye having prolamine and with barley or oats may offer an innovative treatment option that is cost effective and goal targeted. Many researchers have suggested repurposing and replacing gluten-free cereal products, like donuts, cakes, biscuits, bread, and cookies (Karimi et al. 2018).

Also, OC can be used in clinical trials such as managing and reducing nematodes. Due to economic issues, using OC in clinical studies can also help control economic problems. We know that chemical substances used to kill pests, parasites, and nematodes generally harm human health. Using OCs and other natural by-products allows us to reduce the harmful effects of these substances on human health (Tiyagi et al. 2011; Sadighara et al. 2022). One of these cakes is MSC. Nematodes reduce agricultural output and therefore increase food production costs. For example, studies have shown that due to the growth of nematodes, tomato production yields in India have decreased by 27.24 percent (Nilam and Patel 2019). Other agriculture products like yam (*Dioscorea* spp.) are affected by *Scutellonema bradys* (yam nematode), *Pratylenchus sudanensis*, *Meloidogyne arenaria*, *Meloidogyne enterolobii*, *Meloidogyne incognita*, and *Meloidogyne javanica*. Maize is affected by *Pratylenchus brachyurus*, *Pratylenchus zeae*, *Pratylenchus delattrei*, *Pratylenchus hexincisus*, *Pratylenchus penetrans*, *Pratylenchus sefaensis*, and *Pratylenchus scribneri*. Bananas and plantains (*Musa* Spp.) are affected by *Radopholus similis* (burrowing nematode), *Hoplolaimus pararobustus* (lance nematode), and *Rotylenchulus reniformis* (reniform nematode). Potatoes are affected by *Meloidogyne* species, such as *Meloidogyne hapla, Meloidogyne incognita*, *Meloidogyne javanica*, *Ditylenchus* spp., *Paratrichodorus* spp., *Scutellonema bradys*, and *Trichodorus* spp (Coyne et al. 2018). For this reason, crop losses due to nematodes are estimated from US$80 billion to US$157 billion per year. Compared to pests, nematode crop losses (US$10 billion) are higher (US$6.6 billion; Nicol et al. 2011; Coyne et al. 2018).

In examining the amino acid profile of MSC, the amino acid with the highest amount in mustard cake is threonine, and the one with the lowest amount is histidine (Abedini et al. 2022b). Also, its antioxidant compounds are sinapic acid, ferulic acid, caffeic acid, and p-Coumaric acid (Ancuța and Sonia 2020). In addition, secondary metabolites of this cake are gluconapine, sinapine, 3-methylsulfinyl propyl isothiocyanate, 1-methoxybrassitin, rutalexin, 5-methylthiopentylglucosinolate, methyl oleate, and other compounds that can be effective in reducing nematodes (Nadeem et al. 2021). These compounds are more effective when efficient methods are used; for example, ultrasonic applications can lead to better extraction of these materials (Bhargava et al. 2021). MSC has nematicidal potential against plant-parasitic nematodes (*M. incognita, Rotylenchulus reniformis, Tylenchorhynchus brasscae*, *Helicotylenchus indicus*), *Heterodera cajani*, *Meloidogyne incognita*, *Rotylenchulus reniformis*, and *Meloidogyne javanica* (Sumbul et al. 2015). For example, Nadeem et al. (2021) investigated the effects of MSC on reducing the nematode *Meloidogyne javanica*. They concluded that ultrasonic applications can be used to extract effective compounds from MSC and use them to reduce nematodes. This effect on nematode reduction will be much greater with ultrasonic applications (Nadeem et al. 2021). This chapter aims to focus on the significance of MSC and investigate its uses against nematodes.

3.2 MSC AND PHYSICOCHEMICAL PROPERTIES

Mustard seeds are the small, spherical seeds of three plant species: black mustard (*Brassica nigra*), brown Chinese or Indian mustard (*Brassica juncea*), and white mustard (*Sinapis alba* L.). In traditional folk medicine, mustards have been used as a stimulants, diuretics, and purgatives to treat a variety of ailments, including peritonitis and neuralgia (Chan et al. 2014).

Yellow or white mustard seeds are used as seasonings. White mustard mucilage's neutral component is mostly made up of glucose (39.3 percent), arabinose (25.4 percent), galactose (17.9 percent), mannose (5.4 percent), rhamnose (4.0 percent), and xylose (7.0 percent) (w/w), according to previous research. The water-soluble component of yellow mustard mucilage is composed of glucose (22.3 percent), galactose (15.2 percent), mannose (6.3 percent), rhamnose (3.9 percent), arabinose (3.2 percent), xylose (1.8 percent), and uronic acid (18.7 percent) (w/w). Both pectic and non-pectic polysaccharides make up its structure. Most pectic polysaccharides have the rhamnogalacturonan-II structure, which comprises 1,4- and 1,2-linked D- and L-rhamnopyranosyl residues. Non-pectic polysaccharides contain a D-glucosidic backbone that is 1,4-linked and only sometimes has branches at the O-2, O-3, and O-6 positions. The synergistic interactions of galactoglucomannans are due to the non-pectic portion. Water-soluble yellow mustard mucilage produces weak gels or viscous solutions depending on the concentration. It may be used as a stabilizer and thickener and has a high surface activity (Cui et al. 2013). Figure 3.2 shows that the crude protein and fiber levels in yellow and black mustards ranged from 28.8 to 38.17 and 14.8 to 12.17, respectively (Sarker et al. 2015). The minerals included in mustard seeds include iron, magnesium, zinc, calcium, and phosphorus. A mineral deficit, which may have potentially severe side effects, can be avoided by consuming enough of these minerals, which are required for the regular operation of several biological and biochemical processes in the body. As shown in Figure 3.3, according to the United States Department of Agriculture (USDA) National Nutrient Database for Standard Reference, pulverized mustard seed contains varying proportions of essential minerals (Divakaran and Babu 2016).

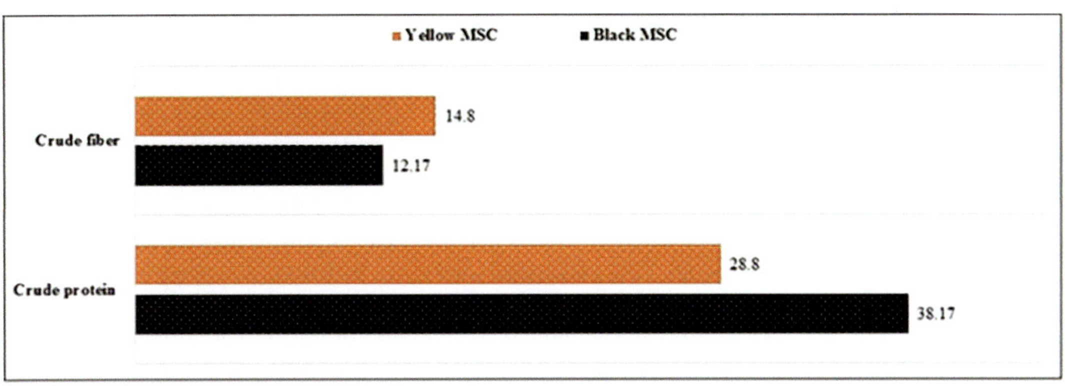

FIGURE 3.2 Chemical compositions of mustard oilseed cakes.

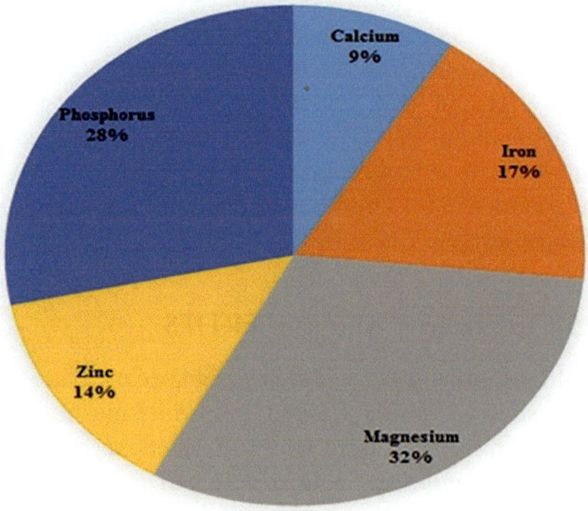

FIGURE 3.3 Various minerals in ground mustard seeds.

As demonstrated in Figure 3.4, using the dry matter in black and yellow mustards as a basis, histidine has the lowest amino acid level (0.43 and 0.90, respectively). Furthermore, threonine had the greatest quantity of amino acid in black and yellow mustards (19.17 and 14.31 percent, respectively). Sinapic acid, ferulic acid, caffeic acid, and p-Coumaric acid are all antioxidants found in mustard seeds (Teh et al. 2014; Sarker et al. 2015; Teh and Bekhit 2015). The mustard powder does not have a lingering scent. However, taste and pungency are detected when water is present and an enzymatic response releases the flavor and pungency of mustard. As with brown and black mustard, yellow mustard flavor and pungency may only be completely experienced by activating the enzyme myrosinase, which releases them. It is caused by a range of isothiocyanate chemicals that occur as glycosides in mustard tissue. Allyl isothiocyanate (AITC) is the principal pungent component of brown and black mustard. There is a myrosinase enzyme in brown and black mustards; after consumption, a burning sensation is caused due to the activity of the enzyme. The myrosinase enzyme transforms the glycoside (sinal-bin) in yellow mustard or the sinigrin in black or brown mustard into para-hydroxybenzyl isothiocyanate, which gives mustard its particular pungent smell. The odors persist until the enzymes stop functioning (Thomas et al. 2012). Mustard flour serves as a preservative and antioxidant, giving flavor and color. The most essential quality of the spice in salad dressings is its emulsifying ability, which can bind lipophilic and hydrophilic phases; this creates viscosity. The fixed oil in mustard, which accounts for 30–35 percent of its weight, is removed by cold pressing. In India and other Asian nations, oil is utilized in cooking. The scent of mustard oil is derived from the sulfur-containing essential oils that are generated during the hydrolysis of the glucosides in the seeds. The concentration of fatty acids and their proportion in mustard oil determine the quality of the oil. Due to the fact that it has a high AITC concentration, mustard oil is dangerous. The way to recover the essential oil is that when the myrosinase enzyme hydrolyzes the mustard oil (brown seed), AITC is separated from the glucoside, and the essential oil is recovered from the pressed cakes by steam distillation. Due to the press cake's low to nonexistent volatile oil content, the yellow mustard essential oil is extracted by solvent extraction. The press cake must first be hydrolyzed to liberate the end products created by enzyme activity. The press cake is macerated overheated water to produce hydrolysis. The three different types of mustard are often combined to create oleoresin of mustard seed, which imparts a balanced taste. Checking the oil's color has shown that this product's color is yellow to light brown with around 5 ml of volatile oil per 100 g. Mustard flour was the most efficient of the spices, condiments, and herbs tested regarding yeast fermentation

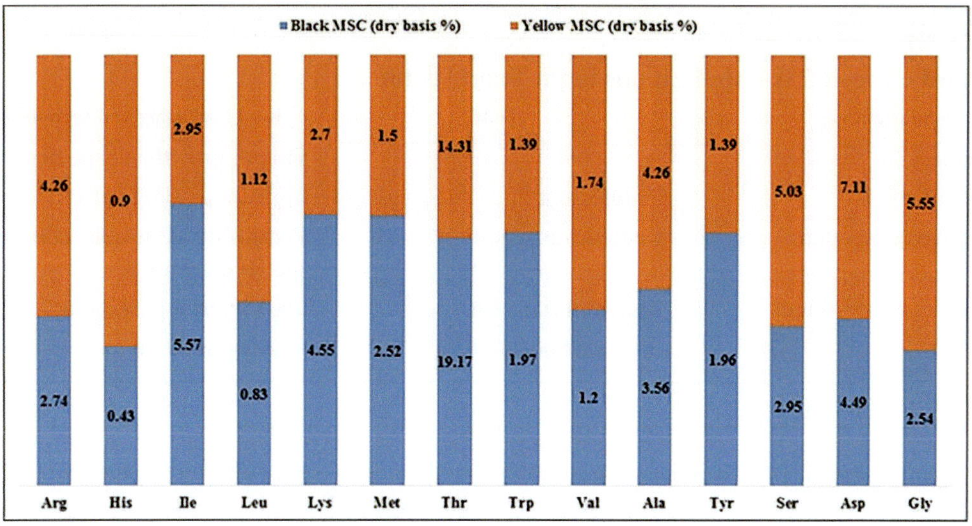

FIGURE 3.4 Amino acid profiles in types of oilseed cakes.

in vinegar and others. It was shown to be more effective than the chemical preservatives tested, sulfur dioxide and benzoic acid. Bactericidal characteristics are found in mustard and its component allyl isocyanate (Thomas et al. 2012).

The chemical AITC is the major pungency component of mustard seed. Isothiocyanates (R – N = C = S) are glucosinolate derivatives found in the cells of Cruciferae or mustard plants (cabbage, kohlrabi, brussels sprouts, cauliflower, broccoli, kale, horseradish, mustard, turnips, rutabaga). When plant tissue is wounded or physically disturbed, the enzyme myrosinase (thioglucoside glucohydrolase EC 3.2.3.1) acts on the glucosinolates, forming these chemicals. Other isothiocyanate side groups include ethyl, methyl, benzyl, and phenyl, in addition to the allyl side group. Antimicrobial activity has been documented for these substances (Davidson and Zivanovic 2003). According to Table 3.1, MSC has a variety of uses, including antimicrobial coatings, enzyme manufacturing, raw sugar decolorization, and antioxidant sources in foods (Singh et al. 2013; Szydlowska-Czerniak et al. 2015; Badhe et al. 2016; Chen and Liu 2016; Kumari et al. 2016). Some research looked at the production of enzymes by bacteria that feed on food waste. This work employed submerged fermentation (SMF) conditions to produce protease by the gram-positive, mesophilic bacterium *B. subtilis*. Because of MSC protein content, cost, and wide application, MSC was employed in this article. Numerous physical and chemical characteristics, including pH and temperature, were also studied. The highest protease activity was attained at pH 11 and 40°C with 4 percent MSC oil. (Badhe et al. 2016). Another study examined how antioxidants are extracted and used in meals and supplements. They observed that a cake without mustard oil had many antioxidants. As a result, it has been promoted for use in dietary supplements, animal nutrition, and food preservation (Kumari et al. 2016). Raw sugar might get discoloroed when exposed to active mustard cake powder. Singh et al. (2013) studied the efficiency of the physical characteristics, molasses, and sugar breakdown. They claim that, although the active powder of MSC has lower colour removal efficacy than chemicals, it presents a fresh approach in this field due to its affordability and environmental friendliness. It was also shown that active mustard cake powder may be utilized on an industrial scale to minimize activated carbon but that it had complicated interactions and reactions (Singh et al. 2013). Chen and Liu (2016) investigated the use of mustard oil and β-glucodextrin in the production of antibacterial films. They experimented with various MSC oil concentrations. Mustard oil enhanced water absorption, decreased Young's modulus, decreased elasticity, and increased elongation but had no effect on

TABLE 3.1
Mustard Oilseed Cake Applications in the Food Sector

Effective Substances	Food	Food Packaging and Technology
Enzyme	–	Fungal growth subspecies
Antioxidants	Supplement	high antioxidant activity
Bioactive compounds	Decoloring sugar	Possibility of discoloration in the food sector
Nutritional compounds	Antibacterial film	antibacterial packaging
	Food and non-food	antioxidant activity

water vapor permeability. Mustard oil has been shown to have good antibacterial properties against *E. coli* and *S. aureus* but not so much against *B. subtilis* and *A. niger*. In addition, they indicated that the application of β-cyclodextrin aids in the preservation of mustard oil for film production, as 7 ml MSC oil remained in the absence of β-cyclodextrin, while 11 mg remained in the film after the inclusion of β-cyclodextrin (Chen and Liu 2016). Breeding factors, including development, ripening, culture, and stressful situations, significantly impact the antioxidant levels in MSC. They used ultrasound to extract nutrients, suggesting that the resulting compounds may have industrial use. It's also important to note that its usage is limited by the presence of erucic acid (25–35%) and glucosinolates (250–350 mol/g defatted seed meal). Recent advances in genetic engineering have enabled the manufacture of mustard oil with higher quantities of antioxidants and fewer glucosinolates and erucic acids (Szydlowska-Czerniak et al. 2015).

MSC has typically been employed in the synthesis of enzymes under SMF conditions. Film production, sugar decolorization, and food enrichment have all been used. MSC is unique in its excellent antioxidant properties, followed by economic effectiveness and environmental friendliness. MSC's use is made safer and more dependable by reducing hazardous chemicals.

3.3 ALLERGENICITY OF MSC

Dietary allergy is characterized as an immune reaction to dietary proteins and has posed a risk to many people's safety and health. According to estimates, roughly 8 percent of children have food allergies, whereas about 5 percent of adults have food allergies. In addition, several studies indicate that in the last 20–30 years, the number of people with food allergies has grown. Studies have shown that small amounts of allergens in food can cause diseases. For example, there are examples of anaphylaxis, rashes, and hives (Chen et al. 2019; Pi et al. 2019; Palmer et al. 2020).

In addition, commonly consumed foods contain 90 percent of allergens in foods. For example, milk, soy, wheat, fish, and peanuts are among the reported examples, according to the Food and Agriculture Organization of the United Nations (FAO; Chen et al. 2019; Crevel et al. 2000). It should be noted that the consumption of animal food is effectively controlled in terms of the number of allergens because the remaining allergens and their transmission in the food cycle to humans can be harmful. The proposed solutions are effective monitoring of animal feed and food labeling. Customers, however, have lost faith in food labels since there is no correlation between cautious labeling and the existence of allergies. In New Zealand and Australia, the Voluntary Incidental Trace Allergen Labelling (VITAL) systems have set allergen limits based on clinical trials that protect 95–99 percent of allergic people (Planque et al. 2017; Madsen et al. 2020). Table 3.2 lists a variety of allergens found in MSC, along with their negative health consequences (FSA 2015; Pałgan et al.

TABLE 3.2
Probable Allergens in MSC

Allergen Compounds	Allergen Thresholds (mg of proteins)	Employment Criteria	Health Effects
• 2S albumin (Sin a 1) • 11S globulin (Sin a 2) • nsLTP (Sin a 3) • Profilin (Sin a 4) for yellow mustard • 2S albumin (Bra j 1) for oriental mustard	• VITAL: 0.05 • EAACI: 0.05 • NVWA: 0.022	This phrase refers to the mustard plant as well as various mustard-related products. Labeling should use the correct terminology. The guidelines are not specific to any mustard species and should be applied to all mustard varieties.	• Anaphylactic shock • Hoarseness • Dyspnea • Angioedema nausea • Vomiting • Chest tightness • Wheezing

2018; L'Hocine et al. 2019). In addition, there is a scarcity of information on allergy thresholds for OCs, which should be investigated further in future research.

3.4 ANTAGONISTIC PROPERTIES OF MSC

Livestock diets, processing processes, personal workplace hygiene, insecticides, good manufacturing practice (GAP) non compliance, and transport procedures have been identified as key contributors to the toxicity of cakes. Recent studies on the mycotoxin content of oil seed cakes revealed that the most prevalent mycotoxins include ochratoxin A, alternariol, and aflatoxin B1, B2, G1, and G2. The results of the mycotoxin production potential of fungi isolate obtained from stored mustard seeds revealed that *Aspergillus flavus*, *Fusarium moniliforme*, and *Penicillium citrinum* were present in all samples and that all of these fungi had the capability of producing numerous mycotoxins (which include Aflatoxin B1, B2 & G1, ZEA, and Cit) in stored mustard seeds (Ahmad and Sinha 2002). The presence of AFB1 and total AF in mustard seed samples revealed that 20 of the 34 imported samples (6.41 ± 2.7 µg/kg for AFB1 and 12.95 ± 3.5 µg/kg for total aflatoxin) were infected with mycotoxins. In addition, 23 out of 46 local samples tested were positive for mycotoxins (8.74 ± 4.5 µg/kg for AFB1 and 13.88 ± 4.6 µg/kg for total aflatoxin). Furthermore, the presence of these toxins in mustard oil samples revealed that 21 of 35 imported samples (5.69 ± 2.4 µg/kg for AFB1 and 11.56 ± 1.9 µg/kg for total aflatoxin) were infected with mycotoxins. In addition, 28 out of 50 local samples tested positive for mycotoxins (7.71 ± 1.8 µg/kg for AFB1 and 13.43 ± 4.6 µg/kg for total aflatoxin; Waqas et al. 2021). In research conducted by Sahay and Prasad (1990) on the incidence of aflatoxins in mustard seed and its derivatives, aflatoxins at concentrations of 750, 87, and 1420 µg/kg were found in 44, 33, and 80 samples out of 100 samples each of mustard seeds, oil, and cakes, respectively. The percentage incidence of the *Aspergillus flavus* group was highest during the monsoon and summer months among the 17 species of fungus isolated from mustard seeds. Twenty-five percent of isolates in the *A. flavus* group (out of 1143 isolates analyzed) were toxigenic, producing varied amounts of aflatoxins (Sahay and Prasad 1990). Sahay and Ranjan studied the dispersion pattern of aflatoxins from the oilseeds of mustard, groundnut, and linseed to their oil and cakes. In oilseed samples, the quantity of aflatoxin B1 produced by *A. flavus* was rather high. It was highest in peanuts (2200 ± 316.22 µg/kg), next in mustard (1839 ± 273.59 µg/kg), and last in linseed (1234 ± 216.28 µg/kg). When oil was extracted, the amounts of the toxin were 786 ± 159.31, 604 ± 96.33, and 420 ± 164.24 µg/kg, respectively. In all cases, leftover cake had a higher concentration of aflatoxin than oil (1122 ± 148.05, 1020 ± 257.29, and 650.5 ± 133.62 µg/kg, respectively; Sahay and Ranjan 1990; Karimi et al. 2019).

3.5 ANTI-NUTRITIONAL FACTORS IN MUSTARD OILSEED CAKE

Anti-nutrient factors are chemicals that negatively impact the digestion and absorption of both micronutrients and macronutrients, as well as the activities of the body's organs (Thakur et al. 2019; Diouf et al. 2019). The reuse of OCs in the food market draws attention to significant challenges, such as methods to lessen or do away with them and their impact on health. Table 3.3 displays the effects of anti-nutrient factors on people's health. α-galactosides, for example, are compounds that a human lacks the digestive enzyme to break down, and bacteria in the large intestine break down and absorb these chemicals there, resulting in flatulence, muscular cramps, and diarrhea (Sreerama et al. 2012). Tannins can attach to carbohydrates, minerals, and proteins, which causes the body to become deficient in these nutrients (Ba et al. 2010). Proteins, carbohydrates, calcium, magnesium, iron, and zinc are more difficult to absorb due to the presence of phytic acid, which also reduces their bioavailability (Abdulwaliyu et al. 2019). The antinutritional compounds inherent in oilseeds are enhanced during the extraction of oil and meal. Therefore, by altering the extraction procedure, these substances may be extracted in smaller quantities, enhancing the nutritious value of food. The methods used to decrease these materials are listed in Table 3.3. For example, the fermentation process creates enzymes used in the food industry and decreases the number of antinutritional factors (Adeyemo and Onilude 2013; Samtiya et al. 2020). Because there are many different kinds of oilseeds and because the food industry makes greater use of waste materials, the nutritional value and safety of these components are

TABLE 3.3
Anti-Nutritional Compounds, Their Impact on Health, and Techniques for Reducing Them

Anti-Nutritional Substances	Effect on the Human Body	Eliminating Procedures and Techniques
Sinapic acid (sinapine, methyl and ethyl sinapate)	Indigestibility and diminishes the flavor	• Extrusion-cooking HTST (high temperature, short time) • Chemical methods (thiols, sulfite, Cu salts) • Soaking procedures (water and salt solutions) • Pressure measurement (application of high temperature) • The blanching technique (application of temperature to reduce enzymatic activity) • Germination of seed (activation of enzymes such as phytase and reduction of phytic acid) • Fermentation process (remove phytic acid and metal complexes) • Enzyme-inhibiting agents (trypsin and chymotrypsin inhibitors) • Types of cooking methods
Erucic acid	Negative health consequences	
Tannis	Reduced protein digestion	
Glucosinolates	Pathologies of the thyroid gland	
Cyanogenic glycosides	Protein absorption disruption	
Trypsin inhibitor MTI-2	Reduce protein digestion	
Phytic acid	Absorption of carbs and proteins	
Oxalates	Development of renal calculi and crystals	
p-Hydroxybenzoic (alkyl esters)	Estrogen production	
Saponins	Reduce vitamin absorption and interfere with acid	
References	Gilani et al. 2005; Alireza et al. 2006; Samtiya et al. 2020; Pojić et al. 2015; EFSA Panel on Contaminants in the Food Chain 2016; Łopusiewicz et al. 2019; Ogidi et al. 2019; Hassan et al. 2020	Fekadu 2014; Diouf et al. 2019; Samtiya et al. 2020

extremely important. Especially when creating OCs, anti-nutrient–decreasing techniques should be employed in accordance with the kind of product. This is because oilseed meal contains a high concentration of anti-nutrients.

3.6 NEMATICIDAL ACTION OF MSC

Nematodes are present in food in different forms. They appear as eggs or second-stage juveniles on plant and soil parts. In terms of the environmental impact, nematodes also have some effects; they can be transported to different places by polluted water. Also, if the soil of other areas becomes infected with nematodes, it can cause the transfer of nematodes in agricultural areas. It should be noted that seedling health is one of the important parameters in crop safety in terms of nematode contamination; if a seedling is contaminated, it can act as a nematode transmitter and increase contamination (Rehman et al. 2012; Kumar et al. 2018).

Because of financial difficulties, the application of by-products in the food business has increased. These by-products have many natural compounds that can enhance the food product's qualities. One of these applications is food safety, like nematicidal, antimicrobial, and antifungal applications. These goals are achieved when the effective ingredients of food by-products are extracted correctly. For example, MSC contains effective compounds to kill nematodes. Compounds include gluconapine, sinapine, 3-methylsulfinyl propyl isothiocyanate, 1-methoxybrassitin, rutalexin, 5-methylthiopentylglucosinolate, and methyl oleate (Nadeem et al. 2021). In addition to these ingredients, oilseed cake may also contain antioxidants, dietary fibre, vitamins, minerals, and proteins (Abedini et al. 2022a). These compounds will be effective when extracted by the correct method because the extraction and storage method will directly impact their nematicidal characteristics. Techniques such as ultrasonication can increase the extraction efficiency of natural MSC materials and their nematicidal properties. The presence of steroid chemicals, terpenoid glycosides, volatile compounds, and nitrogen content all contribute to the oilseed cake's nematicidal mechanism action (Yang et al. 2015; Devi et al. 2019). The nematicidal mechanism of plant active compounds directly depends on the compounds present in the MSC. Plant essential oils' lethal power is due to their constituent compounds' biological activity on insects. Plant compounds have significant toxicity to many pests, including nematodes, due to their constituents, which are mainly monoterpenes (Oro et al. 2020; Oka 2001).

The action of essential oils against nematodes is not very clear. It can be noted that they affect nematode enzymes; for example, plant essential oils affect acetylcholinesterase activity. The effect of plant compounds in essential oil on the nervous system of nematodes and the process of neurotransmission is unclear. Some researchers suspect that essential oils kill nematodes by destroying nematode skin or disrupting nematode cell membranes and altering their permeability (Eloh et al. 2020; Boyko and Brygadyrenko 2021). In general, exposure time, dose, and type of active substance are the most important parameters affecting the nematicidal ability of MSC (Panpatte et al. 2021). Many studies investigated the nematicidal ability of MCS; for example, Oliveira et al. (2011) investigated the nematicidal activity of Brazilian wild mustard tissues against *Meloidogyne incognita* in tomatoes. To analyze the nematicidal effect of mustard seed, the seed was divided into three groups, dry leaf meal (LM), whole seed meal (WSM), and hexane defatted seed meal (DSM). Allyl isothiocyanate (AITC) was the most effective compound for reducing nematodes. The concentration of this effective substance in DSM was higher than in other species. The content of active compounds in the soil could also reduce nematodes. Changing the soil composition with mustard seed-derived compounds was able to minimize galls, egg masses, and eggs of nematodes by up to 90 percent. The compounds used were reported to be 1.6 percent LM, 0.2 percent WSM, or 0.05 percent DSM, respectively. The mustard seeds and meal were expressed as a compound to strengthen the soil and prevent the growth of nematodes. Mustard reduction power was directly related to AITC content (Oliveira et al. 2011).

In conclusion, it can be briefly stated that a direct relationship was observed between increasing the amount of cake and reducing the percentage of nematodes. However, using these wastes improved plant growth indices, especially in non-sterile soil treatment, which indicates the importance of different microorganisms in this type of soil. Also, adding oilseed cake causes biochemical changes in the plant and increases the amount of total phenol in the root, reduces the C:N rate of the soil, and thus better controls the nematode. The effect of adding cake on plant growth factors is significant because if adding cake reduces or does not grow plants, its use will not be appropriate (Radwan et al. 2009; Karimi et al. 2022; Jena et al. 2021).

3.6.1 Application Against Plant-Parasitic Nematodes

Many studies have been performed in recent years on the nematicidal ability of MSC. In the investigations, it can be seen that the type of cake, effective dose, and composition are the most effective nematicidal parameters. For example, Chandrawat et al. (2020) investigated the effect of MSC on *Meloidogyne incognita*. For this experiment, they used tomatoes. In this study, they examined the enzymatic activities (peroxidase [PO], polyphenol oxidase [PPO], phenylalanine ammonia-lyase [PAL], and superoxide dismutase [SOD]) for plant defense parameters against nematodes. The amount used for MSC was reported at 2.5 and 5.0 q/ha. MSC can increase the enzymatic activity of the plant against nematodes and promote plant safety. As observed, when 5.0 q/ha of MSC was used, the production of plant defense enzymes increased. In conclusion, they reported that the combination of cakes could be used to increase the production of enzymes with nematicidal potency (Table 3.4; Chandrawat et al. 2020). Rehman et al. (2011) investigated the effect of MSC on nematode growth in chickpeas. They used two doses of 50 and 100 gr of cake. They reported if the number of cakes increases, the growth of nematodes will decrease. In the end, they reported that to increase the environment's health, natural alternatives such as MSC can be used to maintain plant safety and reduce nematodes (Rehman et al. 2011). Parveen et al. (2019) studied the impacts of MSC on nematode reduction and plant growth. MSC-treated plants (1 percent) showed less nematode penetration in the roots than untreated control plants (Parveen et al. 2019). Sahu et al. (2018) investigated the effect of oilseed cake on the growth of *Meloidogyne incognita*. They used 100 g of MSC to examine nematode growth. In the end, they reported that neem oil cake performed better than other oilseed cakes (Sahu et al. 2018). In another study by Charles et al. (2015), application of nematicidal potency of MSC on *Meloidogyne incognita* was investigated. This study studied the number of glucosinolates and their effect on nematodes. Glucosinolates have a direct impact on nematodes and can reduce plant products. MSC had the highest nematicidal effect in this study compared to other cakes. They suggested that the cake can be used to minimize nematode growth in tomatoes and increase tomato safety (Table 3.4; Charles et al. 2015). Patra et al. (2019) investigated the effect of a different meal on the growth of *Meloidogyne incognita* on *Vigna mungo*. Using 100g/m MSC improved the biological properties of the plant and helped its growth, such as the uptake of phosphorus. In addition, the reduction of nematodes increased significantly. They reported MSC is the best natural nematicidal source for nematode reduction (Patra et al. 2019). Baheti et al. (2019) investigated the impact of different oil cakes (neem, karanj, and mustard) on the management of *Meloidogyne incognita* in okra. For these purposes, they used three different amounts of each oil cake (2, 4, and 6 q/ha). Finally, they reported the neem cake (6 q/ha) was the best nematicidal oil cake for the reduction of nematodes in okra. Kumar (2017) examined the functions of linseed, karanj, groundnut, castor, and mustard on the nematode growth of tomatoes. In contrast to the previous study, MSC was found to be the best effective nematicidal source for reducing nematodes (Table 3.4; Kumar 2017). The study results depend on the dose of MSC and the composition of the soil.

In another study, Jena et al. (2021) investigated the influence of oil cakes on *Meloidogyne incognita* reduction in green germ. As mentioned earlier, plant growth parameters are very important in using cake to control nematodes. This study showed that adding 50 g/m^2 MSC increased plant

TABLE 3.4
Nematicidal Action of Mustard Oilseed Cake against Different Plant-Parasitic Nematodes

Host Plant	Nematode	Reference
Tomato	Plant parasitic nematodes, including *Meloidogyne incognita*, *Meloidogyne javanica*, *Hoplolaimus indicus*, *Tylenchorhynchus brassicae*, and *Helicotylenchus indicus*	Hameed 1968; Alam 1980, 1991; Khan and Saxena 1997; Javed et al. 2007; Singh and Khurma 2007; Charles et al. 2015; Kumar 2017; Sahu 2018; Chandrawat et al. 2020
Okra	Plant parasitic nematodes, including *Rotylenchulus reniformis*, *Hoplolaimus indicus*, *Tylenchulus filiformis*, *Tylenchorynchus brassicae*, and *Meloidogyne incognita*	Siddiqui and Alam 1987; Alam 1991; Anver and Alam 1996; Baheti et al. 2019
Eggplant	Plant parasitic nematodes, including *Meloidogyne incognita*	Husain et al. 1984; Alam 1991
Red pepper	Plant parasitic nematodes, including *Meloidogyne* spp.	Trivedi 1978; Alam et al. 1980; Alam 1991; Chaudhary and Kaul 2013
Cabbage	Plant parasitic nematodes	Alam 1991
Soybean	*Meloidogyne incognita*	Mishra and Gupta 1997
Pigeon pea	*Heterodera cajani*	Singh 2004
Mung bean	PPN including *M. incognita*, *Rotylenchulus reniformis*, *Tylenchorhynchus brasscae*, *Helicotylenchus indicus*)	Tiyagi and Alam 1995
Cucumber	*Meloidogyne incognita*	Devi and Das 2016
Chickpea	*Meloidogyne incognita*	Rehman et al. 2011
Vigna mungo	*Meloidogyne incognita*	Patra et al. 2019
Green garm	*Meloidogyne incognita*	Jena et al. 2021
Lettuce	*Meloidogyne incognita*	Dias-Arieira et al. 2015
Rice nursery	*Meloidogyne incognita*	Devi et al. 2019
Onion	*Meloidogyne incognita*	Faruk 2019
Spinach	*Meloidogyne incognita*	Sahu et al. 2019

growth indices 2 weeks before sowing and significantly reduced nematode growth (Jena et al. 2021). Dias-Arieira et al. (2015) examined the application of organic amendments in the management of *Meloidogyne incognita* in lettuce. They reported the nematicidal efficiency of brown MSC was higher than canola seed cake (Dias-Arieira et al. 2015).

Faruk (2019) studied the effects of organic nematicidal on the growth rate of nematodes in onions. The focus of this study was on the management of soil structure; for these goals, they added mustard oil cake, poultry refuse, rice bran, and sawdust to the soil. They reported that the best way to reduce nematodes was the management of soil ingredients before transplanting. They added these materials three weeks before transplanting. Results showed a significant reduction of nematodes in all three materials (Faruk 2019). In another study by Sahu et al. (2019), the management of *Meloidogyne incognita* in Indian spinach was investigated. They used three doses of MSC (0.5, 1.0, and 1.5 t/ha) for the reduction of nematodes. They studied plant height, root length, fresh shoot weight, fresh root weight, dry shoot weight, dry root weight, root galls, number of egg masses, and root-knot nematode population. In the previously discussed cases, the most effective amount of MSC was reported as 1.5 t/ha. They concluded that MSC could be used to effectively control the root-knot nematode (*M. incognita*) in Indian spinach (Table 3.4; Sahu et al. 2019).

3.7 CONCLUSION

Soil quality is the most crucial component in agriculture; if it is inappropriate, it will reduce plant growth. The soil should be rich in nitrogen and other minerals and properly enriched so that pathogens, including nematodes, cannot grow and reproduce. Enriching the soil with oilseed cake due to its protein, nitrogen, fiber, antioxidants, minerals, and vitamins contents increases plant growth efficiency. Soils enriched with MSC are generally suitable for growing most plants, making them economically viable. These materials also have a high phenolic and nitrogen content that can reduce the growth of nematodes. Due to terpenes, volatile compounds, and anti-nutrients, oilseed cake directly affects nematode's tissue, coating, and permeability and causes their death and growth retardation. Furthermore, oilseed cake has a similar effect on nematode eggs. Nematode infestations are commonly found in potatoes, tomatoes, black gram (*Vigna mungo*), chickpeas (*Cicer arietinum*), cowpeas (*Vigna unguiculata*), French beans, lentils (*Lens culinaris*), mung beans (*Vigna radiata*), pigeon peas (*Cajanus cajan*), soybeans (*Glycine max*), maize (*Zea mays*), rice (*Oryza sativa*), wheat (*Triticum aestivum*), cabbage, cauliflower, chilis, eggplants, okra, pointed gourds, squashes, acid limes, bananas, tangerines (*Citrus reticulata*), papayas (*Carica papaya*), bajra (*Pennisetum typhoides*), betelvines (betel pepper), groundnuts, Japanese mint, subabul, sugarcane, tea (*Camellia sinensis*), tobacco (*Nicotiana tabacum*), and tulsi (*Ocimum basilicum*). MSC reduces and kills nematodes due to its high nitrogen, AITC, and phenolic compounds. In addition, using this oilseed cake also improves soil properties and can increase the safety of plants and fruits by directly affecting nematodes and their eggs.

REFERENCES

Abdulwaliyu, I., S. Arekemase, J. Adudu, M. Batari, M. Egbule, and S. Okoduwa. 2019. Investigation of the medicinal significance of phytic acid as an indispensable anti-nutrient in diseases. *Clinical Nutrition Experimental* 28:42–61. https://doi.org/10.1016/j.yclnex.2019.10.002.

Abedini, A., A. Mahdavi, A. Mirza Alizadeh, E. Hejazi, and H. Hosseini. 2022a. A review on dietary additive, food supplement and exercise effects on the prevention of covid-19. *Nutrition and Food Sciences Research* 9 (1):1–14. https://doi.org/10.52547/nfsr.9.1.1.

Abedini, A., A. Mirza Alizadeh, A. Mahdavi, A. Golzan, M. Salimi, B. Tajdar-Oranj, and H. Hosseini. 2022b. Oilseed cakes in the food industry; a review on applications, challenges, and future perspectives. *Current Nutrition & Food Science* 18 (4):345–364. http://dx.doi.org/10.2174/1573401317666211209150147.

Adeyemo, S. M., and A. A. Onilude. 2013. Enzymatic reduction of anti-nutritional factors in fermenting soybeans by *Lactobacillus plantarum* isolates from fermenting cereals. *Nigerian Food Journal* 31 (2):84–90. https://doi.org/10.1016/S0189-7241(15)30080-1.

Ahmad, M., and K. Sinha. 2002. Mycotoxin contamination of mustard seeds during storage. *Indian Phytopathology* 55 (3):299–302.

Alam, M. 1991. Control of plant-parasitic nematodes with oilseed cakes on some vegetables in field. *Pakistan Jornal Nematology* 9:21–30.

Alam, M., M. Ahmad, and A. M. Khan. 1980. Effect of organic amendments on the growth and chemical composition of tomato, eggplant and chilli and their susceptibility to attack by *Meloidogyne incognita*. *Plant Soil* 57:231–236.

Alireza Sadeghi, M., A. G. Appu Rao, and S. Bhagya. 2006. Evaluation of mustard (*Brassica juncea*) protein isolate prepared by steam injection heating for reduction of antinutritional factors. *LWT-Food Science and Technology* 39 (8):911–917. https://doi.org/10.1016/j.lwt.2005.06.012.

Ancuța, P., and A. Sonia. 2020. Oil press-cakes and meals valorization through circular economy approaches: A review. *Applied Sciences* 10 (21):7432.

Anver, S., and M. Alam. 1996. Effect of some oil-cakes against *Meloidogyne incognita* and *Rotylenchulus reniformis* infecting okra. *Annals of Plant Protection Sciences* 4:176–178.

Ayrilmis, N., A. Kaymakci, and F. Ozdemir. 2013. Sunflower seed cake as reinforcing filler in thermoplastic composites. *Journal of Applied Polymer Science* 129 (3):1170–1178. https://doi.org/10.1002/app.38806.

Ba, K., E. Tine, J. Destain, N. Cisse, and P. Thonart. 2010. Comparative study of phenolic compounds, antioxidant power of various Senegalese sorghum cultivars and amylolytic enzymes of their malt. *Biotechnology, Agronomy and Society and Environment* 14:131–139.

Badhe, P., M. Damale, and R. Adivarekar. 2016. Use of mustard oil cake for protease production by *Bacillus subtilis. International Journal of Current Microbiology and Applied Sciences* 5:845–853. https://doi.org/10.20546/ijcmas.2016.507.097.

Baheti, B., S. Bhati, and H. Singh. 2019. Efficacy of different oil-cakes as soil amendment for the management of root-knot nematode, *Meloidogyne incognita* infecting Okra (*Abelmoschus esculentus* L.). *International Journal of Current Microbiology and Applied Sciences* 8 (12):799–808.

Bello, B., S. Mustafa, J. Tan, T. Ibrahim, Y. Tam, A. Ariff, M. Manap, and S. Abbasiliasi. 2018. Evaluation of the effect of soluble polysaccharides of palm kernel cake as a potential prebiotic on the growth of probiotics. *3 Biotech* 8 (8):346. https://doi.org/10.1007/s13205-018-1362-4.

Bhargava, N., S. Rahul Mor, K. Kumar, and V. Sharanagat. 2021. Advances in application of ultrasound in food processing: A review. *Ultrasonics Sonochemistry* 70:105293. https://doi.org/10.1016/j.ultsonch.2020.105293.

Boyko, O., and V. Brygadyrenko. 2021. Nematicidal activity of essential oils of medicinal plants. *Folia Oecologica* 48 (1):42–48.

Chan, K., H. Zhang, and Z. X. Lin. 2014. Treatments used in complementary and alternative medicine. In *Side Effects of Drugs Annual*, edited by Sidhartha D. Ray, 717–724. Elsevier.

Chandrawat, B., A. Siddiqui, S. Bhati, and V. Saharan. 2020. Response of defence related enzymes in tomato treated with oil-cakes against root-knot nematode, *Meloidogyne incognita. International Journal of Current Microbiology and Applied Sciences* 9 (11):1100–1111.

Charles, K., K. Agathar, M. Ronald, P. Cosmas, M. Ignitius, and M. Blessing. 2015. Nematicidal effects of brassica formulations against root knot nematodes (*Meloidogyne javanica*) in tomatoes (*Solanum lycopersicum* L.). *Pakistan Journal of Phytopathology* 27 (2):109–114.

Chaudhary, K., and R. Kaul. 2013. Efficacy of *Pasteuria penetrans* and various oil seed cakes in management of *Meloidogyne incognita* in chilli pepper (*Capsicum annuum* L.). *Journal of Agricultural Science and Technology* 15:617–626.

Chen, F., H. Ma, Y. Li, H. Wang, A. Samad, J. Zhou, L. Zhu, Y. Zhang, J. He, and X. Fan. 2019. Screening of nanobody specific for peanut major allergen Ara h 3 by phage display. *Journal of Agricultural and Food Chemistry* 67 (40):11219–11229.

Chen, G., and B. Liu. 2016. Cellulose sulfate based film with slow-release antimicrobial properties prepared by incorporation of mustard essential oil and β-cyclodextrin. *Food Hydrocolloids* 55:100–107. https://doi.org/10.1016/j.foodhyd.2015.11.009.

Coyne, D. L., L. Cortada, J. Dalzell, A. Claudius-Cole, S. Haukeland, N. Luambano, and H. Talwana. 2018. Plant-parasitic nematodes and food security in sub-Saharan Africa. *Annual Review of Phytopathology* 56:381–403. https://doi.org/10.1146/annurev-phyto-080417-045833.

Crevel, R., M. Kerkhoff, and M. Koning. 2000. Allergenicity of refined vegetable oils. *Food and Chemical Toxicology* 38 (4):385–93. https://doi.org/10.1016/s0278-6915(99)00158-1.

Cui, S., Y. Wu, and H. Ding. 2013. The range of dietary fibre ingredients and a comparison of their technical functionality. In *Fibre-Rich and Wholegrain Foods*, edited by Jan A. Delcour and Kaisa Poutanen, 96–119. Woodhead Publishing.

Davidson, P., and S. Zivanovic. 2003. The use of natural antimicrobials. In *Food Preservation Techniques*, edited by Peter Zeuthen and Leif Bøgh-Sørensen, 5–30. Woodhead Publishing.

Devi, P., R. Kanwar, and J. Patil. 2019. Effect of oil cakes for the management of *Meloidogyne graminicola* in rice nursery. *Journal of Entomology and Zoology Studies* 7:180–182.

Devi, T., and D. Das. 2016. Effect of organic amendments on root-knot nematode (*Meloidogyne incognita*) in cucumber. *Pest Management in Horticultural Ecosystems* 22 (2):176–181.

Dias-Arieira, C., D. Mattei, H. Puerari, and R. Ribeiro. 2015. Use of organic amendments in the management of root-knot nematode in lettuce. *Horticultura Brasileira* 33:488–492.

Diouf, A., F. Sarr, B. Sene, C. Ndiaye, S. Fall, and N. Ayessou. 2019. Pathways for reducing anti-nutritional factors: Prospects for *Vigna unguiculata. Journal of Nutritional Health & Food Science* 7. https://doi.org/10.15226/jnhfs.2019.001157.

Divakaran, M., and K. Babu. 2016. Mustard. In *Encyclopedia of Food and Health*, edited by Benjamin Caballero, Paul M. Finglas and Fidel Toldrá, 9–19. Oxford: Academic Press.

EFSA Panel on Contaminants in the Food Chain, H. Knutsen, J. Alexander, L. Barregård, M. Bignami, B. Brüschweiler, S. Ceccatelli, M. Dinovi, L. Edler, and B. Grasl-Kraupp. 2016. Erucic acid in feed and food. *EFSA Journal* 14 (11):e04593.

Eloh, K., K. Kpegba, N. Sasanelli, H. Koumaglo, and P. Caboni. 2020. Nematicidal activity of some essential plant oils from tropical West Africa. *International Journal of Pest Management* 66 (2):131–141.

Faruk, I. 2019. Efficacy of organic soil amendments and a nematicide for management of root-knot nematode *Meloidogyne* Spp. of Onion. *American Journal of Bioscience and Bioengineering* 7 (4):57–64.

Fekadu Gemede, H. 2014. Antinutritional factors in plant foods: Potential health benefits and adverse effects. *International Journal of Nutrition and Food Sciences* 3:284. https://doi.org/10.11648/j.ijnfs.20140304.18.

Food Standards Agency. 2015. *Food Allergen Labelling and Information Requirements under the EU Food Information for Consumers*. Food Standards Agency (FSA).

Gilani, G., K. Cockell, and E. Sepehr. 2005. Effects of antinutritional factors on protein digestibility and amino acid availability in foods. *Journal of AOAC International* 88 (3):967–987.

Gültekin Subaşı, B., B. Vahapoğlu, E. Capanoglu, and M. Mohammadifar. 2021. A review on protein extracts from sunflower cake: Techno-functional properties and promising modification methods. *Critical Reviews in Food Science and Nutrition* 1–16. https://doi.org/10.1080/10408398.2021.1904821.

Hameed, S. F. 1968. Relative efficacy of different oil cakes on the incidences of root-knot nematodes. *Madras Agricultural Journal* 55:168–171.

Hassan, S., A. Hagrassi, O. Hammam, A. Soliman, E. Ezzeldin, and W. Aziz. 2020. *Brassica juncea* L. (mustard) extract silver nanoparticles and knocking off oxidative stress, proinflammatory cytokine and reverse DNA genotoxicity. *Biomolecules* 10 (12):1650.

Husain, S. I., R. Kumar, T. Khan, and A. Titov. 1984. Effect of root dip treatment of eggplant seedlings with plants extracts, nematicides, oil-cake extracts and anthelmintic drugs on plant growth and root knot development. *Pakistan Journal Nematology* 2:79–83.

Javed, N., S. Gowen, M. Inam-ul-Haq, K. Abdullah, and F. Shahina. 2007. Systemic and persistent effect of neem (*Azadirachta indica*) formulations against root-knot nematodes, *Meloidogyne javanica* and their storage life. *Crop Protection* 26 (7):911–916. https://doi.org/10.1016/j.cropro.2006.08.011.

Jena, P., N. Sahoo, and J. Mahalik. 2021. Effect of oil cakes and bio-agents each alone and in combination for management of root knot nematode (*Meloidogyne incognita*) in green gram. *International Journal of Bio-Resource & Stress Management* 12 (4):286–294.

Karimi, M., Z. Sheikholeslami, B. Sahraiyan, M. Ghiyafeh Davoodi, and F. Naghipour. 2018. Using sesame meal flour in free gluten French bread (rice-corn) containing guar and CMC gums to produce functional food. *Food Science and Technology* 14 (73):1–2.

Karimi, S., Z. Azami-Sardooei, and F. Fekraf. 2022. Effects of the grapefruit wastes and sesame oil cake to control of the root-knot nematode *Meloidogyne javanica* on tomato plants. *Iranian Journal of Plant Protection Science* 52 (2).

Karimi, Z., A. Mirza Alizadeh, J. Ezzati Nazhad Dolatabadi, and P. Dehghan. 2019. Nigella sativaand its derivatives as food toxicity protectant agents. *Advanced Pharmaceutical Bulletin* 9 (1):22–37. https://doi.org/10.15171/apb.2019.004.

Khan, T., and S. Saxena. 1997. Integrated management of root knot nematode *Meloidogyne javanica* infecting tomato using organic materials and *Paecilomyces lilacinus*. *Bioresource Technology* 61 (3):247–250. https://doi.org/10.1016/S0960-8524(97)00024-2.

Kumar, A., K. Khilari, S. Jain, A. Yadav, and A. Mishra. 2018. Efficacy of different oil cakes on management of root knot nematode (*Meloidogyne graminicola*) in rice crop in Western Uttar Pradesh, India. *International Journal of Current Microbiology and Applied Sciences* 7 (12):3524–3530.

Kumar, M. 2017. Efficacy of different oil-cakes on the incidence of root-knot nematodes (*Meloidogyne Javanica*) on tomato. *The Journal of Middle East and North Africa Sciences* 3 (10):41–43.

Kumari, N., R. Avtar, B. Sharma, and N. Thakral. 2016. Antioxidant potential in seed meal of different Indian mustard genotypes. *Journal of Oilseed Brassica* 1 (1):63–67.

L'Hocine, L., M. Pitre, and A. Achouri. 2019. Detection and identification of allergens from Canadian mustard varieties of Sinapis alba and *Brassica juncea*. *Biomolecules* 9 (9):489.

Łopusiewicz, Ł., E. Drozłowska, P. Siedlecka, M. Mężyńska, A. Bartkowiak, M. Sienkiewicz, H. Zielińska-Bliźniewska, and P. Kwiatkowski. 2019. Development, characterization, and bioactivity of non-dairy kefir-like fermented beverage based on flaxseed oil cake. *Foods* 8 (11):544.

Madsen, C., M. van den Dungen, S. Cochrane, G. Houben, R. Knibb, A. Knulst, S. Ronsmans, R. Yarham, S. Schnadt, and P. Turner. 2020. Can we define a level of protection for allergic consumers that everyone can accept? *Regulatory Toxicology and Pharmacology* 117:104751.

Milani, E., N. Hashemi, S. A. Mortazavi, and F. Tabatabaie Yazdi. 2017. Investigating the effect of extrusion process conditions and formulations on some physicochemical properties of bulk snack based on almond meal (*Amygdalus communis* L.) and corn semolina. *Innovative Food Technologies* 5 (1):123–140.

Mishra, S., and P. Gupta. 1997. Effect of some oil cakes and saw dust application on root-knot (*Meloidogyne incognita*) infection and root nodulation of soybean. *Indian Phytopathology* 50:131–132.

Nadeem, H., P. Malan, A. Khan, M. Asif, M. Ahmad Siddiqui, S. Tuhafeni Angombe, and F. Ahmad. 2021. New insights on the utilization of ultrasonicated mustard seed cake: Chemical composition and antagonistic potential for root-knot nematode, *Meloidogyne javanica*. *Journal of Zhejiang University, Science B* 22 (7):563–574. https://doi.org/10.1631/jzus.B2000746.

Nicol, J., S. Turner, D. Coyne, L. Nijs, S. Hockland, and Z. Maafi. 2011. Current nematode threats to world agriculture. In *Genomics and Molecular Genetics of Plant-Nematode Interactions*, edited by John Jones, Godelieve Gheysen, and Carmen Fenoll, 21–43. Springer.

Nilam, D., and A. D. Patel. 2019. Management of root-knot nematodes (*Meloidogyne* spp.) using different chemicals in tomato nursery. *International Journal of Current Microbiology and Applied Sciences* 8 (3):2047–2051.

OECD/FAO. 2020. Oilseeds and oilseed products. In *OECD-FAO Agricultural Outlook 2020–2029*. Paris; Rome: OECD Publishing.

Ogidi, I., O. Omu, and P. Ezeagba. 2019. Ethno pharmacologically active components of *Brassica juncea* (brown mustard) seeds. *International Journal of Pharmaceutical Research and Development*, 1 (1):9–13.

Oka, Y. 2001. Nematicidal activity of essential oil components against the root-knot nematode *Meloidogyne javanica*. *Nematology* 3:159–164. https://doi.org/10.1163/156854101750236286.

Oliveira, R. D. L., O. D. Dhingra, A. O. Lima, G. N. Jham, M. A. Berhow, R. K. Holloway, S. F. Vaughn. 2011. Glucosinolate content and nematicidal activity of Brazilian wild mustard tissues against *Meloidogyne incognita* in tomato. *Plant Soil* 341:155–164. https://doi.org/10.1007/s11104-010-0631-8.

Oro, V., S. Krnjajic, M. Tabakovic, J. Stanojevic, and S. Ilic-Stojanovic. 2020. Nematicidal activity of essential oils on a *Psychrophilic panagrolaimus* sp. (Nematoda: Panagrolaimidae). *Plants* 9 (11):1588.

Pałgan, K., M. Żbikowska-Gotz, and Z. Bartuzi. 2018. Dangerous anaphylactic reaction to mustard. *Archives of Medical Science: AMS* 14 (2):477–479. https://doi.org/10.5114/aoms.2016.60580.

Palmer, L., J. Marsh, J. Baumert, and P. Johnson. 2020. Persistence of peanut allergen-derived peptides throughout excessive dry thermal processing. *LWT* 132:109903.

Panpatte, D., H. Shelat, Y. Jhala, and R. Vyas. 2021. Fortified bacterial consortium–a novel approach to control root knot nematode in cucumber (*cucumis sativum*). *Biological Control* 155:104528.

Parveen, G., R. Noreen, H. Shafique, V. Sultana, S. Ehteshamul-Haque, and M. Athar. 2019. Role of rhizobia in suppressing the root diseases of soybean under soil amendment. *Planta Daninha* 37.

Patra, M., C. Dash, and N. Sahoo. 2019. Effect of AM fungus and oilcakes with different doses of fertilizers on root knot nematode (*Meloidogyne incognita*) infecting black gram. *Journal of Entomology and Zoology Studies* 7:603–606.

Pi, X., Y. Wan, Y. Yang, R. Li, X. Wu, M. Xie, X. Li, and G. Fu. 2019. Research progress in peanut allergens and their allergenicity reduction. *Trends in Food Science & Technology* 93:212–220.

Planque, M., T. Arnould, and N. Gillard. 2017. Food allergen analysis: Detection, quantification and validation by mass spectrometry. *Allergen* 7–41.

Pojić, M., T. Dapčević Hadnađev, M. Hadnađev, S. Rakita, and T. Brlek. 2015. Bread supplementation with hemp seed cake: A by-product of hemp oil processing. *Journal of Food Quality* 38 (6):431–440. https://doi.org/10.1111/jfq.12159.

Purnawan, A., Y. Yopi, and T. T. Irawadi. 2017. Production of manooligomannan from palm kernel cake by mannanase produced from Streptomyces cyaenus. *Biosaintifika: Journal of Biology & Biology Education* 9 (1):73–80.

Radwan, M., E. El-Maadawy, S. Kassem, and M. Abu-Elamayem. 2009. Oil cakes soil amendment effects on *Meloidogyne incognita*, root-knot nematode infecting tomato. *Archives of Phytopathology and Plant Protection* 42 (1):58–64.

Rani, R., and L. Badwaik. 2021. Functional properties of oilseed cakes and defatted meals of mustard, soybean and flaxseed. *Waste and Biomass Valorization* 12: 5639–5647. https://doi.org/10.1007/s12649-021-01407-z.

Rehman, B., M. Ganai, K. Parihar, M. Siddiqui, and A. Usman. 2012. Management of root knot nematode, *Meloidogyne incognita* affecting chickpea, Cicer arietinum for sustainable production. *Biosciences International* 1 (1):1–5.

Rehman, B., K. Parihar, M. Ganai, A. Usman, and M. Siddiqui. 2011. Biomanagement of root knot nematode through non edible oil seed cakes infesting *Cicer arietinum* L. *International Journal of Applied Biology and Pharmaceutical Technology* 2 (4):10.

Sadighara, P., M. Safta, I. Limam, K. Ghanati, Z. Nazari, M. Karami, and A. Abedini. 2022. Association between food additives and prevalence of allergic reactions in children: A systematic review. *Reviews on Environmental Health*. https://doi.org/10.1515/reveh-2021-0158.

Sahay, S., and T. Prasad. 1990. The occurrence of aflatoxins in mustard and mustard products. *Food Addit Contam* 7 (4):509–513. https://doi.org/10.1080/02652039009373912.

Sahay, S., and K. Ranjan. 1990. Distribution of aflatoxin in oil seeds to their oil and cakes. *Geobios (Jodhpur)* 17 (1):47–48.

Sahu, A., J. Mahalik, and S. Behera. 2019. Management of root knot nematode, *Meloidogyne incognita* in Indian spinach (*Basella alba* L.) through organic amendments. *Journal of Entomology and Zoology Studies* 7 (6):428–431.

Sahu, S., M. Patra, and B. Dash. 2018. Management of root knot nematode (*Meloidogyne incognita*) in tomato (cv. Pusa ruby) using different oil cakes. *International Journal of Current Microbiology Applied Science* 7 (3):2527–2532.

Samtiya, M., R. Aluko, and T. Dhewa. 2020. Plant food anti-nutritional factors and their reduction strategies: An overview. *Food Production, Processing and Nutrition* 2 (1):6. https://doi.org/10.1186/s43014-020-0020-5.

Sarker, A., D. Saha, H. Begum, A. Zaman, and M. Rahman. 2015. Comparison of cake compositions, pepsin digestibility and amino acids concentration of proteins isolated from black mustard and yellow mustard cakes. *AMB Express* 5:22–22. https://doi.org/10.1186/s13568-015-0110-y.

Siddiqui, M., and M. Alam. 1987. Utilization of marigold plant wastes for the control of plant parasitic nematodes. *Biological Waste*, 21:221–229.

Singh, K., R. Bharose, S. Verma, and V. Singh. 2013. Potential of powdered activated mustard cake for decolorising raw sugar. *Journal of the Science of Food and Agriculture* 93. https://doi.org/10.1002/jsfa.5744.

Singh, R., S. Langyan, S. Sangwan, B. Rohtagi, A. Khandelwal, and M. Shrivastava. 2022. Protein for human consumption from oilseed cakes: A review. *Frontiers in Sustainable Food Systems* 101.

Singh, S., and U. Khurma. 2007. Susceptibility of six tomato cultivars to the root-knot nematode, *Meloidogyne incognita*. *The South Pacific Journal of Natural Science* 25:73–77. https://doi.org/10.1071/SP07013.

Singh, V. 2004. Management of heterodera cajani on pigeonpea with nematicides and organic amendments. *Indian Journal of Nematology* 34:213–214.

Sreerama, Y., V. Sashikala, V. Pratape, and V. Singh. 2012. Nutrients and antinutrients in cowpea and horse gram flours in comparison to chickpea flour: Evaluation of their flour functionality. *Food Chemistry* 131 (2):462–468. https://doi.org/10.1016/j.foodchem.2011.09.008.

Sumbul, A., R. Rizvi, I. Mahmood, and R. Ansari. 2015. Oil-cake amendments: Useful tools for the management of phytonematodes. *Asian Journal of Plant Pathology* 9:91–111. https://doi.org/10.3923/ajppaj.2015.91.111.

Sunil, L., P. Appaiah, P. Prasanth Kumar, and A. Gopala Krishna. 2015. Preparation of food supplements from oilseed cakes. *Journal of Food Science and Technology* 52 (5):2998–3005. https://doi.org/10.1007/s13197-014-1386-7.

Szydlowska-Czerniak, A., A. Tulodziecka, G. Karlovits, and E. Szlyk. 2015. Optimisation of ultrasound-assisted extraction of natural antioxidants from mustard seed cultivars. *Journal of the Science of Food and Agriculture* 95 (7):1445–53. https://doi.org/10.1002/jsfa.6840.

Teh, S., and A. Bekhit. 2015. Utilization of oilseed cakes for human nutrition and health benefits. In *Agricultural Biomass Based Potential Materials*, edited by Hakeem, K., Jawaid, M., Y. Alothman, O. Springer. Cham. https://doi.org/10.1007/978-3-319-13847-3_10.

Teh, S., A. Bekhit, and J. Birch. 2014. Antioxidative polyphenols from defatted oilseed cakes: Effect of solvents. *Antioxidants (Basel, Switzerland)* 3 (1):67–80. https://doi.org/10.3390/antiox3010067.

Thakur, A., V. Sharma, A. Thakur, V. Correspondence, and V. Sharma. 2019. An overview of anti-nutritional factors in food. *International Journal of Chemical Studies* 7 (1):2472–2479.

Thomas, J., K. Kuruvilla, and T. Hrideek. 2012. Mustard. In *Handbook of Herbs and Spices* (2nd edition), edited by K. V. Peter, 388–398. Woodhead Publishing.

Tiyagi, S., and M. Alam. 1995. Efficacy of oil-seed cakes against plant-parasitic nematodes and soil-inhabiting fungi on mungbean and chickpea. *Bioresource Technology* 51:233–239.

Tiyagi, S., I. Mahmood, Z. Khan, and H. Ahmad. 2011. Biological control of soil-pathogenic nematodes infecting mungbean using *Pseudomonas fluorescens*. *Archives of Phytopathology and Plant Protection* 44 (18):1770–1778. https://doi.org/10.1080/03235401003633840.

Trivedi, P., A. Bhatnagar, and B. Tiagi. 1978. Control of root-knot nematodes on capsicum annum by application of oil-cake. *Indian Phytopathogy* 31:75–76.

Waqas, M., S. Iqbal, A. Abdull Razis, W. Pervaiz, T. Ahmad, S. Usman, N. Ali, and M. Asi. 2021. Occurrence of aflatoxins in edible vegetable seeds and oil samples available in Pakistani retail markets and estimation of dietary intake in consumers. *International Journal of Environmental Research and Public Health* 18 (15):8015.

Yang, X., X. Wang, K. Wang, L. Su, H. Li, R. Li, and Q. Shen. 2015. The nematicidal effect of camellia seed cake on root-knot nematode, *Meloidogyne javanica* of banana. *PLoS One* 10 (4):e0119700. https://doi.org/10.1371/journal.pone.0119700.

Zhao, X., J. Chen, and F. Du. 2012. Potential use of peanut by-products in food processing: A review. *Journal of Food Science and Technology* 49:521–529.

4 Cotton Oilseed Cake
Chemical Composition and Nematicidal Potential

Raman Tikoria, Nandni Sharma, Sandeep Kour, Deepak Kumar, Mohd. Ali, Roohi Sharma, Parkirti and Puja Ohri

CONTENTS

4.1 INTRODUCTION

Oilseed cakes and oil meals are the remnants of oil extraction from seeds. There are two kinds of oil cakes: edible and non-edible. Edible oil cakes have a high nutritional value, specifically in protein content ranging from 15–50% (Ostbring et al. 2019). The nutrient content varies according to variety, growth conditions and extraction procedures. Due to their high protein content, these oilseed cakes are used as animal feed (Fekria et al. 2012). Because of the high NPK content, non-edible oil cakes such as castor cake, cotton cake, karanja cake and neem cake are employed as organic nitrogenous fertilisers (Rodriguez-Miranda et al. 2012). Some oil cakes have been documented to boost plant nitrogen absorption by slowing soil nitrification (Ramachandran et al. 2007). Also, oilseed cakes stimulate fungus growth and promote the formation of aflatoxins due to their high moisture level and water activity (Pettersson 2012). They also shield the crops from soil nematodes, insects and parasites, providing them with high resilience to infection (Radwan et al. 2009). Over the next decade, the yearly increase in cotton seed cake output is expected to average 2.3%. Seventeen million tons of the overall growth in oil meal output of 23 million tons comes from developing nations such as India, Brazil and Argentina (Jha et al. 2012). The demand for adequate animal feed resources emphasised pressed oilseed cakes as a feed option, a residue created after the oil has been removed (Chakraborty et al. 2022). In India, the oilseed crops are primarily farmed in Rajasthan, Haryana, Madhya Pradesh, West Bengal, Uttar Pradesh and Gujarat, with an estimated production of 7.1 million tons accounting for around 93% of the total output in the country (Sharma 2017).

According to FAOSTAT (2020), India and China have the highest cottonseed production, producing about 10 million tons of cottonseed per annum. Cottonseed oil is procured from the seeds of several plant species, especially *Gossypium hirsutum* and *G. herbaceum* (Mirpoor et al. 2021). Also, cotton seed cake (CSC) is gaining huge popularity because of its high mineral content. Cotton

DOI: 10.1201/9781003319252-4

(*G. barbadense*) seeds are used to make oilseed cakes containing various chemical components. When employing a cold press process, high-grade oils may be extracted from cotton seeds under less severe circumstances and at a lower temperature. As a result, using these cold-pressed oilseed cakes as feed could be a cost-effective way in the agro-industrial sector (Chakraborty et al. 2022). Moreover, it also contains a good amount of amino acids such as lysine, methionine, tryptophan and threonine (Swick 1999). Cotton oilseed cakes promote N absorption by plants and reduce soil/urea nitrification; oil cakes with 2–7% protein improve plant health and decrease the population of soil nematodes (Tiyagi et al. 2002). Oilcake extracts are also used as a seed coating material to protect crops from nematodes (El-Sherif et al. 2010).

Root-knot nematodes (RKNs) are the most common and damaging type of plant-parasitic nematodes (PPNs; Jones et al. 2013). They cause severe diseases in tomatoes, brinjal, chilli, capsicum and okra globally, thus limiting the fruit yield. RKNs significantly impact plant development and make plants vulnerable to additional pathogen infections (Williamson 1998). The control of these pathogens has traditionally relied on soil treatment alternatives using synthetic nematicides, but rising demand for the safety of the environment and humans has resulted in a progressive removal of these products (Mire et al. 2016; Tikoria et al. 2022). In conjunction with crop rotation, biocontrol agents are being utilised to control worm infestations (Abrar et al. 2020). However, researchers worldwide are still working to standardise alternate nematode management systems that use non-chemical and environmentally benign approaches to maintain crop output (Sivakumar and Gunasekaran 2011). CSC is effective against a wide range of nematodes and may be used on many crops (Rehman et al. 2011; Sharma et al. 2016; Kumar et al. 2018). It is proposed that chemical substances present in the CSC have a negative impact on nematode egg hatching, juvenile viability and root penetration. CSCs are incredibly efficient against *Meloidogyne*, *Pratylenchus*, *Tylenchorhynchus* and other PPNs (Parihar et al. 2015; Sharma et al. 2016). CSC as an aqueous solution effectively controls RKNs infesting soybean crops (Prasad and Chawla 1994). Furthermore, organic the amendment of CSC along with nematophagous fungi *Pochonia chlamidospora* is effective in the management of nematodes (Yasmin et al. 2020). Keeping the facts previously mentioned in mind, this chapter focuses primarily on cotton oilseed cakes, their chemical composition, their physicochemical properties and their potential in managing nematodes with its possible mechanism.

4.2 PREPARATION OF CSC

Cottonseed consists of about 15–20% oil and 30–38% seed kernel (Sekhar and Rao 2011). The cotton seeds are cleaned prior to oil extraction (Shahidi et al. 2020). After removing the cotton fibre, the seed still contains short cellulose linters that must be removed to improve the oil yield. If left on seed, these linter fibres absorb the oil during extraction and affect the quality and yield of cotton oil. Mechanical de-linting is performed in commercial mills. After lint removal, cottonseeds are dehulled; that is, their outer covering is separated from the seed kernels in hullers. The dehulled kernels are flaked or diminished in size to facilitate oil removal. For flaking, the seeds need to have a good moisture content, and if the seeds are too dry, their moisture content is increased by 11% by conditioning (Alderks 1948). A series of five stacked-crushing rolls are mostly employed for flaking cottonseeds as they give thinner flakes (Shahidi et al. 2020). Before oil extraction, the flakes are heated or cooked because this helps break the cell wall to allow easy escape of oil, controls moisture content, kills microorganisms and inactivates enzymes, reduces oil viscosity, coagulates proteins and, to some extent, detoxifies Gossypol attached to the protein (Ward 1976). Overcooking is avoided as it darkens the oil and decreases the meal's nutritional value.

After cleaning the seeds, cottonseed oil is extracted either mechanically using a hydraulic press or solvent extraction that uses hexane to remove oil. Mechanical pressing is employed in the case of oilseeds having a high content of extractable oil such as hull-free cottonseed kernels. The solvent extraction process uses a non-polar solvent like hexane, which only dissolves the oil and does not remove proteins and other compounds. In most cases, the obtained cottonseed oil is refined and

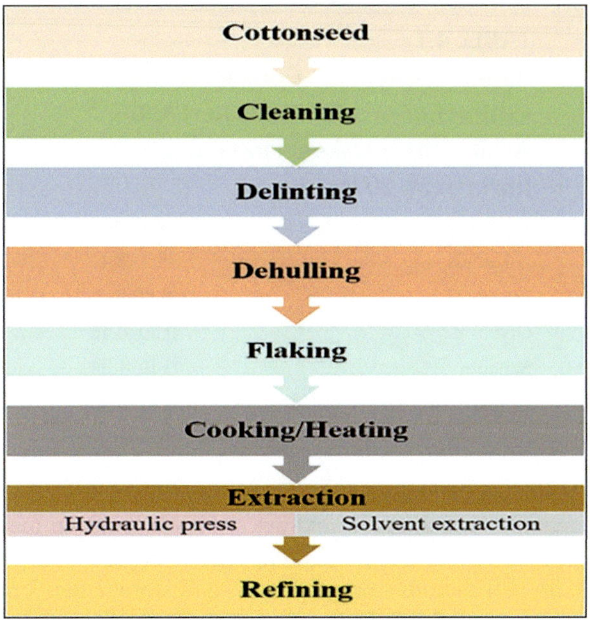

FIGURE 4.1 Process of cottonseed oil extraction.

purified to get rid of free fatty acids, gossypol, phosphatides and other impurities in cottonseed oil. The refining process can be coupled with solvent extraction to obtain better-quality cottonseed oil. The cotton meal's compacted mixture left after the cotton oil extraction is referred to as cotton oilseed cake (Sumbul et al. 2015; Figure 4.1).

4.3 CHEMICAL CONSTITUENTS

Various techniques such as calorimeter, flame photometer (AOAC 2000) and atomic absorption spectrophotometer (AAS) have been reported to calculate the proximal chemical composition of cotton oilseed cake (Aziz et al. 2018). Studies suggest that these cakes contain several volatile organic compounds, including crude protein (CP), crude fibre (CF), calcium, phosphorus, flavonoids, polysaccharides, terpene and phenol and their derivatives. These cakes also contain glycerol-1 alkanoates, bikojic acid, limonene, cellulose, hexadecanoic acid, vicilin and tetradecanoic acid (Egbuta et al. 2017). Gossypol is found in the cotton glands of *Gossypium* L. (Altman et al. 1990). It is a triterpenoid aldehyde ($C_{30}H_{30}O_8$), and its derivatives are present in cotton glands as the most common secondary metabolites (Benbouza et al. 2002). The mechanically extracted cotton oilseed cake method contains 0.06% gossypol, and the expander solvent CSC method includes 0.14% free gossypol (Osti et al. 2006). Cotton oilseed cake (sometimes used as CSM, cottonseed meal) containing gossypol comes in two forms, toxic and bound, after the commercial processors remove the oil (Gadelha et al. 2014). The proteins in the form of amino acids chain are present such as alanine (3.9%), arginine (11.1%), cystine (1.6%), glutamic acid (18.2%), methionine (1.4%) and proline (3.6%), threonine (3.3%) and histidine (2.9%). The most critical amino acid is lysine, which varies from 14–38% (Pousga et al. 2007; Nadeem et al. 2005; Nagalakshmi et al. 2007). It also contains flavonoids, alkaloids, saponins, tannins and glycosides, whose value varies considerably (Jibrin et al. 2020). Further, the fraction of CP is found in the range of 20–26% (Jibrin et al. 2020; Thakur et al. 2019). The percentage of CF is found in the range of 14–17% (Ashiru 2010), and about 7.9–16% is decorticated, whereas 26.9% is in un-decorticated cotton seed (Jibrin et al. 2020). The variation

TABLE 4.1
Range of Chemical Constituents in Cotton Oilseed Cake (Hastert 1986; Ashiru 2010; Oskoueian et al. 2011; Jibrin et al. 2020)

Compounds	Percentage (%)
Alkaloids	2.1–2.4
Flavanoids	0.80–1.2
Glycoside	0.39–0.46
Saponin	0.40–1.38
Tannin	0.87–0.96

of fibre content also depends on the non-dehulled (25%) to fully dehulled (5%) form of crude fibre (Morgan 2015). The content of nitrogen-free extract (NFE) is found in the range of 47–50%. The nutrients, minerals and physical and chemical composition of CSCs vary with climatic, soil and temperature conditions, as well as with storage process and storage temperature and the extraction and dehulling process (Foley et al. 2007; Oskoueian et al. 2011). The protein content also depends on genetic variations (Elmore et al. 1979; Scott et al. 1984; Yu et al. 2012). Besides the previously mentioned chemical constituents, various secondary metabolites, such as hemigossypolone, heliocide H1, heliocide H2, heliocide H3, heliocide H4, terpenes glycerol-1 alkanoates, bikojic acid, limonene, cellulose, hexadecanoic acid, vicilin, tetradecanoic acid and thymol, are also found in CSC (Soler-Serratosa et al. 1996). Elements like calcium, magnesium and phosphorus are present in lower amounts than major elements like potassium and sodium, which are abundant in CSC (Aziz et al. 2018). Besides these, some other chemicals mentioned in Table 4.1 also add to the chemical composition of CSC.

4.4 PHYSICOCHEMICAL PROPERTIES OF CSC

For obtaining the maximum activity of CSC, the pH of cakes should be slightly acidic to neutral. Several reports suggest that pH in the range of 6–8 provides more accurate results (Egbuta et al. 2017). The density of the CSC varies from 0.5 to 0.9 g/cc depending on the extraction process (Okonkwo and Okafor 2016). It was reported that oilseed cakes prepared from cotton possess higher water-absorbing capacity due to dry matter (Rhee et al. 1988). The average moisture content in oilseed cake prepared from cotton is reported to be 3–12%. The ash content is in the range of 5–7%, and lipids account for about 3–28% (Blasi and Drouillard 2002; Okonkwo and Okafor 2016; Jibrin et al. 2020). The water-insoluble portion in cotton oilseed cake is 90.00%, and the water-soluble portion is 4% (Aziz et al. 2018).

4.5 EFFECT OF CSC ON PPNS

The efficacy of numerous oilseed cakes from numerous plants has been reported against PPNs (Goswami et al. 1993; Ashraf and Khan 2007). Oilseed cake is a condensed meal mixture after processing vegetable oils from seeds like cotton, neem, *Brassica*, peanut and others. These are high in protein and minerals, making them ideal feed for poultry and other livestock. Some seeds, like castor beans and tung nuts, produce oilseed cakes that are toxic to various pathogenic microorganisms and are utilised as fertilisers rather than feed. Food and vegetable growers use cotton oilseed cakes to source plant nutrients and manage nematode problems. These cakes are either sprinkled or pelleted straight into the soil, watered and left to degrade. Their nematicidal action rises as the breakdown time increases. Plant seeds can also be treated with oil cake extracts before sowing to protect the crop from worms. Nematologists have recently become more interested in this trend because it effectively manages the target nematode while minimising pollution.

Pot experiments have been conducted to evaluate the effect of different oilseed cakes in the management of root-knot nematode *M. graminicola* in rice crops (Kumar et al. 2018). CSCs (25gm/5kg soil) resulted in about a 60% decrease in the number of galls per plant while doubling the amount of oilseed cake further decreased the number of galls up to 70%. Additionally, an increase of 54% and 64% was observed in shoot length and root length, respectively. Furthermore, the efficacy of CSC was also enhanced when the fungus *Pochonia chlamydosporia* was used in the combination resulting in a 42.4% increase in plant length. In contrast, only a 23.9% increase was observed in the individual application of CSC (Parihar et al. 2015). Sharma et al. (2016) also reported the effect of cotton oilseed cake on three PPNs *M. incognita*, *Helicotylenchus dihystera* and *Tylenchorhynchus mashhoodi* that resulted in decreased nematode population and root galling, leading to an increase in crop yield. Thus, cotton oilseed cake can be a useful eco-friendly management strategy for PPNs. Several studies related to this are summarised in Table 4.2.

TABLE 4.2
Effect of Cotton Oilseed Cake on Plant Parasitic Nematodes

S. No.	Nematode species	Test Crop	Effect	Reference
1.	*Meloidogyne arenaria*	*Cucurbita pepo*	Decreased gall formation and increased shoot and root length along with fresh plant weight	Mian and Rodriguez-Kabana 1982
2.	*Rotylenchulus reniformis*	*Vigna unguiculata*	Inhibited nematode reproduction and increased plant growth	Yassin and Ismail 1994
3.	*M. incognita*	–	Extract of oilseed cake increased larval mortality	Prasad and Chawla 1994
4.	*Heterodera cajani*	*Cajanus cajan*	Reduced nematode population leading to better plant growth	Meena et al. 2009
5.	*M. incognita*	*Solanum lycopersicum*	Reduced juvenile population and gall formation and enhanced plant growth in terms of length and fresh weight	Radwan et al. 2009
6.	Sugarcane nematode	*Saccharum officinarum*	Reduced nematode population to 60.84%	Mohan 2011
7.	*M. incognita*	*Cicer arietinum*	Decreased gall development and enhanced plant growth	Rehman et al. 2011
8.	*M. javanica*	*Solanum melongena*	Reduced nematode reproduction rate and egg masses and enhanced chlorophyll and carotenoid content	Parihar et al. 2015
9.	*M. incognita, Helicotylenchus dihystera and Tylenchorhynchus mashhoodi*	*Solanum lycopersicum*	Reduced nematode populations, root galling and increased crop yield	Sharma et al. 2016
10.	*M. graminicola*	*Oryza sativa*	Reduced gall formation and increased plant height and weight	Kumar et al. 2018
11.	*M. incognita*	*Abelmoschus esculentus*	Reduced nematode infestation and enhanced plant growth and yield	Tariq et al. 2018

4.6 MECHANISM OF ACTION

Cotton oilseed cake is made from the seeds of various cotton plants, primarily *G. hirsutum* and *G. herbaceum*. Cotton fibre and animal feed are the main reasons for its cultivation. It is categorised as polyunsaturated oil (18% monounsaturated and 52% polyunsaturated). It comprises 20–25% palmitic acid, 2–7% stearic acid, 18–30% oleic acid and 40–55% linoleic acid (Fennema 1985). The cotton plant's roots, leaves and seeds contain gossypol, a triterpene aldehyde. The terpenoids desoxyhemigossypol and hemigossypol can also be found in the roots. Cotton leaves, flower buds and seeds also contain so-called gossypol-like terpenoid chemicals such as hemigossypolone, heliocide H1, heliocide H2, heliocide H3 and heliocide H4, and because of their toxicity to *Heliothis*, these chemicals are known as heliocides (Nazarova et al. 1981). In addition, oil cakes contain several volatile organic compounds, including CP, CF, calcium, phosphorus, terpenes and phenols responsible for nematicidal effects. Furthermore, these cakes also contain glycerol-1 alkanoates, bikojic acid, limonene, cellulose, hexadecanoic acid, vicilin and tetradecanoic acid, which have adverse effects on nematodes. Thymol, a phenolic monoterpene found in the essential oils of various plant families, was found to have potent antinematodal activity against PPNs *M. arenaria, Heterodera glycines*, *Paratri chodorus* minor and free-living *Dorylaimid* nematodes. The active compounds such as thymol glycerol-1 alkanoates, bikojic acid and limonene found in the CSC act against nematodes through different mechanisms. However, the mechanism of CSC against nematodes is still unclear. Moreover, it has been reported that several constituents of CSC prevent the action of acetylcholinesterase in other pests such as insects (Ryan and Byrne 1988). Furthermore, CSC is lipophilic; it quickly passes through cell walls and can accumulate in the cell membrane of RKNs, causing structural disruption and the release of important cellular components (Nazzaro et al. 2013). Many simple phenols are volatile and can be found in plant essential oils with anti-nematodal activity. Limonene, the most abundant terpene in CSC, has shown anti-hatching activity (Oka et al. 2000; Figure 4.2).

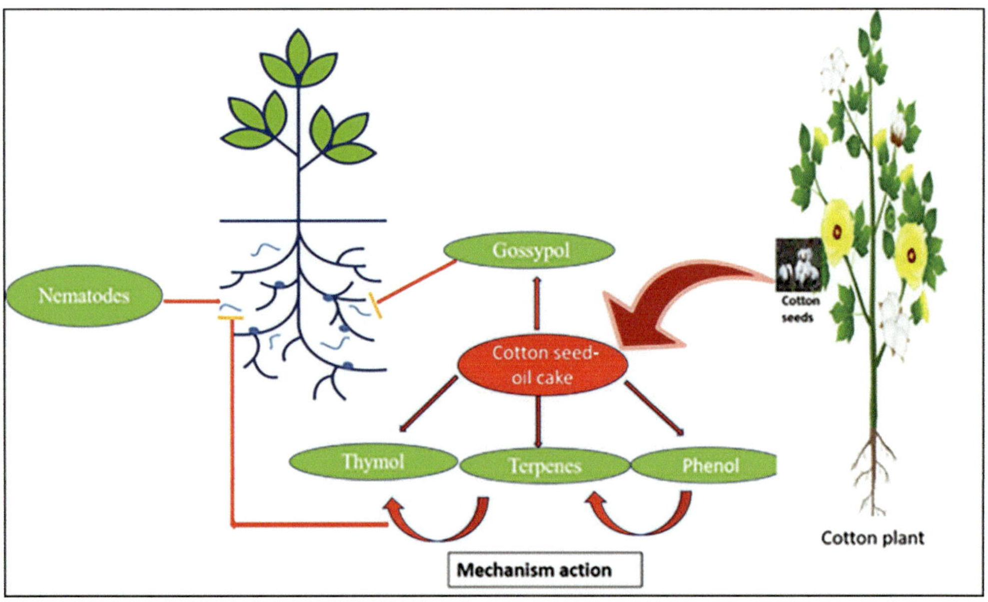

FIGURE 4.2 Generalised mechanism of action of cotton oilseed cake.

4.7 ADVANTAGES AND DRAWBACKS

The use of CSCs is much more beneficial as they act as good nematicides due to several volatile compounds (Estupinan-Lopez et al. 2017). Furthermore, its improvement of soil health and nutrient-rich manure for plants make it much more favourable for use in agriculture (Singh et al. 2015). Additionally, these cakes act as an antagonist for nematodes and can be used for various crops for nematode management. Among modes of applications, it can be used directly in the form of powder or pellet or crude form as well as in an aqueous or alcoholic extract form (Sumbul et al. 2015). These cakes can also be used in combination with other compounds for purpose of enhancing the antagonist potential. CSC and its combination with mustard cake, carbofuran, sodium azide and D-D (mixture of 1, 2-dichloropropene) enhance plant growth (Hossain et al. 1989). Once applied in the field, they gradually release their nutrients, making them effective for the long term and for replacing chemical fertilisers (Mohan 2011).

However, excessive use of CSC at higher rates can lead to phytotoxicity in plants due to the aggregation of salts and ionic species or due to the accumulation of plant pathogenic microorganisms (Radwan et al. 2009). The compound in CSC is gossypol, which can be toxic to consumers and lead to various acute diseases (Gadelha et al. 2014).

4.8 FUTURE PERSPECTIVES

Organic supplements used as cotton oilseed cakes in modern agriculture encounter enormous barriers, such as financial issues, high application rates, labour demand and environmental disruptions. The amount of cotton oilseed cake needed is usually rather large. In large-scale agriculture, all these pose issues with access, transportation and application. However, in developing nations, the production of these wastes is enormous. Government institutions, research centres and others must work to mobilise them in forms that farmers can use more easily. Second, using cotton oilseed cakes takes a lot of effort and energy. However, this is not a major issue in developing countries because abundant labour is available for this purpose (Sumbul et al. 2015). It is a prevalent misperception that using soil amendments to increase oilseed cake levels is always a good idea. An excessive amount

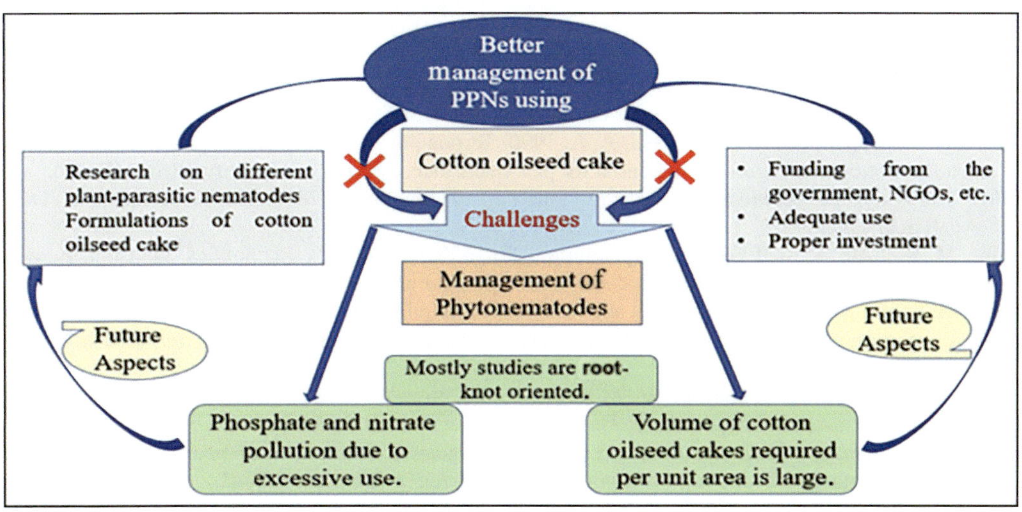

FIGURE 4.3 Future perspectives of using cotton oilseed cake for the management of plant-parasitic nematodes.

of amendment increases the concentration of micronutrients in the soil, resulting in phosphorus and nitrate pollution of surface and groundwater (Diacono and Montemurro 2011).

Furthermore, due to a lack of funding from the government, non-governmental organisations and other sources, research and the formulation of oilseed cakes in organic farming are restricted in developing nations. Nonetheless, eliminating production subsidies that have negative economic, social and environmental consequences might improve organic farming in many developing countries. But the demand for healthy food will undoubtedly increase day by day as people's health consciousness grows. Food produced with fewer inorganic fertilisers and chemical nematicides will play an essential part in safeguarding people's health in this approach (Prasad et al. 2022). Management of PPNs using CSC in crops will undoubtedly play a vital role in reducing yield losses due to nematodes. The studies for nematode management using CSC revolve around the control of RKNs. Therefore, further studies will also be required to depict the role of CSC in mitigating stress caused by other nematodes, for example, cyst and root lesion nematodes. This will emerge as a great tool in providing environment-friendly sustainable management of nematodes aided by increased crop yield (Figure 4.3).

4.9 CONCLUSION

CSCs are effective against various species of plant parasitic nematodes. Their crude, aqueous and powdered extracts have nematicidal potential and have been tested in both *in vitro* and *in vivo* studies. Because of their high nitrogen, phosphorus, and potassium (NPK) content and various volatile organic compounds, polyphenols like gossypol and other chemical constituents like limonene, cellulose, vicilin and others, several vegetable and food growers utilise CSCs to reduce populations of PPNs. When applied in the soil, the oil cakes are rapidly decomposed by microorganisms in the rhizosphere, increasing their nematicidal effect. However, further studies are being continued to understand the mechanism behind their nematicidal effect and incorporation in the integrated management of PPNs.

REFERENCES

Abrar, S., Seid, A. and Dejene, M. 2020. Integrated management of *Meloidogyne incognita* in tomato (*Solanum lycopersicum*) through botanical and intercropping. *African Journal of Agricultural Research* 15(4): 492–501.

Alderks, O. H. 1948. Cooking of meats and recovery of oil. In *Cottonseed and Cottonseed Products: Their Chemistry and Chemical Technology*, ed. A. E. Bailey, 615–650. New York: Interscience Publisher, Inc.

Altman, D. W., Stipanovic, R. D. and Bell, A. A. 1990. Terpenoids in foliar pigment glands of A, D, and AD genome cottons: Introgression potential for pest resistance. *Journal of Heredity* 81(6): 447–454.

AOAC. 2000. *Official Methods of Analysis* (17th ed.). Gaithersburg, MD: The Association of Official Analytical Chemists.

Ashiru, M. S. 2010. Animal Feeds Compendium. AER-656, Commodity Economics, Division, Economic Research Service, USDA 2010. https://www.feedipedia.org/node/6578.

Ashraf, M. S. and Khan, T. A. 2007. Efficacy of *Gliocladium virens* and *Talaromyces flavus* with and without organic amendments against *Meloidogyne javanica* infecting eggplant. *Asian Journal of Plant Pathology* 1(1): 18–21.

Aziz, S., Siddique Md, A. B. and Begum, M. H. A. 2018. Cotton seed oil cake as a valuable source of plant nutrients for sustainable agriculture. *The Pharmaceutical and Chemistry Journal* 5(3): 39–45.

Benbouza, H., Lognay, G., Palm, R., Baudoin, J. P. and Mergeai, G. 2002. Development of a visual method to quantify the gossypol content in cotton seeds. *Crop Science* 42(6): 1937–1942.

Blasi, D. and Drouillard, J. 2002. Cottonseed feed products for beef cattle, composition and feeding value. *Kansas State University*, 22p. www.ksre.ksu.edu/bookstore/Item.aspx?catId=562&pubId=1054.

Chakraborty, D., Sarkar, N., Agrahari, P., Dutta, R., Singh, A. A. and Jacob, S. 2022. Strategic nutrient enhancement of mustard oil seed cake by briquetting and koji fermentation for ruminants feed. *Waste and Biomass Valorisation* 13(2): 1207–1221.

Diacono, M. and Montemurro, F. 2011. Long-term effects of organic amendments on soil fertility. In *Sustainable agriculture volume*, eds E. Lichtfouse, M. Hamelin, M. Navarrete, and P. Debaeke, 761–786. Dordrecht: Springer.

Egbuta, M. A., McIntosh, S., Waters, D. L., Vancov, T. and Liu, L. 2017. Biological importance of cotton by-products relative to chemical constituents of the cotton plant. *Molecules* 22(1): 93.

Elmore, C. D., Spurgeon, W. I. and Thom, W. O. 1979. Nitrogen fertilisation increases N and alters amino acid concentration of cottonseed. *Agronomy Journal* 71(5): 713–716.

El-Sherif, A. G., Refaei., A. R., El-Nagar, M. E. and Salem, M. M. H. 2010. Impact of certain oilseed cakes or powder in comparison with oxamyl or urea on *Meloidogyne incognita* infecting eggplant. *Archives Phytopathology and Plant Protection* 43: 88–94.

Estupinan-Lopez, L., Campos, V. P., Silva, A. P., Barros, A. F., Pedroso, M. P., Silva, J. C. and Terra, W. C. 2017. Volatile organic compounds from cottonseed meal are toxic to *Meloidogyne incognita*. *Tropical Plant Pathology* 42: 443–450.

FAOSTAT, F. 2020. Statistics, Food and Agriculture Organization of the United Nations, Rome. https://www.fao.org/documents/card/en/c/cb1329en

Fekria, A. M., Isam, A. M. A., Suha, O. A. and Elfadil, E. B. 2012. Nutritional and functional characterization of defatted seed cakeflour of two Sudanese groundnut (*Arachishy pogaea*) cultivars. *International Food Research Journal (Malaysia)* 19(2): 629–673.

Fennema, O. R. and Rol, L. 1985. Industrial gum: Polysaccharides and their derivates. In *Food Chemistry* (2nd ed., revised and expanded). New York: Marcell Dekker Inc.

Foley, J. A., Asner, G. P., Costa, M. H., Coe, M. T., DeFries, R., Gibbs, H. K., Howard, E. A., Olson, S., Patz, J., Ramankutty, N. and Snyder, P. 2007. Amazonia revealed: Forest degradation and loss of ecosystem goods and services in the Amazon Basin. *Frontiers in Ecology and the Environment* 5(1): 25–32.

Gadelha, I. C. N., Fonseca, N. B. S., Oloris, S. C. S., Melo, M. M. and Soto-Blanco, B. 2014. Gossypol toxicity from cottonseed products. *The Scientific World Journal* 231635.

Goswami, B. K., Rao, U. and Manton, M. 1993. Efficacy of decomposed products of some oilcakes and ABCD collected at different time intervals on hatching and mortality of *Meloidogyne incognita*. *Current Nematology* 4: 215–217.

Hastert, R. C. 1986. Hydrogenation. In *Proceedings of world conference on emerging technologies in the fats and oils industry*, ed. A. R.Baldwin, Champaign: American Oil Chemists' Society.

Hossain, S., Mian, I. H. and Tsuno, K. 1989. Efficacy of three nematicides and two oilcakes for control of root knot nematode (*Meloidogyne incognita*) in potato seedlings. *Journal Faculty of Agriculture, Kyushu University* 34(1/2): 115–121.

Jha, G. K., Pal, S., Mathur, V. C., Bisaria, G., Anbukkani, P., Burman, R. R. and Dubey, S. K. 2012. *Edible Oilseeds Supply and Demand Scenario in India: Implications for Policy.* New Delhi: Division of Agricultural Economics, Indian Agricultural Research Institute.

Jibrin, H., Gusau, I. A., Musa, M., Shuaibu, A. A., Abubakar, M., Muhammad, U. and Omatah, C. U. 2020. Nutritive and phytochemical assessment of cotton (*Gossypium* spp.) seed meal for fish feed. *International Journal of Fisheries and Aquatic Studies* 8(5): 380–385.

Jones, T. J., Haegeman, A., Danchin, E. G., Gaur, H. S., Helder, J., Jones, M. G. K., Kikuchi, T., Manzanilla-Lopez, R., Palomares-Rius, J. E., Wesemael, W. M. L. and Perry, R. N. 2013. Top 10 plant-parasitic nematodes in molecular plant pathology. *Molecular Plant Pathology*14(9): 946–961.

Kumar, A., Khilari, K., Jain, S. K., Yadav, A. K. and Mishra, A. K. 2018. Efficacy of different oil cakes on management of root knot nematode (*Meloidogyne graminicola*) in rice crop in Western Uttar Pradesh, India. *International Journal of Current Microbiology and Applied Sciences* 7(12): 3524–3530.

Meena, P., Nehra, S. and Trivedi, P. C. 2009. Efficacy of decomposed organic cakes against *Heteroderacajani* infecting *Cajanus cajan*. *Asian Journal of Experimental Science* 23: 181–184.

Mian, I. H. and Rodriguez-Kabana, R. 1982. Soil amendments with oil cakes and chicken litter for control of Meloidogyne arenaria. *Nematropica* 12(12): 205–220.

Mire, G. L., Nguyen., M. L., Fassotte, B., Jardin, P. D., Verheggen, F., Delaplace, P. M. and Jijakli, H. M. 2016. Review: Implementing plant biostimulants and biocontrol strategies in the agroecological management of cultivated ecosystems. *Biotechnologie, Agronomie, Société et Environnement* 20(S): 299–313. https://hdl.handle.net/2268/188662.

Mirpoor, S. F., Giosafatto, C. V. L. and Porta, R. 2021. Biorefining of seed oil cakes as industrial co-streams for production of innovative bioplastics. A review. *Trends in Food Science & Technology* 109: 259–270.

Mohan, K. 2011. Comparison of inorganic and organic nematicides on the population of soil nematodes in hybrids of Saccharum species. *Journal of Biopesticides* 4(2): 201.

Morgan, S. E. 2015. Gossypol toxicity in livestock. *Oklahoma Cooperative Extension Service*. 1–2. http://osufacts.okstate.edu. (accessed June 14, 2022).

Nadeem, K. and Azhar, F. M. 2005. Genetic analysis of fibre length and strength of *Gossypium hirsutum* L. *International Journal of Agriculture and Biology* 7: 263–265.

Nagalakshmi, D., Rao, S. V. R., Panda, A. K. and Sastry, V. R. 2007. Cottonseed meal in poultry diets: A review. *The Journal of Poultry Science* 44(2): 119–134.

Nazarova, I. P., Glushenkova, A. I.and Umarov, A. U. 1981. Gossypol-like compounds of the cotton plant. Methods of determining Gossypol. *Chemistry of Natural Compounds* 17(2): 87–102.

Nazzaro, F., Fratianni, F., De Martino, L., Coppola, R. and De Feo, V. 2013. Effect of essential oils on pathogenic bacteria. *Pharmaceuticals* 6(12): 1451–1474.

Oka, Y., Nacar, S., Putievsky, E., Ravid, U., Yaniv, Z. and Spiegel, Y. 2000. Nematicidal activity of essential oils and their components against the root-knot nematode. *Phytopathology* 90(7): 710–715.

Okonkwo, S. I. and Okafor, E. C. 2016. Determination of the proximate composition, physicochemical analysis and characterization of fatty acid on the seed and oil of *Gossypiumhirsutum*. *International Journal of Chemistry* 8(3): 57–61.

Oskoueian, E., Abdullah, N., Hendra, R. and Karimi, E. 2011. Bioactive compounds, antioxidant, xanthine oxidase inhibitory, tyrosinase inhibitory and anti-inflammatory activities of selected agro-industrial by-products. *International Journal of Molecular Sciences* 12(12): 8610–8625.

Ostbring, K., Malmqvist, E., Nilsson, K., Rosenlind, I. and Rayner, M. 2019. The effects of oil extraction methods on recovery yield and emulsifying properties of proteins from rapeseed meal and press cake. *Foods* 9(1): 19.

Osti, N. P. and Pandey, S. B. 2006. Use of whole cotton seed and cotton seed meal as a protein source in the diet of ruminant animals: Prevailing situation and opportunity. In *Proceedings of the 6th national workshop on livestock and fisheries research* 111–119.

Parihar, K., Rehman, B., Ganai, M. A., Asif, M. and Siddiqui, M. A. 2015. Role of oil cakes and *Pochonia chlamydosporia* for the management of *Meloidogyne javanica* attacking *Solanum melongena* L. *International Journal of Plant Pathology* 6: 12–20.

Pettersson, H. 2012. Mycotoxin contamination of animal feed. In *Animal Feed Contamination*, ed. J. Fink-Gremmels, 233–285. Oxford: Woodhead Publishing.

Pousga, S., Boly, H., Lindberg, J. E. and Ogle, B. 2007. Evaluation of traditional sorghum (*Sorghum bicolor*) beer residue, shea-nut (*Vitellaria paradoxa*) cake and cottonseed (*Gossypium* Spp.) cake for poultry in Burkina Faso: Availability and amino acid digestibility. *International Journal of Poultry Science* 6(9): 666–672.

Prasad, B. K., Singh, G. and Sharma, A. K. 2022. Bio-management of diseases and insect pests in vegetable crops. *The Pharma Innovation Journal* 11(1): 153–155.

Prasad, D. and Chawla, M. L. 1994. Cotton seed oil cake extract for the control of root-knot nematode, *Meloidogyne incognita* on soybean. *Indian Journal of Nematology* 24(2): 191–194.

Radwan, M. A., El-Maadawy, E. K., Kassem, S. I. and Abu-Elamayem, M. M. 2009. Oil cakes soil amendment effects on *Meloidogyne incognita*, root-knot nematode infecting tomato. *Archives of Phytopathology and Plant Protection* 42(1): 58–64.

Ramachandran, S., Singh, S. K., Larroche, C., Soccol, C. R. and Pandey, A. 2007. Oil cakes and their biotechnological applications–a review. *Bioresource Technology* 98(10): 2000–2009.

Rehman, B., Parihar, K., Ganai, M. A., Usman, A. and Siddiqui, M. A. 2011. Biomanagement of root knot nematode through non edible oil seed cakes infesting *Cicer arietinum* L. *International journal of Applied Biology and Pharmaceutical Technology* 2(4): 411–414.

Rhee K.C., Lusas E. W., Watkins L. R., and Kinard D. H. 1988. Proceedings 37th Oilseed Processing Clinic. New Orleans, LA. March 14–15.

Rodriguez-Miranda, J., Hernandez-Santos, B., Herman-Lara, E., Vivar-Vera, M. A., Carmona-García, R., Gómez-Aldapa, C. A. and Martinez-Sanchez, C. E. 2012. Physicochemical and functional properties of whole and defatted meals from Mexican (*Cucurbita pepo*) pumpkin seeds. *International Journal of Food Science & Technology* 47(11): 2297–2303.

Ryan, M. F. and Byrne, O. 1988. Plant-insect coevolution and inhibition of acetylcholinesterase. *Journal of Chemical Ecology* 14(10): 1965–1975.

Scott, S. J., Jones, R. A. and Williams, W. 1984. Review of data analysis methods for seed germination 1. *Crop Science* 24(6): 1192–1199.

Sekhar, S. C. and Rao, B. V. K. 2011. Cottonseed oil as health oil. *Pertanika Journal of Tropical Agricultural Science* 34 (1): 17–24.

Shahidi, F., Oh, W. Y., Wan, P. J. and Wakelyn, P. J. 2020. Cottonseed oil. In *Bailey's Industrial Oil and Fat Products* (7th ed.), ed. F. Shahidi, 1–95. New Jersey: John Wiley & Sons, Ltd.

Sharma, G. C., Gupta, M., Singh, P. and Kumari, S. 2016. Management of plant parasitic nematodes in tomato through chemicals and organic amendments. *International Journal of Farm Sciences* 6(4): 273–278.

Sharma, V. P. 2017. Overview of oilseeds sector: Current status and growth behaviour. In *Oilseed Production in India*, 17–54. New Delhi: Springer.

Singh, S., Singh, B. and Singh, A. P. 2015. Nematodes: A threat to sustainability of agriculture. *Procedia Environmental Sciences* 29: 215–216.

Sivakumar, M. and Gunasekaran, K. 2011. Management of root knot nematodes in tomato, chilli and brinjal by neem oil formulations. *Journal of Biopesticides* 4(2): 198–200.

Soler-Serratosa, A., Kokalis-Burelle, N., Rodríguez-Kabana, R., Weaver, C. F. and King, P. S. 1996. Allelochemicals for control of plant-parasitic nematodes. 1. *In vivo* nematicidal efficacy of thymol and thymol/benzaldehyde combinations. *Nematropica* 57–71.

Sumbul, A., Rizvi, R., Mahmood, I. and Ansari, R. A. 2015. Oilcake amendments: Useful tools for the management of phytonematodes. *Asian Journal of Plant Pathology* 9(3): 91–111.

Swick, R. A. 1999. Considerations in using protein meals for poultry and swine. *ASA Technical Bulletin* 21: 1–11.

Tariq, M., Amir, K. and Siddiqui, M. A. 2018. Comparative efficacy of different oil cakes for sustainable management of root knot nematode, *Meloidogyne incognita* affecting *Abelmoschus esculentus* L. *Trends in Biosciences* 11(2): 131–135.

Thakur, A., Katoch, S. and Mane, B. G. 2019. Cottonseed meal as a protein replacer in broiler chicken. *Veterinary Research* 7(1): 42–50.

Tikoria, R., Kaur, A. and Ohri, P. 2022. Potential of vermicompost extract in enhancing the biomass and bioactive components along with mitigation of *Meloidogyne incognita*-induced stress in tomato. *Environmental Science and Pollution Research* 29: 56023–56036.

Tiyagi, S. A., Khan, A. V. and Alam, M. M. 2002. Biodegradable effect of oil seed cakes on plant-parasitic nematodes and soil inhabiting fungi infesting *Trigonella foenumgreacum* and *Phaseolus aureus*. *Indian Journal of Nematology* 32: 47–57.

Ward, J. A. 1976. Processing high oil content seeds in continuous screw presses. *Journal of the American Oil Chemists Society* 53: 261–264.

Williamson, V. M. 1998. Root-knot nematode resistance genes in tomato and their potential for future use. *Annual Review of Phytopathology* 36: 277–293.

Yasmin, S., Hashem, M. A. and Hoque, T. S. 2020. Influence of integrated use of cotton seed oilcake and mustard oilcake with chemical fertilisers on growth, yield and nutritional quality of rice. *Progressive Agriculture* 31(3): 144–153.

Yassin, M. Y. and Ismail, A. E. 1994. Effect of some oilseed cakes as soil amendments and Aldicarb on cowpea plants infected with *Rotylenchulusreniformis* Lin. & Ol. in relation to soil type. *Anzeiger für Schädlingskunde, Pflanzenschutz, Umweltschutz* 67(8): 176–178.

Yu, J., Yu, S., Fan, S., Song, M., Zhai, H., Li, X. and Zhang, J. 2012. Mapping quantitative trait loci for cottonseed oil, protein and Gossypol content in a *Gossypium hirsutum*, *Gossypium barbadense* backcross inbred line population. *Euphytica* 187(2): 191–201.

5 Castor Oilseed Cake

Chemical Compositions and Nematicidal Potential

Venkadesh G, Karthik R and Parthiban M

CONTENTS

5.1 INTRODUCTION

Plant-parasitic nematodes (PPNs) are a significant biological barrier for commercially important economic crops found worldwide, resulting in substantial losses. PPNs are thought to cause an annual loss in horticulture crops of 8.8–14.6% of overall crop production and a monetary cost of 100–157 billion in US dollars worldwide (Sumbul et al. 2015; Adomako et al. 2017). The underground sections of plants, such as their roots, bulbs, corms, and rhizomes, are the major targets of phytonematodes, which impede the uptake and movement of water. Some PPNs can also infect things that are aboveground, like leaves, buds, and flowers. In addition to having a detrimental effect on plant growth, phytonematodes also hinder nodulation, nitrogen fixation, and the overall yield of plants. Also, they act as a vector for various virus diseases in plants (Sumbul et al. 2015; Galbieri et al. 2021). Today, the chemical application has greatly grown to satisfy the expanding demand for aggressive farming, leading to contamination of ground and surface water and interference with soil biota. In this contest, numerous common nematicides, like dibromochloropropane and ethylene dibromide, have been outlawed completely. Methyl bromide, the most popular and efficient fumigant for soil-borne illnesses and weeds, including nematodes, has already been banned in certain nations, and international accords call for its complete removal from the market in the majority of nations (Oka et al. 2000; Abdel-Rahman et al. 2008). Due to the growing demand for safe and healthy food and long-term sustainability, as well as the concern about environmental contamination brought on by the indiscriminate use of pesticides, organic farming has drawn many producers' attention (Galbieri et al. 2021). Ultimately, the application of organic manure is paramount for

DOI: 10.1201/9781003319252-5

managing plant diseases caused by parasitic nematodes (Sumbul et al. 2015). Oil cakes, chopped plant parts, and seed dressing prepared from plant extracts are only a few examples of organic additives of plant origin that have been utilized to combat nematodes.

Among the organic amendments, some plant oilcake has been discovered to be useful against phytonematodes. A compacted mixture of the meal left over after vegetable oils have been extracted from seeds like cottonseed, groundnut, and castor, among others, is referred to as oilseed cake. These are excellent as chicken and other animal feed because they are high in protein and mineral content. Certain seeds, including castor beans and tung nuts, produce oil cakes that are toxic to different dangerous microorganisms and are utilized as fertilizers rather than food. Oilseed cakes can be sprinkled or placed in pellet form on the soil, watered, and let to break down. The addition of organic amendment caused a high concentration of nitrates and ammonia to accumulate, which was extremely harmful to various nematodes in the soil. In addition to being toxic to nematodes such *Hoplolaimus indicus*, *Rotylenchulus reniformis*, and *Tylenchorynchus brassicae*, water-soluble fractions of oilseed cakes derived from neem, mahua, groundnut, and castor also prevented *M. incognita* larvae from hatching (Varaprasad et al. 2005; Mishra and Pandre 2017). Their nematicidal influence grows as the decomposition time grows. Before planting, plant seeds can also be treated with oilcake extracts to nematode-proof the crop. The literature claims that the application of castor oilseed cake (Moraes and Lordello 1977; Santos et al. 2013; Pedroso et al. 2019) or plant extract (Katooli et al. 2010; Adomako and Kwoseh 2013) significantly suppresses the nematode population in various crop and enhance the crop growth. Globally, 1.29 million hectares of castor were planted in 2018, yielding 1.39 million tonnes of castor oilseed. India, China, and Brazil have the highest production levels (FAO 2020). A significant amount of "by-product" is produced during the industrial process of extracting oil from the castor seed, with castor seed cake accounting for about 53% of the grain weight. Nearly 1.13 tonnes of castor seed cake are made from every ton of castor oil (Lima et al. 2011). Nematologists are becoming more interested in this trend because it offers efficient management against the target nematode and prevents environmental degradation. In this chapter, we discuss the chemical composition, mode of action of castor oilseed cake on phytonematodes and various formulations, and the merits and demerits of castor oilseed cake in nematode management.

5.2 PRODUCTION OF CASTOR OILSEED CAKE

Castor oilseed cake is a common source of organic fertilizer. Many industries exist for the production of castor oilseed cake commercially. It contains more nitrogen (N); (Oliveira et al. 2010), which helps improve crop growth. The stepwise process for the production of castor oilseed cake, as described by Deshamukh et al. (2018), is depicted in Figure 5.1.

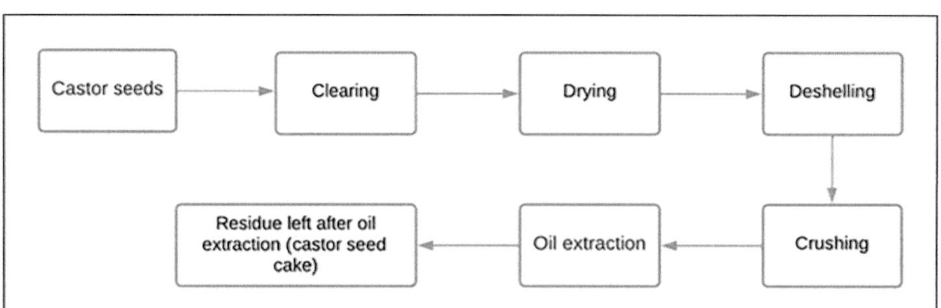

FIGURE 5.1 Stepwise process of castor oilseed cake production.

5.3 CHEMICAL COMPOUNDS IN CASTOR OILSEED CAKE

5.3.1 Mineral Composition in Castor Oilseed Cake

Carbon is an important source for nematode control, where carbon helps improve the biocontrol agent's ability and increases urea's activity against nematodes (Rodriguez- Kabana and King 1980). As the castor seed cake possesses about 45% of carbon (de Castro et al. 2016), it probably helps to control nematodes by various mechanisms. The nitrogen content of castor seed cake ranges from 3.14 to 8.7% (de Castro et al. 2016; Pedroso et al. 2019), which assists in reducing the incidence of plant parasitic nematodes (Zhou et al. 2021). Phosphorus (P) is also essential for plant growth, and P supplementation reduces the nematode population (Zhou et al. 2014; Olatunji et al. 2019). However, the amount of phosphorus in castor seed cake is lower (0.62%) (Pedroso et al. 2019), and the P in castor seed cake may also be involved in nematode control. Other macronutrients such as potassium (K), magnesium (Mg), calcium (Ca), and sulfur (S) present in castor seed cake are 0.78%, 0.30%, 0.79%, and 0.27%, respectively (Pedroso et al. 2019). These macronutrients also contribute to controlling nematodes. For example, K inhibits the population of *Heterodera glycines* in soybeans (Barbosa et al. 2010), and Ca controls *Meloidogyne incognita* in *Cucurbita pepo* (Mohamed and Youssef 2009). Besides these important macronutrient components in castor seed cake, it contains 6.5% and 32% hydrogen and oxygen, respectively.

The different micronutrients in castor seed cake are 945.51, 458.54, 115.64, 24.63, and 100.51 ppm of iron, manganese, copper, boron, and zinc, respectively (Pedroso et al. 2019). These micronutrients in castor seed cake provide additional benefits as they also prevent nematode development. For instance, a few studies proved the role of iron, zinc, manganese, and boron on nematode development (Haque and Mukhopadhyaya 1983; Couto et al. 2016; Mansourabad et al. 2016). Apart from nematode control, the mineral composition in castor seed cake also promotes plant growth by various mechanisms, making castor seed cake a suitable organic manure for various crops.

5.3.2 Biochemical Composition of Castor Oilseed Cake

The composition of various biochemical components present in the castor seed cake includes ash (6.25%), moisture (8.79%), carbohydrate (13.72%), lipids (8.72%), protein (43.54%), insoluble neutral detergent fiber (20.99%), insoluble acid detergent fiber (16.64%), cellulose (6.79%), hemicellulose (4.36%), and lignin (9.85%); (de Castro et al. 2016). Annongu and Joseph (2008) reported biochemical composition in a decorticated and un-decorticated seed cake. In the case of un-decorticated seed cake, the biochemical composition is 31.06%, 24.69%, and 19.40% of protein, carbohydrate, and fat, respectively. Compared to un-decorticated seed cake, the decorticated seed cake has a higher amount of biochemical composition as 35.43% protein and 24.88% carbohydrate. Jones (1947) also proved that the decorticated castor seed cake contains more protein (46.37%) than un-decorticated whole castor seed cake (20.44%). The amino acid composition of castor seed cake also varies from study to study. In general, amino acids such as lysine, isoleucine, and tryptophan are deficient in castor seed cake (Annongu and Joseph 2008).

Ricin is a poisonous lectin compound (a carbohydrate-protein bound compound) in castor seed cake. In the case of ricin, the various amino acids, namely alanine (1%), humin (1.8%), phenylalanine (0.4%), cysteine (1%), tyrosine (2.7%), proline (4.8%), aspartic acid (2%), glutamic acid (20%), arginine (11.7%), lysine (6.3%), and tryptophan (0.4%), are present. In contrast, amino acids, namely glycine, histidine, and hydroxyproline, are absent (Jones 1947). Ricin is majorly present in the endosperm of castor seeds and has ribosome-inactivating protein (RIP) (Fernandes and Machado 2012). Ricin in castor seed cake prevents nematodes by decreasing their movement/motility (Rich et al. 1989) and reduces their reproduction (Roldi et al. 2013). Besides ricin, hydro cyanides and other allergens are also potent toxic proteins present in castor seed cake. Ricinine is another protein reported in castor seed cake, but it is relatively harmless compared to any other toxic proteins present in the castor seed cake (Annongu and Joseph 2008). Ricinoleic acid is the dominant fatty acid present in the castor seed cake. Apart from ricinoleic acid, isoricinoleic acid, stearic acid, and dihydroxy stearic acids are also prevalent in castor seed cake (Herrera et al. 2021).

5.3.3 Volatile Organic Compounds in Castor Oilseed Cake

Among all the biochemical compositions of castor seed cake, volatile organic compounds (VOCs) play a major role in nematode control. VOCs present in castor seed cake mostly belong to different classes of chemicals such as alcohols, terpenes, ketones, phenols, and sulfur compounds. Perhaps, the most predominant VOC class in castor seed cake is ketones.

In the case of ketone compounds in castor seed cake, 2-pentanone (ketone), 6-methyl-2-heptanone (ketone), phenyl ethyl ketone (ketone), 2-butanone (ketone), 2-hexanone (ketone), ketone (ketone), p-methyl acetophenone (ketone), methyl-isobutyl ketone (ketone), 2-nonanone (ketone), 2-heptanone (ketone), 2-undecanone (ketone), and 2-octanone (ketone) are the predominant ones. The different alcohol compounds present in castor seed cake are 2-octanol (alcohol), phenol (phenol), phenethyl alcohol (alcohol), and 4-methyl-phenol (phenol). Dimethyl disulfide (sulfurate), dimethyl tetrasulfide (sulfurate), methanethiol (sulfurate), and dimethyl trisulfide (sulfurate) are the common sulfurate compounds in castor seed cake. In the case of terpene compounds, pinocarveol (terpene), verborene (terpene), nopinone (terpene), cuparene (terpene), isopinocanphone (terpene), and borneol (terpene) are the usual ones and release aromatic compounds against nematodes. The other minor VOCs identified in castor seed cake are 2-methy-propanoic acid (carboxylic acid), skatole (amine), p -methylanisole (ether), carbon dioxide (oxide), γ-decalactone (ester), and 2-methylbutanoic acid (carboxylic acid; Pedroso et al. 2019). Among all the volatile compounds identified, 2-octanone, 4-methyl-phenol, γ-decalactone, phenol, and skatole are quantified at a higher intensity and proved effective against root-knot nematode *Meloidogyne incognita* (Pedroso et al. 2020). Thus, the application of VOCs could serve as an alternative for nematicides and provide effective control against PPNs.

5.4 CASTOR OILSEED CAKE MECHANISM/MODE OF ACTION IN NEMATODE MANAGEMENT

Castor oil cakes are applied in several formulations like crude oilseed cake, aqueous extract, and powdered oil cake for PPN management. The application of castor oil cake as a soil amendment releases essential plant nutrients that promote root growth and thus help plants escape nematode attack (Pandey and Singh 1990). The application of castor oil cake significantly decreases the mobility and egg hatching and increases the mortality of phytonematodes (Pedroso et al. 2019; Galbieri et al. 2021). The nematicidal properties of castor bean cake are mainly due to the production of volatile organic compounds in the soil, such as y-decalactone, skatole, phenol, and 4-methyl-phenol, which are highly toxic for nematode growth and development (Figure 5.2). The nematicidal properties are increased with castor oil cake's decomposition time (Galbieri et al. 2021). In addition, castor oil cake improves microbial growth and is a good source of nutrients and minerals for different soil types (Capuani et al. 2012).

Crude castor bean and oil cake aqueous extract applied in various concentrations to soil cause a direct effect on the eggs and juveniles of nematodes. This is mainly attributed to the toxic chemical ricin and other secondary metabolites in castor, which have been described as having ovicidal and larvicidal properties to nematodes (Audi et al. 2005; Adegbite and Adesiyan 2005; Adomako and Kwoseh 2013).

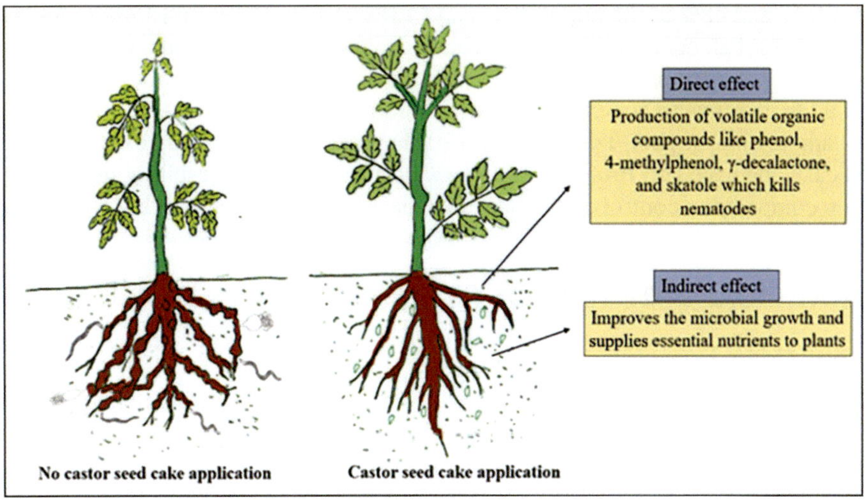

FIGURE 5.2 Mode of action of castor oilseed cake application in nematode management.

5.5 PHYTOTOXICITY OF CASTOR OILSEED CAKE

However, the application of castor oilseed cake significantly suppresses the nematode population, and an increase in concentration shows phytotoxicity in plants. Moraes and Lordello (1977) recorded 3% concentration of castor seed cake causing phytotoxicity in coffee seedlings. Santos et al. (2013) recorded phytotoxicity in banana plants. Galbieri et al. (2021) found that applying castor seed cake above 100kg/ha causes phytotoxicity to cotton. The phytotoxic effect of castor cake has been attributed to its low C/N ratio (7:1), which would lead to fast nitrogen mineralization and accumulation of ammonium and nitrate in the soil (Lima et al. 2011; Santos et al. 2013). Therefore, it is essential to standardize the optimum quantity/concentration of castor oilseed cake, aqueous extract, and the amendment to be applied in soil for successful nematode management and enhancing plant growth.

5.6 NEMATICIDAL POTENTIAL OF CASTOR OILSEED CAKE

For producing 1 ton of castor oil from castor, 1.31 tons of husk and 1.13 tons of castor oilseed cake (meal) are produced as a by-product (Lima et al. 2011). The castor seed cake applied as organic amendments in soil improves plant growth and yield and has an antagonistic effect against nematodes (Table 5.1).

Pedroso et al. (2019) found that castor seed cake application in soil resulted in immobility, death, and reduction in the hatching of *M. incognita* at >97.3%, >96.9%, and >74.3%, respectively, in tomato plants. Furthermore, they identified 32 VOCs from castor seed cake application in soil of which phenol, 4-methylphenol, γ-decalactone, and skatole are important VOCs with nematicidal potential. Devi et al. (2019) conducted an experiment to find the efficacy of neem cake, mustard cake, and castor cake at 5g and 10g per kg of soil against *Meloidogyne graminicola* inoculated in a rice nursery. Results suggest that castor cake reduces nematode galls, and reproduction further improves plant growth.

Prasad et al. (2005) conducted an experiment on the efficacy of different oil cakes against *Meloidogyne graminicola* in rice in which castor cake at 5g/kg soil increased the plant growth even in the presence of rice root-knot nematode. Similarly, castor seed cake amended in soil reduced eggs per plant by 18% and 48% at the rate of 0.5% and 1.0% in tomatoes against *Meloidogyne javanica* (Lopes et al. 2009). Patel et al. (2005) evaluated organic cakes and nematicides alone or in combination against *M. javanica* in fennel and they were found that castor cake one ton per ha + phorate 1 kg per ha as an effective treatment, with a minimum root-knot index of 1.72.

For easy application in pots, Izidoro et al. (2021) prepared an aqueous extract with 10% from castor seed cake and tested on *Pratylenchus brachyurus* in soybeans, resulting in 65% mortality for 20% dilution compared to control and a 40% increase in the shoot dry weight of soybeans.

5.7 NEMATICIDAL POTENTIAL OF OTHER PARTS OF CASTOR PLANT

5.7.1 Plant Extract

Parts of the plant and its extract have been used to manage plant parasitic nematodes (Khan et al. 2011, Mousa et al. 2011) (Table 5.1). Castor leaves are more readily available than castor seeds, so their extraction is much easier than castor cake or seed extract.

TABLE 5.1
Compilation of Different Castor-Based Formulations against Plant-Parasitic Nematodes

Application	Experiment Site	Nematode spp.	Crop	Dose	% Larval Mortality	Reference
Castor Oilseed Cake	Laboratory	*Meloidogyne incognita*	Tomato	0–6% (v/v)	>96.9%	Pedroso et al. 2019
		Meloidogyne graminicola	Rice	5g/kg and 10 g/kg soil	–	Devi et al. 2019
		Meloidogyne graminicola	Rice	2.5g/kg and 5g/kg soil	–	Prasad et al. 2005
		Meloidogyne javanica	Tomato	1%	–	Lopes et al. 2009
	Field	*Meloidogyne javanica*	Fennel	1 ton/ha	–	Patel et al. 2005
Aqueous Extract of Castor Cake	Laboratory	*Pratylenchus brachyurus*	Soybean	20%	65%	Izidoro et al. 2021
	Laboratory and field	*Meloidogyne incognita*	Tomato	10–40%	–	Adomako and Kwoseh 2013
	Laboratory	*Meloidogyne javanica*	Tomato	100%	–	Gardiano et al. 2009
	Laboratory	*Meloidogyne incognita*	Carrot	100%	–	Baldin et al. 2012
	Laboratory and greenhouse	*Meloidogyne incognita*	Cucumber	0–1000 ppm	–	Katooli et al. 2010
	Laboratory	*Meloidogyne* spp.	-	5% and 10%	26% and 81%	Salim et al. 2016
	Laboratory and field	*Meloidogyne* spp.	Tomato	10%, 20%, 30%, and 40%	–	Adomako and Kwoseh 2013
Castor Root Extract	Laboratory	*Meloidogyne incognita*	Soybean	100%	62.1%	Adegbite and Adesiyan 2006

Gardiano et al. (2009) used 20 different plant extracts to control parasitic plant nematode *Meloidogyne javanica* in tomatoes, of which *Ricinus communis* extract reduced the number of galls to 54.4% and the number of eggs to 56.6% compared to the control. Baldin et al. (2012) also investigated the effectiveness of various plant extracts against *M. incognita* in carrot *(Dacus carota). Ricinus communis* extract, which included the leaves, branches, and fruits, had the highest root weight and diameter of all of them.

Katooli et al. (2010) did an experiment to manage *M. incognita* in cucumbers using different plant extracts with different concentrations both in the laboratory and greenhouse, resulting in the immobility of 61.33% in the castor leaf extract. Furthermore, castor plant extract at 5% and 10% concentrations indicated mortality of 26% and 81%, respectively, when Salim et al. (2016) tested several plant extracts on juveniles of *Meloidogyne* spp. in the lab. Adomako and Kwoseh (2013) tested the different concentrations of aqueous extracts of castor seed against root-knot nematodes in both in vitro and field experiments on tomato and They found inhibition in egg hatching. Juvenile mortality increased with an increased concentration of aqueous extracts of 10%, 20%, 30%, and 40%.

5.7.2 Root Extract

Castor root secretes root metabolites such as palmitic acid and linoleic acid, which are repellent to plant parasitic nematodes in soil (Dong et al. 2018), so castor root extracts, when applied in soil, reduce the nematode population (Table 5.1). Root extracts of castor at 100% concentration inhibit egg hatching and increase larval mortality by 93% and 62.1% after 48 hr in soybeans against *Meloidogyne incognita* (Adegbite and Adesiyan 2006).

5.8 CONCLUSION AND FUTURE PERSPECTIVES

In order to maintain a sustainable ecosystem and not harm the environment, organic management practices should be followed to manage nematodes. Although nematicides are preferred over organic management practices in providing quicker results, nematodes can also be effectively managed by applying components like castor seed cake. The application of castor seed cake in the soil acts as a nematicide and organic manure, which has a dual role and a synergistic effect that directly helps in nematode management, indirectly supplies essential nutrients for plant growth and development, and improves the soil microbiota. Castor seed cake helps reduce the cost of cultivation for farmers and improves soil health. One detriment of castor seed cake is phytotoxicity if excessively applied. Although the chemical composition and volatile organic compounds in castor seed cake are very effective against PPNs, studies are limited. However, castor seed cake is essential for nematode management for the conservation of agriculture and sustainable farming. Emphasis should be given to different application methods of castor seed cake into the soil and application at different rates and during different crop stages against nematodes. Studies on the synthesis and commercialization of the volatile chemical compounds found in castor seed cake have provided a option for controlling nematodes. The most challenging aspect of castor crop cultivation is a lack of adequate and loss-free mechanical harvest. However, it is essential in farm mechanization to increase the area and production of castor cultivation. Developing a commercial formulation to meet farmers' needs is also important. To overcome the phytotoxicity of castor seed cake, application quantities for individual crops are needed. Ready-to-use commercial products with castor seed cake should also be developed for a better future. Still, more research must be conducted to find the efficacy of castor seed cake with biocontrol agents for effective control of nematodes in field conditions. Most of the research has been conducted on aqueous extracts from the castor plant, but only a little research has been conducted on castor seed cake. So, future research should focus on castor seed cake extracts and their formulation for easy application *via* drip irrigation, sprinkler irrigation, drone-based spraying, and so on, as agriculture moves toward mechanization, smart agriculture, and precision farming.

REFERENCES

Abdel-Rahman, F. H., Clark, S., and Saleh, M. A. 2008. Natural organic compounds as alternative to methyl bromide for nematodes control. *Journal of Environmental Science and Health* 43(8):680–685.

Adegbite, A. A., and Adesiyan, S. O. 2005. Root extracts of plants to control root knot nematode on edible soybean. *World Journal of Agricultural Science* 1(1):18–21.

Adegbite, A. A., and Adesiyan, S. O. 2006. Root extracts of plants to control root-knot nematode on edible soybean. *Journal of Vegetable Science* 12(2):5–12.

Adomako, J., and Kwoseh, C. K. 2013. Effect of castor bean (*Ricinus communis* L.) aqueous extracts on the performance of root-knot nematodes (*Meloideogyne* spp.) on tomato (*Solanum lycopersicum* L.). *Journal of Science and Technology (Ghana)* 33(1):1–11.

Adomako, J., Osei, K., Kankam, F., and Danso, Y. 2017. Limitations to peri-urban vegetable farming in Ghana: An overview of root knot nematodes contribution. *International Journal of Sciences: Basic and Applied Research* 36(3):75–85.

Annongu, A. A., and Joseph, J. K. 2008. Proximate analysis of castor seeds and cake. *Journal of Applied Sciences and Environmental Management* 12(1):39–41.

Audi, J., Belson, M., Patel, M., Schier, J., and Osterloh, J. 2005. Ricin poisoning: A com-prehensive review. *Journal of American Medical Association* 294(18):2342–2351

Baldin, E. L. L., Wilcken, S. R. S., Pannuti, L. E. D. R., Schlick-Souza, E. C., and Vanzei, F. P. 2012. Use of botanical extracts, cassava wastewater and nematicide for the control of root-knot nematode on carrot. *Summa Phytopathologica* 38:36–41.

Barbosa, K. A. G., Garcia, R. Á., Santos, L. C., Teixeira, R. A., Araújo, F. G. D., Rocha, M. R. D., and Lima, F. S. D. O. 2010. Effect of potassium fertilization on *Heterodera glycines* reproduction in susceptible and resistant soybean cultivars. *Nematologia Brasileira Piracicaba (SP) Brasil* 34(3):150–158.

Capuani, S., Rigon, J. P. G., Beltrão, N. E. M., and Brito Neto, J. F. 2012. Atividade microbiana em solos, influenciada por resíduos de algodão e torta de mamona. *Revista Brasileira de Engenharia Agrícola e Ambiental* 16(12):1269–1274.

Couto, E. A. A., Dias-Arieira, C. R., Kath, J., Homiak, J. A., and Puerari, H. H. 2016. Boron and zinc inhibit embryonic development, hatching and reproduction of *Meloidogyne incognita. Acta Agriculturae Scandinavica, Section B-Soil and Plant Science* 66(4):346–352.

De Castro, A. M., Dos Reis Castilho, L., and Freire, D. M. G. 2016. Characterization of babassu, canola, castor seed and sunflower residual cakes for use as raw materials for fermentation processes. *Industrial Crops and Products* 83:140–148.

Deshamukh, M., Shinde, A., Nirmal, S., and Nair, H. 2018. Synthesis and characterization of castor oil and deoiled cake as fertilizer. *International Journal of Chemical Engineering Research* 10(2):95–103.

Devi, P., Kanwar, R. S., and Patil, J. A. 2019. Effect of oil cakes for the management of *Meloidogyne graminicola* in rice nursery. *Journal of Entomology and Zoology Studies* 7:180–182.

Dong, L., Li, X., Huang, C., Lu, Q., Li, B., Yao, Y., Liu, T., and Zuo, Y. 2018. Reduced *Meloidogyne incognita* infection of tomato in the presence of castor and the involvement of fatty acids. *Scientia Horticulturae* 237:169–175.

FAO. 2020. FAOSTAT. Rome: Food and Agriculture Organization of the United Nations. [accessed 2020 Jul]. Available from: www.fao.org/faostat/en/#data/QC/visualize.

Fernandes, K. V., and Machado, O. L. T. 2012. Approaches for the detection of toxic compounds in castor and physic nut seeds and cakes. In *Biodiesel-Feedstocks, Production and Applications.* IntechOpen.

Galbieri, R., Kobayasti, L., Albuquerque, M. C., de Sá, R. O., Dutra, S. G., Boldt, A. S., and Timper, P. 2021. Castor bean as an option for *Meloidogyne incognita* management in cotton. *International Journal of Pest Management*, 1–9.

Gardiano, C. G., Ferraz, S., Lopes, E. A., Ferreira, P. A., Amora, D. X., and de Freitas, L. G. 2009. Avaliação de extratos aquosos de várias espécies vegetais, aplicados ao solo, sobre Meloidogyne javanica (Treub, 1885) Chitwood, 1949. *Semina: Ciências Agrárias* 30(3):551–556.

Haque, M. S., and Mukhopadhyaya, M. C. 1983. Influence of some micro-nutrients on *Rotylenchulus reniformis. Indian Journal of Nematology* 13(1):115–116.

Herrera, M. D., Jiménez-Ocampo, R., Rosales-Serna, R., Rodríguez-González, J. A., Santos-De la Cruz, J. L., Cruz-Bravo, R. K., Aguilar, P. A., Gamboa-Gómez, C. I., and Guerrero-Romero, F. 2021. A simple thermal-detoxified method for castor bean (*Ricinus communis* L.) cake, and its potential nutraceutical properties. *Industrial Crops and Products* 174:114–151.

Izidoro Jr, A., Silva, E. J., Tarini, G., Bordin, J. C., Silva, B. A., Ambrosano, L., and Dias-Arieira, C. R. 2021. Aqueous extract of castor bean seed cake for the control of *pratylenchus brachyurus* in soybean. *Nematropica* 51(1):1–8.

Jones, B. D. 1947. Proteins of the castor bean–Their preparation, properties, and utilization. *Journal of the American Oil Chemists' Society* 24(7):247–251.

Katooli, N., Moghadam, E. M., Taheri, A., and Nasrollahnejad, S. 2010. Management of root-knot nematode (*Meloidogyne incognita*) on cucumber with the extract and oil of nematicidal plants. *International Journal of Agricultural Research* 5(8):582–586.

Khan, S. A., Javed, N., Khan, M. A., Haq, I. U., and Safdar, A. 2011. Use of plant extracts as bare dip root treatment for the management of *Meloidogyne incognita. Pakistan Journal of Phytopathology* 23(1):9–13.

Lima, R. L. S., Severino, L. S., Sampaio, L. R., Sofiatti, V., Gomes, J. A., and Beltrão, N. E. M. 2011. Blends of castor meal and castor husks for optimized use as organic fertilizer. *Industrial Crops and Products* 33(2):364–368.

Lopes, E. A., Ferraz, S., Dhingra, O. D., Ferreira, A., and Freitas, L. G. 2009. Soil amendment with castor bean cake and jack bean seed powder to control *Meloidogyne javanica* on tomato roots. *Nematologia Brasileira* 33(1):106–109.

Mansourabad, A. M., Bideh, K. A., and Abdollahi, M. 2016. Effects of some micronutrients and macronutrients on the root-knot nematode, *Meloidogyne incognita*, in greenhouse cucumber (*Cucumis sativus* cv. *Negin*). *Journal of Crop Protection* 5(4):507–517.

Mishra, K. K., and Pandre, P. K. 2017. Comparative efficacy of different oil cakes as a soil amendments against *M. incognita* on the yield of soyabean. *International Journal of Advanced Research in Chemical Science* 4(2):1–5.

Mohamed, M. M., and Youssef, M. M. A. 2009. Efficacy of calcium carbide for managing *Meloidogyne incognita* infesting squash. *Egyptian Journal of Agronematology* 19:229–231.

Moraes, M. V., and Lordello, L. G. 1977. Uso de torta de mamona no controle de nematóides em solo para viveiro de café. *Sociedade Brasileira de Nematologia* 2.

Mousa, E. M., Mahdy, M. E., and Younis Dalia, M. 2011. Evaluation of some plant extracts to control root knot nematodes, *Meloidogyne* spp. on tomato plants. *Egyptian Journal of Agronematology* 10(1):1–14.

Oka, Y., Koltai, H., Bar-Eyal, M., Mor, M., Sharon, E., Chet, I., and Spiegel, Y. 2000. New strategies for the control of plant-parasitic nematodes. *Pest Management Science* 56(11):983–988.

Olatunji, O. A., Gong, S., Tariq, A., Pan, K., Sun, X., Chen, W., and Tan, X. 2019. The effect of phosphorus addition, soil moisture, and plant type on soil nematode abundance and community composition. *Journal of Soils and Sediments* 19(3):1139–1150.

Oliveira, D. A. S., Campos, J. M. S., Oliveira, M. R. C., Brito, A. F., Valadares Filho, S. C., Detmann, E., Valadares, R. F. D., Souza, D.M.S., and Machado, O. L. T. 2010. Nutrient digestibility, nitrogen metabolism and hepatic function of sheep fed diets containing solvent or expeller castorseed meal treated with calcium hydroxide. *Animal Feed Science and Technology* 158:15–28.

Pandey, G., and Singh, K. P. 1990. Effect of organic amendments on soil microflora and nematode fauna with special reference to *M. incognita* in soyabean. *New Agriculturist* 1:65–70.

Patel, S. K., Patel, H. V., and Patel, D. J. 2005. Integrated management of root-knot nematode, *Meloidogyne javanica* (Treub) chitwood in fennel (*Foeniculum vulgare* mill.) *Journal of Spices and Aromatic Plants* 14(2):152–154.

Pedroso, L. A., Campos, V. P., Barros, A. F., Justino, J. C., and de Paula, L. L. 2020. Activity against *Meloidogyne incognita* of volatile compounds produced during amendment of soil with castor bean cake. *Nematology* 22(5):505–514.

Pedroso, L. A., Campos, V. P., Pedroso, M. P., Barros, A. F., Freire, E. S., and Resende, F. M. 2019. Volatile organic compounds produced by castor bean cake incorporated into the soil exhibit toxic activity against *Meloidogyne incognita. Pest Management Science* 75(2):476–483.

Prasad, J. S., Varaprasad, K. S., Rao, Y. R., Rao, E. S., and Sankar, M. 2005. Comparative efficacy of some oil seed cakes and extracts against root-knot nematode (*Meloidogyne graminicola*) infection in rice. *Nematologia Mediterranea* 33(2):191–194.

Rich, J. R., Rahi, G. S., Opperman, C. H., and Davis, E. L. 1989. Influence of the castor bean (*Ricinus communis*) lectin (ricin) on motility of *Meloidogyne incognita. Nematropica* 19(1):99–103.

Rodriguez-Kabana, R., and King, P. S. 1980. Use of mixtures of urea and blackstrap molasses for control of root-knot nematodes in soil. *Nematropica* 10(1):38–44.

Roldi, M., Dias-Arieira, C. R., Severino, J. J., de Melo Santana, S., Santo Dadazio, T., Marini, P. M., and Mattei, D. 2013. Use of organic amendments to control *Meloidogyne incognita* on tomatoes. *Nematropica* 43(1):49–55.

Salim, H. A., Salman, I. S., Majeed, I. I., and Hussein, H. H. 2016. Evaluation of some plant extracts for their nematicidal properties against root-knot nematode, *Meloidogyne* sp. *Journal of Genetic and Environmental Resources Conservation* 4(3):241–244.

Santos, B. H. C., Ribeiro, R. C. F., Xavier, A. A., Santos Neto, J. A., and Mota, V. J. G. 2013. Controle de *Meloidogyne javanica* em mudas de bananeira 'prata-anã' por composto orgânico. *Revista Brasileira de Fruticultura* 35(2):650–656.

Sumbul, A., Rizvi, R., Mahmood, I., and Ansari, R. A. 2015. Oilcake amendments: Useful tools for the management of phytonematodes. *Asian Journal of Plant Pathology* 9(3):91–111.

Varaprasad, K. S., Prasad, J. S., Rao, Y. R., Rao, E. S., and Sankar, M. 2005. Comparative efficacy of some oil cakes and extracts against root-knot nematode in tomato and brinjal. *Indian Journal of Plant Protection* 33(2):268–272.

Zhou, J., Wang, F., Li, J., Zou, B., Wang, X., Li, Z., and Fu, S. 2014. Effects of experimental nitrogen and/or phosphorus additions on soil nematode communities in a secondary tropical forest. *Soil Biology and Biochemistry* 75:1–10.

Zhou, Q., Xiang, Y., Li, D., Luo, X., and Wu, J. 2021. Global patterns and controls of soil nematode responses to nitrogen enrichment: A meta-analysis. *Soil Biology and Biochemistry* 163:108433.

6 Mahua Oilseed Cake
Chemical Compounds and Nematicidal Potential

Dania Ahmed, Sheila Shahab, Muneer Ahmad and Neha Pathak

CONTENTS

6.1 INTRODUCTION

Excessive utilization of chemical fertilizers and pesticides is hazardous to the environment and causes toxicity to animals and human beings. Organic farming has become a crucial part of agriculture as an alternative to conventional practices due to the rise in the demand for chemical-free and healthy food. The rising global population has led to increased cultivation, resulting in an enormous increase in the application of chemicals which have adverse outcomes like pollution of water sources and a decline in beneficial soil microflora. Ecological concerns and the safety of life forms have resulted in many popular pesticides like ethylene dibromide and dibromochloropropane (DBCP) being banned. Methyl bromide, one of the most popular and effective pesticides to control plant pathogenic organisms like nematodes and fungi in the soil, has already been restricted. The unavailability of many of these agrochemicals has paved the path for developing alternative approaches. In certain conditions, the employment of traditional chemicals becomes inevitable in agriculture, but still, there are examples of organic productions being suitable for many economically important plants and surrounding habitats.

Organic matter is an important source of macro- and micronutrients and increases the soil's nitrogen, carbon, potassium, calcium and magnesium content. It also improves the water-holding capacity and fertility of soil, leading to better plant growth and yield. Also, incorporating organic additives in soil has shown promising results in effectively managing plant diseases. The mechanism of inhibition of plant pathogens can be direct or indirect. The release of allelochemicals or compounds toxic to nematodes that may be already present or produced during the decomposition of organic amendments like certain fatty acids and ammonia is a direct suppression mechanism. Indirect mechanisms result in pathogen suppression by enhancing the population or introducing diverse antagonistic microorganisms by increasing disease resistance in plants and altering soil structure (Oka 2010). Plant-based materials like oilseed cakes, plant parts either chopped or in powdered form and

DOI: 10.1201/9781003319252-6

plant extracts have been used to control disease-causing organisms, especially nematodes in plants (Ahmad et al. 2010; Nadeem 2022).

Oilseed cake is the coarse residue left after the oil is extracted from the seeds of plants and thus is a by-product of the oil extraction process. Oilseed cakes are mainly categorized into edible and inedible forms. Neem, castor, karanja, jatropha, simarouba and mahua are important inedible cakes, while the edible ones include groundnut, soya bean, mustard, sunflower, coconut, linseed and sesame seed. Edible oilseed cakes are mainly utilized as feed for livestock and fish farming. In contrast, non-edible ones have been used in biofuel industry and for the fermentation production of mushrooms, enzymes, vitamins and antioxidants. Non-edible oil cakes are mainly utilized as organic fertilizer in farming. Oilseed cakes have been used as fertilizer to increase crop yield and enhance soil fertility for a long time. They have a low carbon-to-nitrogen ratio and thus have a slower decomposition rate and nutrient release into the soil, making the plant utilize the nutrients for a longer time. Therefore, this chapter discusses the importance of mahua oilseed cake, particularly emphasizing its chemical composition and nematicidal potential.

6.2 MAHUA OILSEED CAKE

6.2.1 Origin

Mahua cake is a derivative of *Madhuca indica* J.F. Gmel syn. *Madhuca longifolia* (L.) J.F. Macbr. Mahua belongs to the family Sapotaceae and is a midsized deciduous tree, about 12–15 m tall. It has a stubby trunk bearing a shady round crown with many branches. The leaves have an alternate arrangement with a simple leaf blade about 10–25 cm long and 6–12 cm broad, thick and broadly elliptic in shape, and young leaves are pinkish with a woolly lower surface and contain a milky sap. Flowers are small and present in bunches, with each bunch having scented pale white flowers. The fruit is an ovoid, fleshy berry consisting of 1–4 shiny brown elliptical seeds which are 3–5 cm long. The mahua tree is native to India, mainly growing in Uttar Pradesh, Madhya Pradesh, Gujarat, Bihar, Chhattisgarh, Orissa, Maharashtra and certain parts of Rajasthan. It also grows well in countries like Nepal, Myanmar and Sri Lanka. Mahua is a hardy tree and can thrive in tropical and subtropical climates in deep loamy or sandy loam soils or sometimes in calcareous soils with an annual rainfall of 500–1500 mm and temperatures in the range of 2–46°C. This tree's leaves, flowers and fruits are edible and consumed in south Asia, including India. Seeds of mahua contain 35–50% oil on average. Oil extracted from mahua seeds is used in soap and candle making and is also used as a preserving seed from pests (Orwa et al. 2009). Mahua oil is reported to have potential in biodiesel production. After the removal of oil from seeds, the residue oil cake is utilized as an organic fertilizer and bio-pesticide to control nematodes and other plant pathogens (Gupta et al. 2013; Orwa et al. 2009).

6.2.2 Chemical Composition

The cake has a high sugar and protein content. The percentage of carbohydrates is 42.8% in de-oiled seed cake, along with about 30% protein, 8.6% fibre and 9.8% saponins. It also contains some ash (6%) and tannins (1%; Singh and Singh 1991). The average nitrogen, phosphorus and potassium content of mahua oilseed cake is 2.5%, 0.8% and 1.2%, respectively. The phytochemical analysis of oil cake extracts revealed alkaloids, flavonoids, saponins, glycosides, lignins, sterols, tannins, triterpenoids, quinones and anthraquinones (Singh et al. 2018; Vasudha Udupa et al. 2021). Gas chromatography-mass spectrometry (GCMS) analysis of extracts of mahua oilseed cake has revealed different phytochemicals with important biologic activities. Vasudha Udupa et al. (2021) identified other compounds in acetone extracts of mahua oilseed cake. These include glycerine, diacetone, cis-13-eicosenoic acid, thymol, 7-methyl-Z-tetradecen-1-ol acetate, chloramben, methyl ester, stearic acid, linoleic acid, 2-phenyl-4-anilino-6[1H]-pyrimidinone, palmitic acid, methyl ester, linoleic acid, methyl ester, elaidic acid, methyl ester, diisooctyl phthalate, α-Monoacetin and oleic

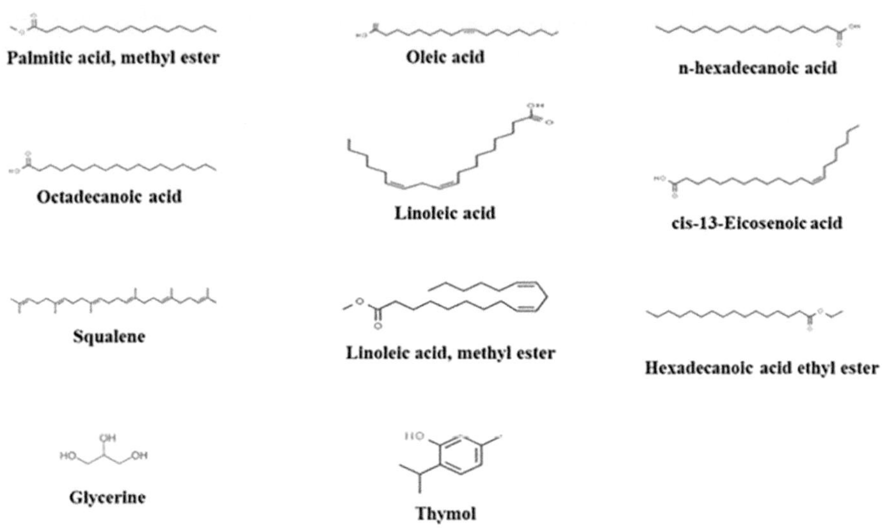

FIGURE 6.1 Structures of some important bioactive compounds identified in mahua oilseed cake.

TABLE 6.1
Some Important Bioactive Compounds Identified in Mahua Oilseed Cake

Phytochemical	Biological Activity
Octadecanoic acid	Bactericidal activity (Ivanova et al. 2017), antimicrobial action (Jubie et al. 2012), fungicidal (Frahm 2004)
Squalene	Antibacterial, antioxidant, antitumor properties (Rajeshwari et al. 2012), fungicidal activity (Elewski 1993)
Thymol	Anti-nematodal (Oka et al. 2000; Ohri and Pannu 2009), antibacterial and antifungal (Memar et al. 2017, Falcone et al. 2005), antioxidant, antiseptic, antiproliferative and anti-inflammatory (Marchese et al. 2016)
n-Hexadecanoic acid	Fungicidal and bactericidal properties (Akpuaka et al. 2013) anti-inflammatory, nematicide (Kumar et al. 2010), larvicidal activity (Rahuman et al. 2000)
Glycerine	Antibacterial (Nalawade et al. 2015), improvement in plant disease resistance (Li et al. 2016), enhancement of nematicidal activity of *Pseudomonas aeruginosa* (Siddiqui and Shaukat 2002)
Hexadecanoic acid, ethyl ester	Antifungal and antibacterial (Agoramoorthy et al. 2007)
Diacetone	Bactericidal and antiproliferative properties (Güvensen et al. 2019)
Linoleic acid	Nematicidal activity (Stadler et al. 1994), antioxidant and antibacterial properties (Pinto et al. 2017), fungicidal (Zheng et al. 2005)
Cis-13-eicosenoic acid	Antibacterial (Gupta and Kumar 2017), Antifungal (Ahsan et al. 2017) and anti-inflammatory properties (Azhar et al. 2016)
Palmitic acid, methyl ester	Antibacterial and antioxidant (Davoodbasha et al 2018), fungicidal properties (Agoramoorthy et al. 2007)
9-Octadecenoic acid, (E)	Nematicidal and antimicrobial (Chandrasekaran et al. 2008)
Linoleic acid, methyl ester	Antifungal and antibacterial (Lima et al. 2011) antioxidant (Pinto et al. 2017)
Oleic Acid	Anti-inflammatory, nematotoxic activity (Tarraf et al. 2019)

acid. In hexane extracts of mahua oil cake, compounds such as 9,12-Octadecadienoic acid (Z,Z)-; methyl ester, 9,12-Octadecadienoic acid (Z,Z)-; 9-Octadecenoic acid; octadecanoic acid; squalene; hexadecanoic acid; ethyl ester; octadecanoic acid; 2,3 dihydroxypropyl ester; n-hexadecanoic acid; tetrapentacontane,1,54-dibromo-; and methyl stearate have also been identified using GCMS analysis (Ayyandurai et al. 2022). The chemical compounds of oilseed cake with different biological activities are listed in Table 6.1, and their structures are given in Figure 6.1.

6.3 NEMATODE PARASITISM AND ITS MANAGEMENT

6.3.1 Plant-Parasitic Nematode

These nematodes are one of the major groups of pathogens infecting agricultural crops all around the world. About 4300 species of these pathogens have been described and placed under 197 genera, including a small group that is substantially pathogenic to plants, causing economic losses in important plants. These organisms have been reported to cause an estimated yield loss of 12.3% (US$157 billion) in different parts of the world (Hassan et al. 2013). The annual damage is reported to be US$40.3 million in India (Singh et al. 2015). Nematodes parasitizing plants are mostly soil-borne, feeding on plant roots penetrating by a protrusible stylet, although some species can feed on other plant parts like stem, leaves, fruits and seeds. The effector molecules produced in the pharyngeal glands of these organisms facilitate penetration, internal migration and parasitism.

Plant parasitic nematodes can be classified into ectoparasitic (*Xiphinema*, *Helicotylenchus. Longidorus*), semi-endoparasitic (*Rotylenchulus, Tylenchulus*) and endoparasitic (*Pratylenchus*, *Meloidogyne*, *Heterodera* and *Globodera*) based on their feeding habits. However, the most important group that causes heavy yield losses in crops are the root-knot (*Meloidogyne* spp.) and cyst nematodes (*Heterodera* and *Globodera* spp.). These pathogens impact the production and economy of crop plants as they lead to the application of extra fertilizers. Using oilseed cake is one strategy that can reduce the impact of nematodes on agricultural crops. Mahua oilseed cake has been reported to have nematicidal, antifungal, insecticidal and bactericidal activities.

6.3.2 Nematicidal Potential

Several experiments have been conducted from time to time to test the potential of mahua oilseed cake against different nematodes infecting plants, but most of the studies are conducted on the root-knot nematode, *Meloidogyne* spp. Mahua oil cake has been used either in the form of an organic soil amendment in powdered form or in the form of aqueous extracts. When used as a soil amendment the grounded cake is mixed with soil and left for decomposition for some time, while aqueous extracts have been used to study larval mortality and the inhibition of egg hatching. These extracts have also been used as a seed dressing in some instances. Aqueous extracts of mahua, neem, groundnut and castor oil cake were found to be inhibitory to nematodes like *R. reniformis*, *Hoplolaimus indicus*, *M. incognita* and *Tylenchorynchus brassicae* (Khan et al. 1973). Khan et al. (1974) observed inhibition of egg hatching in *M. incognita* in water extracts of mahua and neem cakes. Aqueous extracts of oil cake were found to be inhibitory to *Aphelenchus avenae* and *Ditylenchus dipsaci* (Mukherjee 1983). Yadav et al. (2005) used 10%, 15% and 20% aqueous extracts of oil cakes of mustard, mahua, castor, neem and karanj as seed soaking for control of *M. incognita* in chickpeas. A considerable increase in plant growth and weight in grams and a decline in root-knot and nematode population were recorded in all the treatments. The addition of mahua, castor, mustard, neem and groundnut oil cake into soil either in single treatment or in different combinations were found to be detrimental to plant-parasitic nematodes on cabbage, cauliflower, tomato, brinjal, chilli and okra (Alam 1991). Oil cakes of mustard neem and mahua were evaluated in a pot study on *Heterodera cajani*, *F. udum* and their disease complex causing wilt in pigeon peas, along with carbofuran. All the oil cakes proved detrimental to nematode multiplication, while maximum growth of the plants was recorded in treatment with mahua oil cake. However, neem oil cake proved to be the most efficacious in decreasing wilt incidence caused by *F. udum* alone and in combination with *H. cajani* (Rai and Singh 1996). The efficacy of different oil cakes like mahua, neem, mustard, karanj, linseed, cotton seed and castor has been tested against *M. incognita* in soybean and okra plants, and a decline in root-galling and enhancement in plant growth was recorded (Goswami and Satyendra 1998). An increase in root nodulation in soybean was also observed (Mishra and Gupta 1997).

Mahua oil cake has also been used in integration with other biological control agents and nematicides and resulted in better control of nematode populations. The population of *M. incognita* declined to a greater extent in black gram when the mycorrhizal fungus *Glomus fasciculatum* was employed in combination with oil cakes of groundnut, mahua, castor and neem (Sankaranarayanan and Sundarababu 1997). Fungi like *Paecilomyces lilacinus* or *Cladosporium oxysporum* when used along with the oil cakes of castor, groundnut, mahua, linseed and neem inhibited *M. javanica* populations in eggplant (Ashraf and Khan 2010). Shankar et al. (2016) tested different oil cakes (mahua cake neem cake, groundnut cake, sunflower cake) along with carbosulfan against *Meloidogyne* spp. and found that the treatments not only lowered gall number and nematode density but also improved plant growth and nodulation in lentil plant. Treatment with neem and mahua oil cakes proved to be the most promising in different combinations.

The probable mechanism of nematode control by adding mahua oil cake in the soil can be the increase in the supply of essential nutrients resulting in improved health of the host plant. Alterations in soil and root physiology, induction or increased level of defence-related enzymes, photosynthetic pigments, the production of phenolic compounds and elevated levels of host resistance towards pathogens play an important role in disease reduction and enhancement in plant growth. The presence of bioactive compounds like thymol, linoleic acid and n-hexadecanoic acid in mahua oil cake can be credited for its pesticide potential. Some of these compounds can interfere with the nematode's cuticle or hypodermis, and some may negatively interact with lipophilic regions of the plasma membrane. By acting as a substrate for the mass multiplication of antagonistic organisms, mahua oil cake results in enhanced disease suppression by biocontrol agents. Some examples where mahua oil cake is utilized for its potential against nematodes are summarized in Table 6.2.

TABLE 6.2
Use of Mahua Oilseed Cake for Its Nematicidal Efficacy against Plant-Parasitic Nematodes

Host Plant	Nematode spp.	Nematicidal Effect	Reference
Rice	*M. graminicola*	Reduced root-knot infection with improved plant growth parameters	Kumar et al. 2018
Eggplant, tomato and chilli	*M. incognita*	Low larval penetration with reduced galls, better plant growth and enhancement in total phenol content of plants	Alam et al. 1980
Tomato and okra	*M. incognita*	Root gall indices declined, leading to improved growth of plants	Singh and Sitaramaiah 1971
Citrus sp., eggplant and guava	*Hoplolaimus* sp., *Tylenchorynchus* sp., *R. reniformis Helicotylenchus* sp. and *M. incognita*	Decline in population densities of nematodes	Mobin and Khan 1969
Sweet basil	*M. incognita*	Low root-galling and fewer number of nematodes	Haseeb et al. 1988
Brinjal	*R. reniformis*, *M. incognita* and *P. delattrei*	Reduction in nematode population with boost in plant growth	Poornima and Vadivelu 1993
Groundnut	*M. javanica*	Reduction in nematode population and improvement in crop growth	Joshi and Patel 1995

6.4 CONCLUSION

The use of mahua oilseed cake as an organic soil amendment for the control of nematodes is a beneficial approach, as it not only suppresses or kills the pathogen populations but also enriches the nutrient status of the soil, thereby reducing the use of synthetic fertilizers. Mahua oilseed cake contains bioactive compounds, which have proven to be nemato-toxic in various studies. Easy availability, relatively low costs, a slower rate of decomposition and nutrient-enriching properties of oil cakes make them ideal for use as organic fertilizer and nematicides, leading toward sustainability in agriculture.

REFERENCES

Agoramoorthy, G., M. Chandrasekaran, V. Venkatesalu, and M. J. Hsu. 2007. Antibacterial and antifungal activities of fatty acid methyl esters of the blind-your-eye mangrove from India. *Brazilian Journal of Microbiology*, 38(4): 739–742.

Ahmad, F., M. A. Rather, and M. A. Siddiqui. 2010. Nematicidal activity of leaf extracts from *Lantana camara* L. against *Meloidogyne incognita* (kofoid and white) chitwood and its use to manage roots infection of *Solanum melongena* L. *Brazilian Archives of Biology and Technology*, 53: 543–548.

Ahsan, T., J. Chen, X. Zhao, M. Irfan, and Y. Wu. 2017. Extraction and identification of bioactive compounds (eicosane and dibutyl phthalate) produced by Streptomyces strain KX852460 for the biological control of *Rhizoctonia solani* AG-3 strain KX852461 to control target spot disease in tobacco leaf. *AMB Express*, 7(1): 1–9.

Akpuaka, A., M. M. Ekwenchi, D. A. Dashak, and A. Dildar. 2013. Biological activities of characterized isolates of n-hexane extract of *Azadirachta indica* A. Juss (Neem) leaves. *Nature and Science*, 11(5): 141–147.

Alam, M. M. 1991. Control of plant-parasitic nematodes with oilseed cakes on some vegetables in field. *Pakistan Journal of Nematology*, 9: 21–30.

Alam, M. M., M. Ahmad, and A. M. Khan. 1980. Effect of organic amendments on the growth and chemical composition of tomato, eggplant and chilli and their susceptibility to attack by *Meloidogyne incognita*. *Plant Soil*, 57: 231–236.

Ashraf, M. S., and T. A. Khan. 2010. Integrated approach for the management of *Meloidogyne javanica* on eggplant using oil cakes and biocontrol agents. *Archives of Phytopathology and Plant Protection*, 43: 609–614. http://dx.doi.org/10.1080/03235400801972434

Ayyandurai, M., R. Akila, M. L. Mini, and K. Manonmani. 2022. Gas chromatography–mass spectrometry (GC-MS) analysis of antimicrobial compounds produced by mahua oil cake against the stem rot pathogen-*Sclerotium rolfsii*. *Journal of Applied and Natural Science*, 14(2): 600–606.

Azhar, A. S., H. B. Suhaila, and H. H. Imad. 2016. Analysis of bioactive chemical compounds of *Euphorbia lathyrus* using gas chromatography-mass spectrometry and fourier-transform infrared spectroscopy. *Journal of Pharmacognosy and Phytotherapy*, 8(5): 109–126.

Chandrasekaran, M., K. Kannathasan, and V. Venkatesalu. 2008. Antimicrobial activity of fatty acid methyl esters of some members of Chenopodiaceae. *Zeitschrift für Naturforschung*, 63(5–6): 331–336.

Davoodbasha, M., B. Edachery, T. Nooruddin, S. Y. Lee, and J. W. Kim. 2018. An evidence of C16 fatty acid methyl esters extracted from microalga for effective antimicrobial and antioxidant property. *Microbial Pathogenesis*, 115: 233–238.

Elewski, B. E. 1993. Mechanisms of action of systemic antifungal agents. *Journal of the American Academy of Dermatology*, 28(5): S28–S34.

Falcone, P., B. Speranza, M. A. Del Nobile, M. R. Corbo, and M. Sinigaglia. 2005. A study on the antimicrobial activity of thymol intended as a natural preservative. *Journal of Food Protection*, 68(8): 1664–1670.

Frahm, J. P. 2004. Recent developments of commercial products from bryophytes. *The Bryologist*, 107(3): 277–283.

Goswami, B. K., and S. Satyendra. 1998. Efficacy of some soil mycoflora arising from the oil seed cakes of cottonseed, karanj, mahua and mustard amended soil against *Meloidogyne incognita* infecting okra. *Annals of Agricultural Research*, 19(2): 158–161.

Gupta, A., A. Kumar, S. Sharma, and V. K. Vijay. 2013. Comparative evaluation of raw and detoxified mahua seed cake for biogas production. *Applied Energy*, 102: 1514–1521.

Gupta, D., and M. Kumar. 2017. Evaluation of in vitro antimicrobial potential and GC–MS analysis of *Camellia sinensis* and *Terminalia arjuna*. *Biotechnology Reports*, 13: 19–25.

Güvensen, N. C., D. Keskin, H. Güneş, M. K. Oktay, and H. Yıldırım. 2019. Antimicrobial property and antiproliferative activity of *Centaurea babylonica* (L.) L. on human carcinomas and cervical cancer cell lines. *Annals of Agricultural and Environmental Medicine*, 26(2): 290.

Haseeb, A., R. Pandey, and A. Hussain. 1988. A comparison of nematicides and oil cakes for control of *Meloidogyne incognita* on *Ocimum basilicum*. *Nematropica*, 18(1): 65–69.

Hassan, M. A., T. H. Pham, H. Shi, and J. Zheng. 2013. Nematodes threats to global food security. *Acta Agriculturae Scandinavica, Section B: Soil & Plant Science*, 63(5): 420–425.

Ivanova, E. P., S. H. Nguyen, Y. Guo, V. A. Baulin, H. K. Webb, V. K. Truong, J. V. Wandiyanto, C. J. Garvey, P. J. Mahon, D. E. Mainwaring, and R. J. Crawford. 2017. Bactericidal activity of self-assembled palmitic and stearic fatty acid crystals on highly ordered pyrolytic graphite. *Acta Biomaterialia*, 59: 148–157.

Joshi, P. R., and H. R. Patel. 1995. Organic amendments in management of *Meloidogyne javanica* on groundnut. *Indian Journal of Nematology*, 25: 76–78.

Jubie, S., P. N. Ramesh, P. Dhanabal, R. Kalirajan, N. Muruganantham, and A. S. Antony. 2012. Synthesis, antidepressant and antimicrobial activities of some novel stearic acid analogues. *European Journal of Medicinal Chemistry*, 54: 931–935.

Khan, W. M., A. M. Khan, and S. K. Saxena. 1973. Influence of certain oil cakes amendments on nematodes and fungi in tomato fields. *Acta Botanica Indica*, 1: 49–54.

Khan, W. M., A. M. Khan, and S. K. Saxena. 1974. Rhizosphere fungi and nematodes of eggplant as influenced by oil cake amendments. *Indian Phytopathology*, 27: 480–484.

Kumar, A., K. Khilari, S. K. Jain, A. K. Yadav, and A. K. Mishra. 2018. Efficacy of different oil cakes on management of root knot nematode (*Meloidogyne graminicola*) in rice crop in western Uttar Pradesh, India. *International Journal of Current Microbiology and Applied Sciences*, 7(12): 3524–3530.

Kumar, P. P., S. Kumaravel, and C. Lalitha. 2010. Screening of antioxidant activity, total phenolics and GC-MS study of *Vitex negundo*. *African Journal of Biochemistry Research*, 4(7): 191–195.

Li, Y., N. Song, C. Zhao, F. Li, M. Geng, Y. Wang, and Q. Sun. 2016. Application of glycerol for induced powdery mildew resistance in *Triticum aestivum* L. *Frontiers in Physiology*, 7: 413.

Lima, L. A. R. D. S., S. Johann, P. S. Cisalpino, L. P. S. Pimenta, and M. A. D. Boaventura. 2011. In vitro antifungal activity of fatty acid methyl esters of the seeds of *Annona cornifolia* A. St.-Hil.(Annonaceae) against pathogenic fungus *Paracoccidioides brasiliensis*. *Revista da Sociedade Brasileira de Medicina Tropical*, 44: 777–780.

Marchese, A., I. E. Orhan, S. F. Daglia Nabavi, and S. M. Nabavi. 2016. Antibacterial and antifungal activities of thymol: A brief review of the literature. *Food Chemistry*, 210: 402–414.

Memar, M. Y., P. Raei, N. Alizadeh, M. A. Aghdam, and H. S. Kafil. 2017. Carvacrol and thymol: Strong antimicrobial agents against resistant isolates. *Reviews in Medical Microbiology*, 28(2): 63–68.

Mishra, S. M., and P. Gupta. 1997. Effect of some oil cakes and saw dust application on root-knot (*Meloidogyne incognita*) infection and root nodulation of soybean. *Indian Phytopathology*, 50: 131–132.

Mobin, M. P., and A. M. Khan. 1969. Effect of organic amendment on the population of rhizosphere fungi and nematodes around roots of some fruit trees. *All India nematology symposium*. New Delhi: IARI, pp. 11–13.

Mukherjee, S. K. 1983. ICAR, New Delhi, Pers. Commun. In Michael et al. 1985. Plant species reportedly possessing pest control properties. An EWC/UH DATABASE, University of Hawaii, pp 249.

Nadeem, H. 2022. Studies on the potential of biopesticides for management of root-knot nematode. PhD thesis, Aligarh Muslim University, Aligarh.

Nalawade, T. M., K. Bhat, and S. H. Sogi. 2015. Bactericidal activity of propylene glycol, glycerine, polyethylene glycol 400, and polyethylene glycol 1000 against selected microorganisms. *Journal of International Society of Preventive & Community Dentistry*, 5: 114–119.

Ohri, P., and S. K. Pannu. 2009. Effect of terpenoids on nematodes: A review. *Journal of Environmental Research and Development*, 4(1): 171–177.

Oka, Y. 2010. Mechanisms of nematode suppression by organic soil *amendments–a* review. *Applied Soil Ecology*, 44: 101–115.

Oka, Y., S. Nacar, E. Putievsky, U. Ravid, Z. Yaniv, and Y. Spiegel. 2000. Nematicidal activity of essential oils and their components against the root-knot nematode. *Phytopathology*, 90(7): 710–715.

Orwa, C., A. Mutua, R. Kindt, R. Jamnadass, and S. Anthony. 2009. *Agroforestree Database: A tree reference and selection guide version 4.0*. Kenya: World Agroforestry Centre.

Pinto, M. E. A., S. G. Araújo, M. I. Morais, N. P. Sa, C. M. Lima, and C. A. Rosa. 2017. Antifungal and antioxidant activity of fatty acid methyl esters from vegetable oils. *Anais da Academia Brasileira de Ciências*, 89: 1671–1681.

Poornima, K., and S. Vadivelu. 1993. Comparative efficacy of nematicide, oil cakes and plant extracts in the management of *Meloidogyne incognita*, *Pratylenchus delattrei* and *Rotylenchus reniformis* on brinjal. *Indian Journal of Nematology*, 23(2): 170–173.

Rahuman, A. A., G. Gopalakrishnan, B. S. Ghouse, S. Arumugam, and B. Himalayan. 2000. Effect of *Feronia limonia* on mosquito larvae. *Fitoterapia*, 71(5): 553–555.

Rai, P. K., and K. P. Singh. 1996. Efficacy of certain oilcake amendments on *Heterodera cajani*, *Fusarium udum* and associated wilt of pigeon pea. *International Journal of Tropical Plant Diseases*, 14(1): 51–58.

Rajeshwari, G., M. Murugan, and V. R. Mohan. 2012. GC-MS analysis of bioactive components of *Hugonia mystax* L. (Linaceae). *Research Journal of Pharmaceutical, Biological and Chemical Sciences*, 3(4): 301–308.

Sankaranarayanan, C., and R. Sundarababu. 1997. Effect of oil-cakes and nematicides on the growth of blackgram (*Vigna maungo*) inoculated with VAM fungus (*Glomus fasciculatum*) and root knot nematode (*Meloidogyne incognita*). *Indian Journal of Nematology*, 27: 128–130.

Shankar, P., P. Naresh, S. K. Biswas, S. Kumar, A. Kumar, and Erayya. 2016. Integrated effect of oil cakes, bioagent and nematicide on root knot namatode and root nodulation in lentil. *Journal of Pure and Applied Microbiology*, 10: 2949–2955.

Siddiqui, I. A., and S. S. Shaukat. 2002. Zinc and glycerol enhance the production of nematicidal compounds in vitro and improve the biocontrol of *Meloidogyne javanica* in tomato by fluorescent pseudomonads. *Letters in Applied Microbiology*, 35(3): 212–217.

Singh, A., and I. S. Singh. 1991. Chemical evaluation of mahua (*Madhuca indica*) seed. *Food Chemistry*, 40: 221–228.

Singh, P., A. Goel, and M. Sehgal. 2018. Phytochemical analysis and antimicrobial activity of *Madhuca longifolia* and *Clerodendrum infortunatum*: Medicinal plants for application as textile finishes. *International Journal of Chemical Studies*, 6(5): 2031–2040.

Singh, R. S., and K. Sitaramaiah. 1971. Control of root-knot nematode through organic and inorganic amendments in the soil-effect of oil cakes and saw dusts. *Indian Journal of Mycology and Plant Pathology*, 1(1): 20–29.

Singh, S., B. Singh, and A. P. Singh. 2015. Nematodes: A threat to sustainability of agriculture. *Procedia Environmental Sciences*, 29: 215–216.

Stadler, M., A. Mayer, H. Anke, and O. Sterner. 1994. Fatty acids and other compounds with nematicidal activity from cultures of Basidiomycetes. *Planta Medica*, 60(2): 128–132.

Tarraf, W., S. Laquale, G. De Mastro, and T. D'Addabbo. 2019. The potential of *Citrullus colocynthis* oil as a biocide against phytoparasitic nematodes. *Crop Protection*, 124: 104843.

Vasudha Udupa, A., B. Gowda, B. E. Kumarswammy, and M. B. Shivanna. 2021. The antimicrobial and antioxidant property, GC–MS analysis of non-edible oilseed cakes of neem, madhuca, and simarouba. *Bulletin of the National Research Centre*, 45(1): 1–14.

Yadav, Y. S., A. U. Siddiqui, and A. Parihar. 2005. Management of root-knot nematode *Meloidogyne incognita* infesting gram through oil cakes. *Journal of Phytological Research*, 18: 263–264.

Zheng, C. J., J. S. Yoo, T. G. Lee, H. Y. Cho, Y. H., Kim, and W. G. Kim. 2005. Fatty acid synthesis a target for antibacterial activity of unsaturated fatty acids. *FEBS Letters*, 579(23): 5157–5162.

7 Flaxseed Oil Cake
By-Product of a Superfood with Remarkable Antagonistic Properties

Rishil Gupta, Hera Nadeem and Faheem Ahmad

CONTENTS

7.1 INTRODUCTION

Plant-parasitic nematodes (PPNs) are one of the important agricultural parasites, causing significant economic losses annually (Bernard et al. 2017). There are many soil-inhabiting PPNs; root-knot nematodes are one of the most damaging pathogens that can infect a wide range of host plants like tomatoes, brinjal, okra, wheat, sugar beets, soybeans, and others (Sikandar et al. 2020). Root-knot nematodes are responsible for global agriculture crop damages estimated to be 173 billion US dollars per year (Elling et al. 2013). These nematodes can be managed *via* cultural techniques, organic amendments, nematicides, and resistant cultivars. Research right now is concentrated on controlling PPN by utilizing the organic supplement as one of the most promising substitutes for eradicating PPNs and improving the structure and health of the soil. Agricultural residues such as sawdust, rice bran, seeds of fruits and vegetables, paddy straw, and oil cakes are obtained during various stages of extracting and processing oilseeds, cereals, and pulses. These residues offer advantages in the field of bioprocessing, control of phytopathogens, food industries, pharmaceutical applications, and cosmetics because they encourage sustainable development and zero-waste idea for the environment. In recent years, oilseed crops have gained popularity due to their high potential for producing a wide range of value-added products with both economic and high nutritional value. Widely accessible oilseed cakes are additions that promote the growth of beneficial organisms, reducing the number of nematodes and otherwise enhancing the crop in various ways. Among oilseed crops, flaxseed (*Linum usitatissimum* L.) is a well-known oilseed with a nutrient value cultivated in more than 50 nations worldwide.

DOI: 10.1201/9781003319252-7

Flaxseed is a superfood containing vitamins, proteins, minerals, lipids, carbohydrates, and dietary fibre (Goyal et al. 2014). Various flaxseed-based products have recently been on the market, including whole flaxseeds, lignan extracts, and oil cakes (Shim et al. 2015). According to Crouse (1998), the haphazard use of chemically synthesized substances and antibiotics for regulating humans and diseases poses severe risks to human health and the environment, resulting in the formation of multidrug-resistant microbes and pollutants in the ecosystem's food chain. Oilseed cakes contain toxic and non-toxic phytochemicals that can be isolated and used in various industries. Oil cakes are obtained from the seeds after oil extraction and have a high amount of bioactive components such as phenolic acid, coumarins, flavonoids, and antioxidants (Oomah and Sitter 2009). According to various studies, the dark colour, sour taste, and unsavoury flavour of products may be caused by the presence of free phenolic compounds in oilseeds and their products. However, oilseed cakes have a significant amount of phosphorus, potassium, and nitrogen; hence, they could be utilized as a source of organic manure. The majority of flaxseeds are made up of minerals (3%), lipids (43%), and protein (22%), which demonstrates their nutritional content. Apart from this, flaxseed oil cake contains lignin and fibres. The lignin in flaxseed cake, such as secoisolariciresinol diglucoside (SDG), has numerous health advantages. Bakke and Klosterman (1956) performed the first SDG isolation. The most abundant plant lignans are phytoestrogens in flaxseed, with approximately 75,800 times more lignans than in other crops. In addition, lignan is a source of beneficial phytochemicals and other bioactive molecule components in plant foods. This makes it a functional food. Several researchers noted that the bioactive compounds in flaxseed oil cakes have nematicidal, antifungal, and antibacterial properties. Flaxseed containing bioactive compounds plays a crucial role in crop diseases caused by various soil-borne phytopathogens, including PPNs (Rizvi et al. 2012).

7.1.1 Morphology and Distribution

Flaxseed is a summer herbaceous annual plant that grows from 20 to 150 cm high. Its intended need determines the plant's height and branching habits. Large-seeded genotypes designed to produce oil are branched and shorter than typical flax varieties used for fibre (Gill 1987). Oilseed and fibre flax varieties have been developed separately. It has a dense fibrous root system with numerous lateral roots. The leaves are small, alternate on the stalk, and are lance-shaped; the stem is glabrous and greyish green. The plant's stem provides fibre, which is extremely strong and long-lasting. The five-petalled flowers appear in panicle-like inflorescences and develop into five-celled bolls holding up to 10 seeds. The fruits are dry capsules, round in shape, most often indehiscent, and have a diameter of 5–9 mm, containing 10 seeds in large-seeded oil cultivars and fewer in fibre varieties. Seeds with a branching habit have a lower harvest index than fibre types. The seeds are lenticular, compressed, light brown, shining yellow- and olive-coloured, 4–6 mm in length, and 2–3 mm in width. Testa has a shiny, soft, bright red surface that is minutely pitted, and seeds are oval at one end and obliquely pointed at the other. The test weight of approximately 1000 seeds ranges from 4 to 13 g. The endosperm and cotyledons of flaxseed from oil varieties are highly enriched in oil and protein. The number of capsules per unit area is a significant yield component in determining flaxseed yield and quality.

Despite being more widely adopted and having historically been grown in other parts of the United States, the primarily grown flax is currently in North Dakota and Montana. Flax grows best in soils and climates in which spring cereals are grown. Higher yields are typically obtained by planting flax early in the spring, similar to the suggested planting time for spring-planted cereals. Planting flax first, followed by buckwheat, is recommended to enhance profit potential in regions with a long, sufficient growing season in the south-central United States. Over 30 nations on five continents cultivate flax as an oilseed or fibre crop (Table 7.1). According to the Flax Council of Canada (2014), flaxseeds are grown in almost every country worldwide. China, Canada, Argentina, the United States, Poland, Romania, Ethiopia, and India are major producers. Flax was extensively grown in ancient Egypt, where flowering flax was painted on temple walls, and mummies were wrapped in linen (Sekhri 2011). Flaxseed production in the world was

TABLE 7.1
Flaxseed Production (FAOSTAT 2020)

Country	Production (tonnes)
Kazakhstan	1,058,247
China	330,000
Canada	578,000
Russia	787,923
Total World Production	3.4 million

3.4 million metric tonnes in 2020, with Kazakhstan accounting for 31% of the global total (FAO 2020). Canada produces and sells more than 80% of its flax oilseed output, primarily to Europe, South Korea, Japan, and the United States (Flax Council of Canada 2014). Flaxseed grows better on thick loam soils with more moisture than on dry sandy loam. It not only interacts well with fertilizer but also can withstand moderate salinity. Due to issues with root diseases, the flax crop may be cultivated rotationally with cereal crops but not after sugar beets, potatoes, or a previous flaxseed crop (CFIA 2012).

7.1.2 Extraction Process and Types

Oil extraction from the oilseed crop is a critical process in its industrialization. The extraction method directly impacts the quantity and quality of oils. According to Sharma et al. (2019), solvent extraction and mechanical pressing are the most popular ways to extract flaxseed oil and cakes. After pressing, fresh, unfiltered flaxseed oil has a nutty flavour and an orange to yellow colour. Numerous variables, such as flaxseed pretreatment, cultivars, moisture content, pressing conditions, and so on, might impact the extraction process of flaxseed oil. The vast majority of flax is made into oil, which is then used in industrial products, including paints and floor coverings. The protein-rich meal obtained after extraction is utilized in livestock feeds and is commonly called flaxseed meal. High-quality fresh flaxseeds are used as the raw material. Oilseed dehulling can increase oil extraction productivity and oil quality. Those hulls are occasionally removed before oil is extracted to maintain the quality of the oil and flaxseed meal. The mucilage and crude fibre are the two main components of the flaxseed hull.

Lan et al. (2020) explained that there are several dehulling methods of flaxseed, including dry and wet mechanical, wet, and sprouting processes. Before being dehulled, the cleaned flaxseed is kept over magnets to eliminate unwanted traces of metal and debris. The dehulled flax seeds are ground into a uniform fine meal using hammer mills or mechanical grooved rollers to create a wider surface area for pressing. Flaxseed oil contains many alpha-linolenic acids (ALAs), so it must be pressed at a low temperature to protect the ALA content. The oil extraction process is then accelerated by heating this meal. Some trash and contaminants are gathered with the oil during the pressing process and must be removed before the oil is packaged for use. The heated metal is again progressively pushed into a screw press that increases pressure as the meal progresses through a barrel with slots. The oil cake left over in the screw press after the first oil has been recovered is characterized by using solvent extraction to produce a higher yield. Hexane is a volatile hydrocarbon that dissolves the oil from the oilseed cake, and the oil is then separated by distillation of the solvent out, and then the solvent is transported through the matter to be deposited at the bottom. The methods for extracting flaxseed oil that are most frequently utilized are

1. prepress solvent extraction and
2. expeller press extraction.

Prepress solvent extraction combines mechanical pressure and chemical extraction, the most efficient way to extract oil from oilseeds. It is the favoured approach for most large-scale operations. Before preheating, seeds are cleaned to remove stones, chaff, or other debris to guard against damage or wear on the processing machinery. After that, to burst the cell walls and enhance the surface area for oil extraction, the seed is fed between the rolls and crushed into a fragile flake to decrease the oil's viscosity and allow it to move out of seed storage bodies. The compression effect pulls the oil out of the tiny pores in the wall and collects nearly half of the oil, preparing a cake suitable for solvent extraction. When hexane is injected into the cake, the oil from the flax meal is dissolved and absorbed. After the meal, the solvent is allowed to drain (Figure 7.1).

7.1.3 Nutritional Value

Flaxseed is widely known for its high concentration of chemical compounds with distinct bioactivity and functional properties. It has a unique nutritional profile because it is the richest source of PUFAs (polyunsaturated fatty acids), omega-3 family acids, and linolenic acid, which are essential fatty acids. Flaxseed is a functional dietary constituent because it has many positive health effects and nutrient properties (Eyres 2015). Flaxseed's oil compositional concentration varies significantly and is influenced by various factors, including cultivars and growth conditions. In general, flaxseed contents are oil, protein, lignan, and dietary fibre, with ranges of 38–45%, 20–28%, 0.8–1.3%, and 18–28%, respectively. Flaxseed contains 20–30% total protein, primarily in globulins (80%) and glutelin (20%). As one of the most nutrient-dense plant proteins, soybean protein is thought to have the most similar amino acid composition to flax protein. Flaxseed contains 450 kcal of energy, 41 g of total fat, 23 g of ALA, 20 g of protein, 29 g of total carbohydrates, and 28 g of total dietary fibre per 100 g (Morris 2007).

It is an excellent source of necessary vitamins, minerals, and antioxidants like selenium, zinc, magnesium, sodium, and potassium (Lan et al. 2020). Flaxseed oil meal contains increased levels of two essential amino acids, arginine, tryptophan, vitamin B6, and beta carotene (Table 7.2). According to Ganorkar and Jain (2013), flaxseed oil contains fatty acids such as saturated fatty

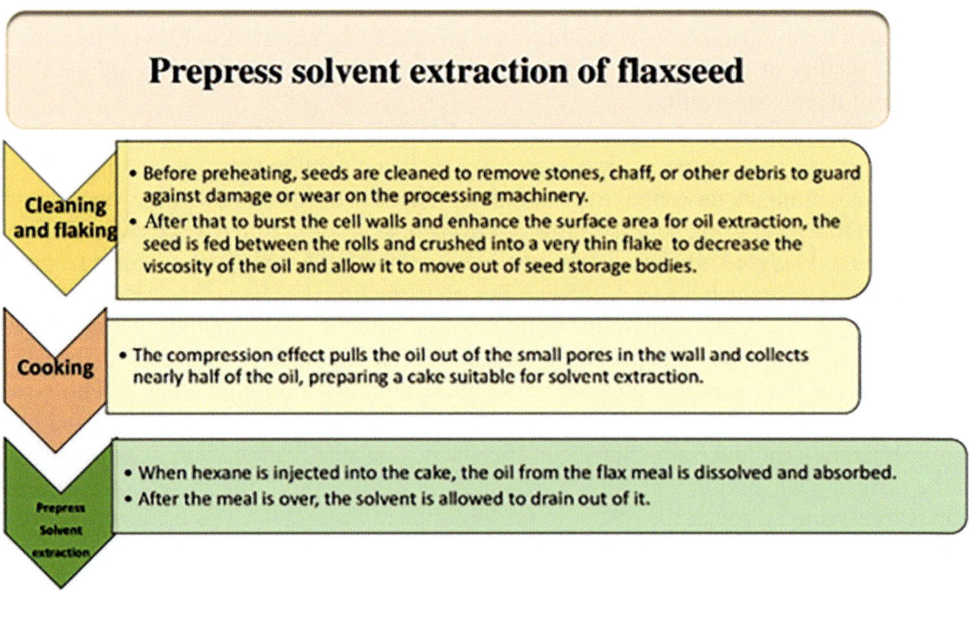

FIGURE 7.1 Solvent extraction of flaxseed oil and cakes.

acid (9%), mono-unsaturated fatty acid (18%), omega-6 fatty acid (16%), and omega-3 fatty acid (57%) that are essential for health (Table 7.3). Compared to other plant sources, flax has the highest concentration of lignan, which has antitumor, antimitotic, and antioxidant properties that lower the risk of cancer and cardiovascular disease. Cyanogenic glycosides (250–550 mg/100 g) are important antinutrients, including whole flaxseed, whereas phytic acid (2.3–3.3%) is present in flaxseed meal, resulting in reduced nutrient absorption (Singh et al. 2011). Protein found in flaxseed seeds ranges from 10% to 30% and is primarily made up of the amino acids, glutamic acid, phenylalanine, leucine, arginine, valine, and tyrosine (Shim et al. 2014). A good n-6:n-3 fatty acid ratio of about 0.3:1 is provided by the 30% of lipids in flaxseed that are composed of linoleic acid (17%), linolenic acid (53%), oleic acid (19%), palmitic acid (5%), and stearic acid (3%). The variety of biomolecules in flaxseed contributes to this plant crop's high nutritional profile, but some of the flaxseed oil cake constituents have also been investigated as food additives due to their specific bioactivity. Because of the mucilage in the seed's outer layers, flaxseed stands out among the oilseeds. Flaxseed mucilage has a strong capacity to bind water, which improves the liquid consistency, viscosity, and stability (Basiri et al. 2018). In flaxseed, there are 800–1000 mg of phenolic acids per 100 g of seeds, which is a significant amount. Herchi et al. (2014) reported various phenolic compounds in the defatted sample, such as gallic acid, chlorogenic acid, and ferulic acid (2.8,7.5 and 10.9 mg/g). Other phenolics, such as p-Coumaric acid glucosides,4-hydroxybenzoic acid, and hydroxycinnamic acid glucosides, are only found in minimal amounts. Even though flax is one of the oldest food-grade oilseed crops, details on how its active ingredients are processed can vary. Furthermore, many processing techniques extract or isolate flaxseed's functional/bioactive ingredients (Figure 7.2).

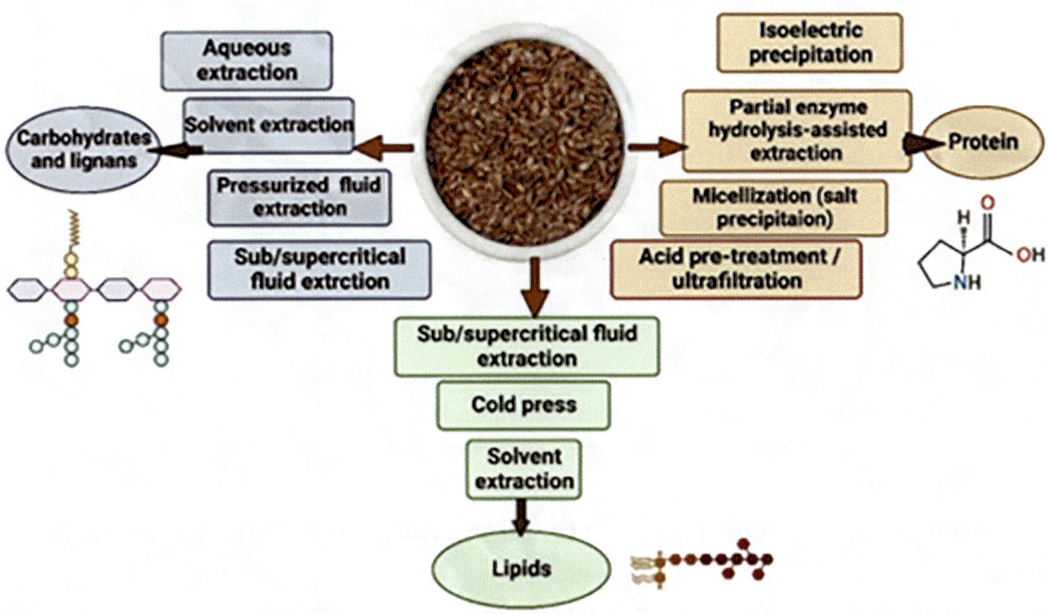

FIGURE 7.2 A simplified diagram of the various methods for extracting flaxseed's active ingredients.

TABLE 7.2
The Fatty Acid Content of Flaxseed Oil (Ganorkar and Jain 2013)

Flaxseed Oil	Saturated fatty acid (9%)
	Mono-unsaturated fatty acid (18%)
	Omega-6 fatty acid (16%)
	Omega-3 fatty acid (57%)
	α-linolenic acid (53%)
	Linoleic acid (17%)
	Oleic acid (19%)
	Stearic acid (3%)
	Palmitic acid (5%)

TABLE 7.3
Nutritional Value of Flaxseed

Nutritional Composition		**Content per 100 g**
Proximate composition (g)	Moisture	68
	Energy	545 kcal
	Protein	18
	Carbohydrates	34.4
	Sugars	1.55
	Total dietary fibre	23.1
	Fat	37.3
Vitamin (mg)	Vitamin C	0.6
	Vitamin E	<0.001
	Niacin	3.76
	Riboflavin (B2)	0.161
	Thiamine (B1)	0.537
	Vitamin B6	0.393
	Pantothenic acid (B5)	0.985
Minerals (mg)	Calcium	230
	Magnesium	372
	Phosphorus	556
	Potassium	793
	Zinc	4.74
	Iron	5.78
	Copper	1.34
	Sodium	37

Source: The United States Department of Agriculture, USDA (2021).

7.2 MORPHOLOGY OF FLAXSEED OIL CAKE AND METABOLITE CONTENT

Tonnes of seed cakes are produced during the extraction of plant seed oils. Oilseed cake polyphenols are trapped in insoluble components such as vacuoles, lipoprotein bilayers, lignin, hulls, and cell walls (Corrales et al. 2008). In plants, phenolic acids exist as esters, glycosides, and bound complexes. These characteristics have made standard polyphenol separation from oilseed cakes complex by traditional extraction methods. Recent research has revealed that ultrasound-assisted extraction (UAE) can improve extraction efficiency *via* acoustic cavitation and specific mechanical effects (Rostagno et al. 2003; Chemat et al. 2004, 2012). Acoustic cavitation can damage cell walls, allowing the solvent to enter the plant material and release intracellular substances. An additional mechanical effect generated by ultrasound is the agitation of the extraction solvent, which increases the contact surface area between the solvent and the desired chemicals by allowing higher solvent penetration into the sample matrix. Shorter extraction periods and lower solvent consumption were also needed for UAE (Zhang et al. 2008; Mason et al. 2011). Hromadkova et al. (1999) revealed that UAE could be performed at a lower temperature, avoiding thermal damage to the extracts and limiting bioactive component loss.

Nadeem (2022) showed the impact of ultrasonication on the morphology of flaxseed oil cake and compared the intensities of metabolites released using the ultrasonication process with the non-sonicated one. Figure 7.3 shows scanning electron microscopic (SEM) images of flax oilseed cakes'

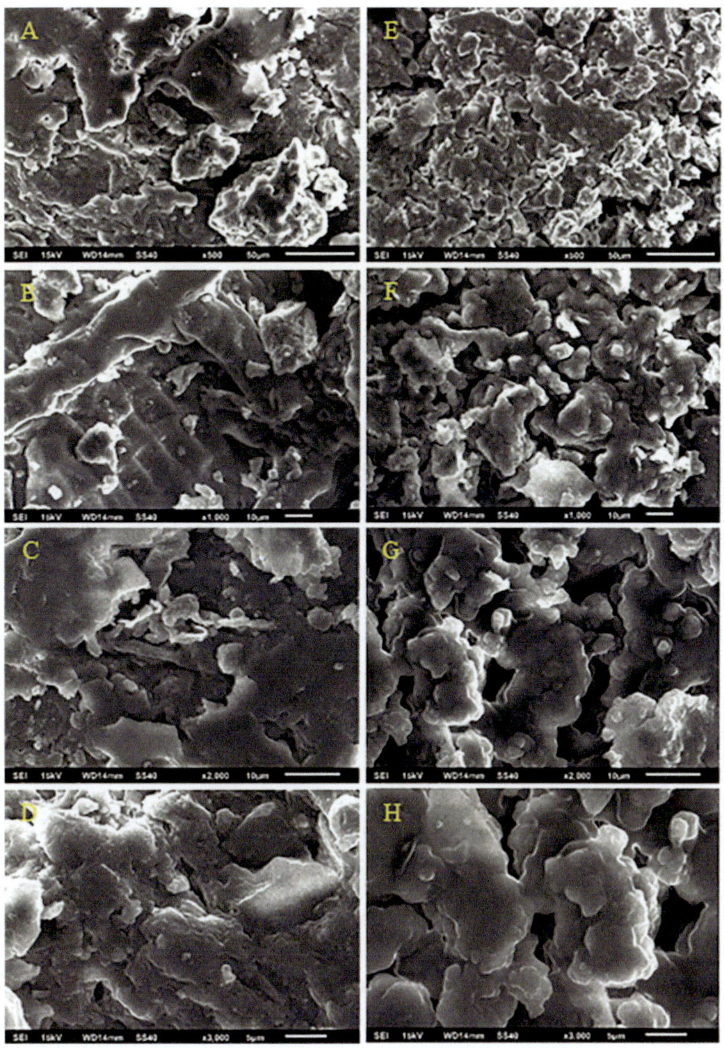

FIGURE 7.3 Scanning electron microscopy (SEM) images of flax oilseed cake at ×500 (A, E), ×1000 (B, F), ×2000 (C, G), and ×3000 (D, H) magnification levels. (A–D) The surface morphology of a non-sonicated extract sample; (E–H) the surface morphology of a sonicated extract sample. The sonicated extract sample images show significant surface morphological alterations, which occurred *via* ultrasonication.

TABLE 7.4
Secondary Metabolites in the Extracted Sample of Flax Oilseed Cake Identified by liquid Chromatography Electrospray Ionization Mass Spectrometry (LC-ESI-MS) in Positive Mode

Rt (min)	Secondary Metabolites	Molecular Ion Mass (m/z)	Adduct Molecular Ion Mass (m/z)	Area% (SE sample)	Area% (NSE sample)
1.71	Neolinustatin	423.4	487 [M+2ACN] $^+$	1.9	PND
2.18	Lotaustralin	261.2	294 [M+CH_3OH+H] $^+$	2.5	2.3
5.85	Linustatin	409.3	450 [M+ACN] $^+$	11.4	9.1
6.35	Linolenic acid	278.4	310 [M+CH_3OH+H] $^+$	9.6	7.9
10.10	Secoisolariciresinol	362.4	380 [M+ NH_4] $^+$	1.5	PND
12.74	Caffeic acid	180.1	213 [M+CH_3OH+H] $^+$	4.7	PND
15.42	Sinapic acid	224.2	246 [M+Na] $^+$	2.3	1.9
18.58	Palmitic acid	256.4	274 [M+NH_4] $^+$	3.8	2.1
23.51	Stearic acid	284.5	302 [M+NH_4] $^+$	1.6	0.5
26.64	Linamarin	247.2	329 [M+2ACN] $^+$	2.8	1.5
34.02	Matairesinol	358.4	391 [M+CH_3OH+H] $^+$	2.1	1.2
34.65	Syringic acid	198.1	414 [M+NH_4] $^+$	1.8	PND

Note: ACN is acetonitrile, PND is a peak not determined, Rt is retention time, SE is sonicated extract, NSE is a non-sonicated extract, m/z is a mass-to-charge ratio, and the m/z value is considered the mass.

sonicated and non-sonicated extract at different magnification levels. These images show smooth surfaces with no pores or fissures and intact surface structures that are tightly connected. The particle size of flaxseed oilcake also decreases in sonicated extract compared to the non-sonicated one (Figures 7.3 A-H). Nadeem (2022), in her study, performed liquid chromatography-mass spectrometry (LC-MS) to visualize the effect of ultrasonication. The following compounds, namely neolinustatin, lotaustralin, linustatin, linolenic acid, secoisolariciresinol, caffeic acid, sinapic acid, palmitic acid, stearic acid, linamarin, matairesinol, and syringic acid, were identified in flaxseed oil cake (Table 7.4). The peak areas of identified compounds in sonicated extract were 11.4% (linustatin), 9.6% (linolenic acid), 4.7% (caffeic acid), 3.8% (palmitic acid), 2.8% (linamarin), and 2.5% (lotaustralin) at the retention times of 5.85, 6.35, 12.74, 18.58, 26.64, and 2.18 min, respectively, in positive mode of LC-ESI-MS. The percentages of compounds were decreased in the non-sonicated extract compared to the sonicated extract.

7.3 ANTAGONISTIC PROPERTIES OF FLAXSEED OIL CAKE

Several chemicals, including chlorine, chlorine dioxide, hydrogen peroxide, and electrolyzed water, are currently used to kill pathogens (Kim et al. 2009; Huang and Chen 2011; Alexandre et al. 2012). But these chemicals harm the environment and human health. Plant by-product extracts containing secondary metabolites have been researched as natural antagonists as an alternative to chemical pesticides (Cowan 1999; Del Carmen Villalobos et al. 2016). Notably, plant by-products (oilseed cake) contain valuable components such as phenolic compounds (Gyawali and Ibrahim 2014); using these compounds has several advantages, including low cost, environmental friendliness, and reduced waste (García-Lomillo et al. 2014). Many studies have linked flaxseed oil cake's antagonistic effects to the presence of natural bioactive compounds, including polyphenols, glucosylated lignans, and aglycones (Barbary et al. 2010; Pag et al. 2014). Furthermore, previous research has shown that, instead of oil and fibres, the residue from the extraction process of flaxseed oil, known as seedcake, can contribute to antagonistic activities because it contains significant levels of phenolic acids that have antimicrobial potential (Zuk et al. 2014).

Tiyagi and Ajaz (2004) reported in their study that flaxseed oilseed cake significantly inhibits the multiplication of phytonematodes. The nematicidal impact was observed on PPNs such as *Hoplolaimus indicus*, *Helicotylenchus indicus*, *Rotylenchulus reniformis*, *Tylenchorhynchus brassicae*, *Tylenchus filiformis*, *Meloidogyne incognita*, *Hemicriconemoides mangiferae*, *Pratylenchus coffeae*, *Dorylaimus* sp., *Longidorus elongatus*, *Xiphinema basiri*, and *Trichodorus mirzai*. Devi and Gupta (1995) also observed dipping egg sacs of cyst nematodes in extracts of flaxseed cakes for 24 and 48 hr suppressed larval development. Flaxseed oil cakes extracted in various solvents resulted in 70–100% nematode mortality (Sebastian and Gupta 1997). In comparison to juveniles, eggs of *M. incognita* were found to be more susceptible to oilcake of flaxseed and fungicide application (Lanjeswar and Shukla 1986). Nadeem (2022) studied the effect of different flaxseed oil cake extracts prepared with or without sonication alone and in combination with *B. subtilis* and *P. putida* at various concentrations on hatching and mortality of *M. javanica* and found it very effective in suppressing this pathogenic nematode (Table 7.5). Flaxseed, compared with synthetic antibiotics such as cefoperazone, also possesses excellent antibacterial activity against several microbial strains, including *Streptococcus agalactiae* (NCIM 2401), *Micrococcus luteus* (ATCC 10240, ITCC9341), *Staphylococcus aureus* (ITCC8531, ATCC 29737), *Bacillus pumilus* (ATCC 14884), *Enterococcus faecalis* (ATCC 51299), *Bacillus subtilis* (ATCC 6633), *Bacillus brevis* (ITCC7096), *Bacillus cereus* (ATCC 11778), and *Escherichia coli* (ATCC 8739) (Kaithwas et al. 2011).

Earlier research found that oligosaccharides derived from flaxseed can inhibit the establishment of pathogens such as *Alternaria alternata* and *Alternaria solani*, as well as the human pathogenic fungus *Candida albicans* (Guilloux et al. 2009). Chitooligosaccharides derived from flax seed cake

TABLE 7.5
Effect of Different Extracts of Flaxseed Oil Cake at 2000-ppm Concentration on Hatching and Mortality of *M. javanica* (Nadeem 2022)

Different Processes of Extract Preparation	Percentage Hatching Inhibition after 5 Days of the Exposure Period	Percentage Mortality after 72 Hr of the Exposure Period
Undecomposed extract	71.9	36.5
Undecomposed extract + *B. subtilis* lipopeptide	86.3	62.5
Undecomposed extract + *P. putida* lipopeptide	84.1	59.7
Sonicated undecomposed extract	72.6	43.0
Sonicated undecomposed extract + sonicated *B. subtilis* lipopeptide	87.8	71.0
Sonicated undecomposed extract + sonicated *P. putida* lipopeptide	85.8	69.0
Decomposed extract	75.5	46.5
Decomposed extract + *B. subtilis* lipopeptide	85.6	70.5
Decomposed extract + *P. putida* lipopeptide	84.8	67.5
Sonicated decomposed extract	79.5	50.0
Sonicated decomposed extract + sonicated *B. subtilis* lipopeptide	89.9	75.2
Sonicated decomposed extract + sonicated *P. putida* lipopeptide	87.1	70.0

exhibited antifungal activity against pathogenic fungi such as *C. albicans*, *Penicillium chrysogenum*, *Fusarium graminearum*, and *Aspergillus flavus* (Xu et al. 2008).

7.3.1 Management of PPNs on Cultivated Crops Using Flaxseed Oil Cake

PPNs in extensive crop production systems have been managed mainly by nematicides and chemical soil fumigants. However, some nematicides have been banned from the marketplace in subsequent years due to ecological and human health concerns. Furthermore, nematicides do not always promote long-term pathogen control. PPNs are more challenging to control than other pests because they dwell in the soil and potentially damage nearly every component of the plant. Using oilseed cakes as organic soil amendments is a conventional cultural strategy for improving soil fertility and structure; it is also known as an approach for combating soil-borne pathogens such as PPNs. Flaxseed oil cakes can prevent damage caused by nematodes both directly and indirectly by boosting plant development, modifying root physiology, and increasing populations of antagonist microorganisms, as well as productivity (Adegbite and Adesiyan 2005, Nadeem 2022).

7.3.1.1 Flaxseed Oil Cake as Organic Soil Amendment

In general, amendments of oilseed cakes in soil decreased PPN populations *via* toxic breakdown products such as phenolics and ammonia or promoted predacious and parasitic activities of soil biota. Flaxseed oilseed cakes not only prevented nematode infestation and galling but also suppressed the nematode population in roots and soil and their egg-laying potential (Nadeem 2022). The incorporation of oilseed cakes significantly affected the multiplication of PPNs, including root-knot nematodes (Radwan et al. 2009; Irshad et al. 2012; Nadeem 2022). Oilseed cakes of flaxseed as soil amendments were also investigated against *M. incognita*–infected tomatoes, and it was reported that root galling was considerably reduced (Radwan et al. 2009). Rehman et al. (2014) conducted a pot study to assess the efficacy of various doses of oilseed cakes against root-knot nematode infecting *Vigna mungo* and reported that flaxseed oil cake at 100 g/pot was effective in controlling *M. incognita* and improving plant growth, pollen fertility, and yield in a pulse crop. Goswami and Swarup (1971) tested flaxseed oil cakes against *M. incognita* on tomato seedlings. Mishra (1972) studied the effect of eight cakes (karanj, neem, mahua, mustard, groundnut, cotton, flaxseed, sesamum) on nematodes attacking wheat, mung beans, and tomatoes. These cakes efficiently lowered the nematode population. Singh and Sitaramaiah (1966, 1969) observed that incorporating oil cakes of flaxseed into infested soil in field plots 3 weeks before growing okra or tomatoes reduced root-knot galls. Nadeem (2022) used flaxseed as different treatments (flaxseed oil cake at 100 g/kg soil and sonicated extract at 20 ml/kg soil) and observed that both treatments effectively control *M. javanica* and improve plant growth and biochemical parameters in fenugreek.

7.4 ROLE OF FLAXSEED OIL CAKE IN INTEGRATED MANAGEMENT

Integrated nematode management is a combination of biocontrol agents with resistant plants, suitable agricultural approaches, and minimal use of pesticides as an integral part of farming around the world and serves as a basis for nematode management in the future (Hillocks 2002; Phipps and Park 2002; Feder et al. 2004). Combining biocontrol agents with oilseed cakes has proved significant in controlling PPNs on vegetable crops (Ashraf and Khan 2010).

Bhatt et al. (2002) conducted a study on betel vine to manage *M. incognita* using oilseed cakes and *Trichoderma viride* and found that three quarterly applications of *T. viride* multiplied on flaxseed oil cake were effective in reducing root-knot incidence in betel vines. The inclusion of the antagonist in oil cake not only improved *T. viride* establishment in the rhizosphere but also increased its rate of multiplication. Combining oilseed cakes of flaxseed in soil with *P. lilacinus* improved plant development and decreased nematode population development and galls on tomatoes, according to Khan and Saxena (1997). Under glasshouse circumstances, Ashraf and Khan (2010) studied the efficacy of biocontrol agents (*P. lilacinus* and *C. oxysporum*) and/or oilseed cakes of flaxseed in the management of root-knot nematode *M. javanica* infecting eggplant. The integration of *Paecilomyces lilacinus* with flaxseed or linseed was found to be highly effective in reducing the multiplication rate of *Meloidogyne* spp. and subsequently increasing the plant growth attributes compared to the plants in which only *P. lilacinus* or oil cake was inoculated (Tiyagi and Ajaz 2004). Combining oil cakes with biocontrol agents including *Trichoderma viride*, *T. harzianum*, *Glomus fasciculatum*, *G. mosseae*, and *Paecilomyces lilacinus* was always efficient in controlling nematodes in soil (Borah and Phukan 2004; Mahanta and Phukan 2004; Borah et al. 2007).

7.5 BIOTECHNOLOGICAL APPROACH

7.5.1 Antibiotic Production

Oil cakes are frequently utilized in the production of industrial enzymes, antibiotics, vitamins, biofertilizers, and other biological products (Figure 7.4). However, with a growing emphasis on industrial process cost reduction and adding value to agro-industrial leftovers, oil cakes may be an essential base of proteinaceous elements as well as a substratum for various biotechnological approaches. Many oilseed cakes, especially edible oilseed cakes, have the potential to be advantageous when used as a substrate for bioprocesses. These have been used to produce antibiotics, enzymes, mushrooms, and more by fermentation. The production of vitamins and antioxidants is another biotechnological use for flaxseed oil cakes. Oilseed cakes can be utilized as a substrate for manufacturing various antimicrobial agents, including antibiotics, which lowers costs. Some of the compounds in flaxseed are converted by intestinal bacteria into lignans, which are responsible for several potential advantages of flaxseed. Because antibiotics destroy these bacteria, lignans do not develop normally. Much research reported the role of flaxseed oilcake extract against pathogenic

FIGURE 7.4 Schematic diagram showing flaxseed oil cake applications.

bacterial strains that were both sensitive and multidrug-resistant (Zuk et al. 2014). They found that seed cake exhibited antibacterial properties because of antibiotic production.

7.5.2 Enzyme Production

Oilseed cakes would be used in the commercial synthesis of enzymes, biopolymers, biogas, antibiotics, and bioelectricity, among other things. Oilseed cakes have the potential to provide benefits in the field of bioprocessing when used as a substrate. Solid-state fermentation (SSF) is a biotechnological technique in which microorganisms thrive in the absence of free-flowing water on a non-soluble substrate or solid substance. This method could turn oilseed cake into a functional product. Several studies have described oil cakes as a source of SSF or a supplement to the production media to produce various enzymes. Oil cakes provide an excellent nutrient supplement in SSF, providing both nitrogen and carbon supplies, and have been reported to be an excellent substrate for enzyme synthesis.

7.6 CONCLUSION

Using flaxseed oil cakes as organic amendments can be a safer alternative to toxic chemicals in managing PPNs in cultivated crops. Flaxseed oil cakes have promising nematicidal action and provide opportunities for crop growers to keep the environment sustainable. Oil cakes are the best substrate for rapidly growing helpful soil microbes, which benefit plants in various ways and inhibit the multiplication of pathogens. Flaxseed oil cakes successfully controlled the nematode population, as proved in multiple published reports. These oil cakes contain phytotoxic components, which exhibit nematicidal activity for nematodes. These compounds also promote plant growth parameters and physiochemical properties.

REFERENCES

Adegbite, A.A., and S.O. Adesiyan. 2005. Root extracts of plants to control root-knot nematode on edible soybean. *World Journal of Agricultural Research* 1:18–21.

Alexandre, E.M., T.R. Brandão, and C.L. Silva. 2012. Assessment of the impact of hydrogen peroxide solutions on microbial loads and quality factors of red bell peppers, strawberries and watercress. *Food Control* 27:362–368.

Ashraf, M.S., and T.A. Khan. 2010. Integrated approach for the management of *Meloidogyne javanica* on eggplant using oil cakes and biocontrol agents. *Archives of Phytopathology and Plant Protection* 43:609–614.

Bakke, J.E., and H.J. Klosterman. 1956. A new diglucoside from flaxseed. *Proceedings of North Dakota Academy of Science* 10:18–22.

Barbary, O.M., S.A. El-Sohaimy, M.A. El-Saadani, and A.M.A. Zeitoun. 2010. Antioxidant, antimicrobial and anti-HCV activities of lignan extracted from flaxseed. *Research Journal of Agriculture and Biological Sciences* 6:247–256.

Basiri, S., N. Haidary, S.S. Shekarforoush, and M. Niakousari. 2018. Flaxseed mucilage: A natural stabilizer in stirred yogurt. *Carbohydrate Polymers* 187:59–65.

Bernard, G.C., M. Egnin, and C. Bonsi. 2017. The impact of plant-parasitic nematodes on agriculture and methods of control. *Nematology-Concepts, Diagnosis and Control* 10:121–151.

Bhatt, J., S.K. Sengupta, and R.K. Chaurasia. 2002. Management of *Meloidogyne incognita* by *Trichoderma viride* in betelvine. *Indian Phytopathology* 55:97–98.

Borah, A., P.P. Neog, and A.K. Sinha. 2007. Efficacy of *Glomus fasciculatum* in comparison with neem cake and carbofuran for management of *Meloidogyne incognita* on blackgram and greengram. *Annals of Plant Protection Sciences* 15:445–448.

Borah, A.P., and N. Phukan. 2004. Comparative efficacy of *Glomus fasciculatum* with neem cake and carbofuran for the management of *Meloidogyne incognita* on brinjal. *Indian Journal of Nematology* 34:129–132.

CFIA. 2012. The biology of *Linum usitatissimum* L. (Flax). https://inspection.canada.ca/plant-varieties/plants-with-novel-traits/applicants/directive-94–08/biology-documents/linum-usitatissimum-l-flax-/eng/1330979709525/1330979779866

Chemat, F., I. Grondin, P. Costes, L. Moutoussamy, A.S.C. Sing, and J. Smadja. 2004. High power ultrasound effects on lipid oxidation of refined sunflower oil. *Ultrasonics Sonochemistry* 11:281–285.

Chemat, F., M.A. Vian, and G. Cravotto. 2012. Green extraction of natural products: Concept and principles. *International Journal of Molecular Sciences* 13:8615–8627.

Corrales, M., S. Toepfl, P. Butz, D. Knorr, and B. Tauscher. 2008. Extraction of anthocyanins from grape by-products assisted by ultrasonics, high hydrostatic pressure or pulsed electric fields: A comparison. *Innovative Food Science and Emerging Technologies* 9:85–91.

Cowan, M.M. 1999. Plant products as antimicrobial agents. *Clinical Microbiology Reviews* 12:564–582.

Crouse, G.D. 1998. *Pesticides: Managing Risks and Optimizing Benefits*. Washington, DC: ACS Symposium Series 734, American Chemical Society. Natural products as leads for new pesticides with reduced risk, pp. 80–95.

Del Carmen Villalobos, M., M.J. Serradilla, A. Martín, S. RuizMoyano, C. Pereira, and M. de Guía Córdoba. 2016. Synergism of defatted soybean meal extract and modified atmosphere packaging to preserve the quality of figs (*Ficus carica* L.). *Postharvest Biology and Technology* 111:264–273.

Devi, S., and P. Gupta. 1995. Larval emergence from egg sacs of *Heterodera cajani* in extracts of cakes in various media and their effect on cowpea. *Indian Journal of Nematology* 25:190–193.

Elling, A.A. 2013. Major emerging problems with minor *Meloidogyne species*. *Phytopathology* 103:1092–1102.

Eyres, L. 2015. Flaxseed fibre–a functional superfood? *Food New Zealand* 15:24.

Feder, G., R. Murgi, and J.B. Quizon. 2004. Sending farmers back to school. The impact of farmer field schools in Indonesia. *Review of Agricultural Economics* 26:45–62.

Flax Council of Canada. 2014. Canada–a flax leader. www.flaxcouncil.ca/

Food and Agriculture Organization FAO. 2020. www.fao.org/statistics/en/

Food and Agriculture Organization, Statistics Division FAOSTAT. 2020. www.fao.org/faostat/en/

Ganorkar, P.M., and R.K. Jain. 2013. Flaxseed–a nutritional punch. *International Food Research Journal* 20:519–525.

García-Lomillo J, M.L. González-SanJosé, R. Del Pino-García, M.D. Rivero-Pérez, and P. Muñiz-Rodríguez. 2014. Antioxidant and antimicrobial properties of wine by-products and their potential uses in the food industry. *Journal of Agriculture and Food Chemistry* 62:12595–12602.

Gill, K. S. 1987. Agronomy. In: *Flaxseed*. New Delhi: Indian Council of Agricultural Research, pp. 241–268.

Goswami, B.K., and G. Swarup. 1971. Effect of oil cake amended soil on the growth of tomato and root-knot nematode population. *Indian Phytopathology* 24:491–494.

Goyal, A., V. Sharma, N. Upadhyay, S. Gill, and M. Sihag. 2014. Flax and flaxseed oil: An ancient medicine & modern functional food. *Journal of Food Science and Technology* 51:1633–1653.

Guilloux, K., I. Gaillard, J. Courtois, B. Courtois, and E. Petit. 2009. Production of arabinoxylanoligosaccharides from flaxseed (*Linum usitatissimum*). *Journal of Agriculture and Food Chemistry* 57:11308–11313.

Gyawali, R., and S.A. Ibrahim. 2014. Natural products as antimicrobial agents. *Food Control* 46:412–429.

Herchi, W., D. Arráez-Román, H. Trabelsi, I. Bouali, S. Boukhchina, H. Kallel, A. Segura-Carretero, and A. Fernández-Gutierrez. 2014. Phenolic compounds in flaxseed: A review of their properties and analytical methods. An overview of the last decade. *Journal of Oleo Science* 63:7–14.

Hillocks, R.J. 2002. Integrated pest management and organic agricultural for small holders in Africa. *International Pest Management Reviews* 7:17–27.

Hromadkova, Z., J. Kováčiková, and A. Ebringerová. 1999. Study of the classical and ultrasound-assisted extraction of the corn cob xylan. *Industrial Crops and Products* 9:101–109.

Huang, Y., and H. Chen. 2011. Effect of organic acids, hydrogen peroxide and mild heat on inactivation of Escherichia coli O157: H7 on baby spinach. *Food Control* 22:1178–1183.

Irshad, M., S.A. Tiyagi, R. Rizvi, and K. Zehra. 2012. Soil amendment with oilseed cakes for the management of phytone-matodes and soil-inhabiting fungi attacking *Capsicum annuum* L. *Journal of Medicinal and Aromatic Plant Sciences* 34:46–52.

Kaithwas, G., A. Mukerjee, P. Kumar, and D.K. Majumdar. 2011. *Linum usitatissimum* (flaxseed/flaxseed) fixed oil: Antimicrobial activity and efficacy in bovine mastitis. *Inflammopharmacology* 19:45–52.

Khan, T.A., and S.K. Saxena. 1997. Effect of root-dip treatment with fungal filtrates on root penetration, development and reproduction of *Meloidogyne javanica* on tomato. *International Journal of Nematology* 7:85–88.

Kim, Y.J., M.H. Kim, and K.B. Song. 2009. Efficacy of aqueous chlorine dioxide and fumaric acid for inactivating pre-existing microorganisms and *Escherichia coli* O157:H7, *Salmonella typhimurium*, and *Listeria monocytogenes* on broccoli sprouts. *Food Control* 20:1002–1005.

Lan, Y., J.B. Ohm, B. Chen, and J. Rao. 2020. Physicochemical properties and aroma profiles of flaxseed proteins extracted from whole flaxseed and flaxseed meal. *Food Hydrocolloids* 104:105731.

Lanjeswar, R.D., and V.N. Shukla. 1986. Vulnerability of larvae and eggs of *Meloidogyne incognita* to some oil cakes and fungicides. *Indian Journal of Nematology* 16:69–73.

Mahanta, B., and P.N. Phukan. 2004. Comparative efficacy of *Glomus fasciculatum*, neem cake and carbofuran for the management of *Meloidogyne incognita. Annals Plant Protection Sciences* 12:377–379.

Mason, T.J., Chemat, F., and M. Vinatoru. 2011. The extraction of natural products using ultrasound or microwaves. *Current Organic Chemistry* 15:237–247.

Mishra, S.D. 1972. Effect of organic matter and inorganic fertilizers on soil and plant nematodes. PhD Thesis, IARI, New Dehli.

Morris, D.H. 2007. *Flax: A Health and Nutrition Primer* (4th edn). Flax Council of Canada, Winnipeg, MB, Canada.

Nadeem, H. 2022. Studies on the potential of biopesticides for the management of root-knot nematode. PhD Thesis, Aligarh Muslim University, Aligarh.

Oomah, B.D., and L. Sitter. 2009. Characteristics of flaxseed hull oil. *Food Chemistry* 114:623–628.

Pag, A.I., D.G. Radu, D. Draganescu, M.I. Popa, and C. Sirghie. 2014. Flaxseed cake–a sustainable source of antioxidant and antibacterial extracts. *Cellulose Chemistry and Technology* 48:265–273.

Phipps, R.H., and J.R. Park. 2002. Environmental benefits of genetically modified crops: Global and European perspective on their ability to reduce pesticide use. *Journal of Animal and Food Sciences* 11:1–18.

Radwan, M. A., E.K. El-Maadawy, S.I. Kassem, and M.M. Abu-Elamayem. 2009. Oil cakes soil amendment effects on *Meloidogyne incognita*, root-knot nematode infecting tomato. *Archives of Phytopathology and Plant Protection* 42:58–64.

Rehman, B., M.A. Ganai, K. Parihar, M. Asif, and M.A. Siddiqui. 2014. Biopotency of oilcakes against *Meloidogyne incognita* affecting *Vigna mungo. European Journal of Applied Sciences* 6:57–63.

Rizvi, R., I. Mahmood, S.A. Tiyagi, and Z. Khan. 2012. Conjoint effect of oilseed cakes and *Pseudomonas fluorescens* on the growth of chickpea in relation to the management of plant-parasitic nematodes. *Brazilian Archives of Biology and Technology* 55:801–808.

Rostagno, M.A., M. Palma, and C.G. Barroso. 2003. Ultrasound-assisted extraction of soy isoflavones. *Journal of Chromatography A* 1012:119–128.

Sebastian, S., and P. Gupta. (1997). Laboratory screening of some nematicides and oilcake extracts against *Pratylenchus thornei. Indian Journal of Nematology* 27:133–134.

Sekhri, S. 2011. *Textbook of Fabric Science: Fundamentals to Finishing.* New Delhi: PHI Learning Private Limited, p. 76.

Sharma, M., K. Dadhwal, Y. Gat, V. Kumar, A. Panghal, R. Prasad, and P. Gat. 2019. A review on newer techniques in extraction of oleaginous flaxseed constituents. *OCL*, 26:14.

Shim, Y.Y., B. Gui, P.G. Arnison, Y. Wang, and M.J.T.Reaney. 2014. Flaxseed (*Linum usitatissimum* L.) bioactive compounds and peptide nomenclature: A review. *Trends Food Science and Technology* 38:5–20.

Shim, Y.Y., B. Gui, Y. Wang, and M.J.T.Reaney. 2015. Flaxseed (*Linum usitatissimum* L.) oil processing and selected products. *Trends in Food Science and Technology* 43:162–177.

Sikandar, A., M.Y. Zhang, Y.Y. Wang, X.F. Zhu, X.Y. Liu, and H.Y. Fan. 2020. Review article: *Meloidogyne incognita* (root-knot nematode) a risk to agriculture. *Applied Ecology and Environmental Research* 18:1679–1690.

Singh, K.K., D. Mridula, J. Rehal, and P. Barnwal. 2011. Flaxseed: A potential source of food, feed and fiber. *Critical Reviews in Food Science and Nutrition* 51:210–222.

Singh, R.S., and K. Sitaramaiah. 1966. Incidence of root-knot of okra and tomato in oil cake amended soil. *Plant Disease Reporter* 50:668–672.

Singh, R.S., and K. Sitaramaiah. 1969. Control of root-knot nematodes through organic and inorganic amendments of soil. III. Effect of sawdust and inorganic nitrogen sources and levels. *First All India Nematology Symposium*, New Dehli, 64.

Tiyagi, S.A., and S. Ajaz. 2004. Biological control of plant parasitic nematodes associated with chickpea using oil cakes and *Paecilomyces lilacinus. Indian Journal of Nematology* 34:44–48.

USDA (U.S Department of Agriculture). 2021. https://fdc.nal.usda.gov/fdc-app.html#/food-details/2262075/nutrients (accessed on 28 September 2022).

Xu, Y., C. Hall 3rd, and C. Wolf-Hall. 2008. Antifungal activity stability of flaxseed protein extract using response surface methodology. *Journal of Food Sciences* 73:M9–M14.

Zhang, Z.S., L.J. Wang, D. Li, S.S. Jiao, X.D. Chen, and Z.H. Mao. 2008. Ultrasound-assisted extraction of oil from flaxseed. *Separation and Purification Technology* 62:192–198.

Zuk, M., A. Dorotkiewicz-Jach, Z. Drulis-Kawa, M. Arendt, A. Kulma, and J.Szopa. 2014. Bactericidal activities of GM flax seedcake extract on pathogenic bacteria clinical strains. *BMC Biotechnology* 14:1–15.

8 Sesame Oilseed Cake
Chemical Compounds and Nematicidal Potential

Hera Nadeem, Ananya Prova, Rishil Gupta,
Md. Saeed Sultan and Faheem Ahmad

CONTENTS

8.1 INTRODUCTION

Plant-parasitic nematodes (PPNs) are significant biological limitations that create severe yield losses in economically essential crops distributed worldwide. In terms of money, the yearly global losses in horticulture crops owing to PPNs are anticipated to be 8.8–14.6% of crop yields and US$100–157 billion (Nicol et al. 2002; Abad et al. 2008). PPNs are obligatory pathogens that target the belowground sections of plants, such as roots, bulbs, and rhizomes, causing water uptake and movement to be hampered. Certain PPNs can invade aboveground tissues like leaves, buds, and flowers as well. Phytonematodes affect plant growth and impede nodulation and nitrogen fixation, limiting crop productivity. The root-knot nematode (RKN) *Meloidogyne* genus is found globally and is responsible for significant crop losses (Wesemael et al. 2011). *Meloidogyne javanica*, *M. incognita*, *M. arenaria*, and *M. hapla* are economically significant among the 70 identified *Meloidogyne* spp. The severity of damage caused by *Meloidogyne* spp. is generally species specific and varies according to species of plants, crop rotation, seasons, and type of soil (Potter and Olthof 1993). The primary and most common symptom is the development of root galls within the root systems of infected plants, where the *Meloidogyne* spp. are rooted. As a result, the nematode-infested host plants suffer severe damage, making them vulnerable to wilting, impairment in growth, and vulnerability to other plant pathogens. These parasites have a wide geographic spread, a high potential for damage, and limited treatment measures, making agricultural production in nematode-infested farms challenging to preserve. Phyto-nematode damages are sometimes confused with nutritional deficiencies, making it difficult to diagnose and treat. Even though they coexist with other advantageous soil microorganisms, scholars and practitioners face significant difficulty controlling PPNs (Akhtar 1997).

Cultural methods, chemical nematicides, and resistance cultivars can all be used to control these nematodes. Among them, for decades, chemical nematicides are currently the most common

DOI: 10.1201/9781003319252-8

method of nematode-induced disease control (Whitehead 1998). However, organic nematicides are beneficial because of their eco-friendly nature (Noling and Becker 1994). Research focusing on regulating PPNs by using organic supplements is a promising alternative for suppressing phytonematodes and enhancing soil health and structure (Faruk et al. 2001). Oilseed cakes are widely available additives that stimulate the establishment of beneficial organisms, lowering the number of nematodes and improving the productivity of crops in diverse methods (Sumbul et al. 2015). Oilseed cakes are packed with micronutrients that can be incorporated straight into the soil either in powdered or pellet form, watered, and left to decompose (Youssef and El-Nagdi 2010). It is proposed that the combative nature of oilseed cake on fungi and nematodes is because of unique fungi toxic and nematicidal compounds released because of decomposition or nitrogen deficiency caused by saprophytes ability to multiply or sensitivity of pathogenic fungi to CO_2 liberation. It is also reported that the amendments of oil cakes also increase the natural competitors' population (Chavarria-Carvajal and Rodriguez-Kabana 1998). Manurial components such as nitrogen, phosphate, and potash are commonly found in oilseed cakes. The decomposition of various oilseed cakes takes about 1–2 weeks, resulting in a prolonged supply of micronutrients to the plants, finally suppressing the PPN population. Oilseed cakes are more effective when supplemented with damp soils than when mixed with dry soils. Volatile fatty acids, such as formaldehyde, acetic, propionic, amino acids butyric, ammonia, hydrogen sulphide, phenols, and formic acid, are liberated during oil cake decomposition and are lethal to phytonematodes (Sumbul et al. 2015; Alam et al. 1980).

Sesame (*Sesamum indicum* L.) is amongst the world's oldest and most widely cultivated oilseeds. According to Food and Agriculture Organization Corporate Statistical Database (FAOSTAT), their annual production in 2019 exceeded 6.5 million tons, with Sudan, Myanmar, and India taking the lead (FAOSTAT 2020). The extraction of sesame oil results in the production of a defatted by-product known as sesame cake. This by-product of the sesame oil business is typically utilized as livestock fodder or compost production. Sesame seeds' popularity may be explained by their high concentration of sesame lignans, including sesamin, sesamolin, and sesamol. Sesame oil has antioxidant and health-promoting properties due to its increased level of lignan, phytosterol, tocopherol, and other components (Afroz et al. 2019). Defatted sesame oil cakes or meals could be an excellent support matrix source for biotechnological applications like synthesizing enzymes, antibiotics, vitamins, antioxidants, and food products. Some studies have found that sesame cake contains a variety of biologically active substances, including sesamin, amino acids, and protein (Suja et al. 2004). The protein content of sesame cake is 35.6%, the crude fibre content is 7.6%, the ash content is 11.8%, the calcium content is 2.45%, and the phosphorus content is 1.11% (Plaitho et al. 2017).

In squash plants, Youssef and El-Nagdi (2004) report that sesame seed cake lessened root knots, egg masses, and nematode development. Sesame has the potential to reduce nematodes which is used as a rotation crop and has been investigated in various tropical and subtropical farming techniques. To control pathogenic nematodes, ground sesame stalks, sesame chaff, and sesame cake are put into the soil before seeding or spreading as a mulch around growing crops. *M. incognita* penetration was significantly reduced in okra (*Abelmoschus esculentus*) intercropped with sesame compared to a monoculture control (Tanda and Atwal 1988).

8.2 ORGANIC PRODUCTS AND THEIR POTENTIAL

Organic soil additions have been used for a long time (Muller and Gooch 1982; D'Addabbo 1995; Rodriguez-Kabana et al. 1995; Akhtar and Malik 2000). It has been established that their chemical properties determine the effectiveness of organic amendments against nematodes. Beneficial microorganisms colonize the rhizosphere because of the organic substrates. The colonized microflora helps to reduce the PPNs present in the soil. These approaches are ecologically sustainable, less expensive, and easily accessible with minimum environmental impact. Nitrogen and phosphorus are generally available in reasonable amounts in oilseed cakes (Akhtar and Alam 1990).

Numerous antibacterial by-products (hydrogen sulfide, phenols, tannins) are produced or generated by microorganisms that indulge in organic amendment decomposition (Rodriguez-Kabana et al. 1995). The use of sesame chaff as a nematicide has been proved, but its mechanism of action against nematodes does not appear to be biocidal, and one source suggested that it should be called a root protectant (McKenry 1994).

Due to their high nitrogen, phosphorus and potassium content, non-edible oil cakes are utilized as organic nitrogenous fertilizers. Some oilseed cakes have been proved to boost plant nitrogen absorption by slowing soil nitrification. Additionally, they defend the plants against phytonematodes, insects, and parasites in the soil. In contrast, edible oilseed cakes are employed as cattle and protein hydrolysate, whereas non-edible cakes are used as biocontrol agents. They have advantages when employed as substrates in the development of bioprocesses for producing organic compounds and biomolecules. Research on their use in the synthesis of industrial enzymes has yielded encouraging results. For such purposes, mixed substrate fermentation has proved to be more beneficial. Furthermore, oil cakes are an excellent alternative to traditional applications by being used to manufacture environmentally acceptable green biofuel. Another essential aspect is that the bioprocess utilizing oil cakes is appealing owing to the relatively affordable supply for the whole year, creating further alluring when economics are considered.

8.3 SESAME OILSEED CAKE

Antioxidants extracted and purified from natural sources, mainly by-products, are of enormous importance to the market sector because they can be employed in functional foods, food additives, nutraceuticals, and pharmaceuticals. The sesame seed contains about 50% oil, typically extracted through mechanical pressing. Defatted sesame oilseed cake, the final product of the pressing, could retain 18–22% residual oil (Hwang 2005), which can be subsequently refined utilizing solvents, enhancing the oil extraction yield and retrieving phytoconstituents and minerals (Sarkis et al. 2015). Hexane, a petroleum isomer mixture, has traditionally been used as a vegetable oil extraction solvent. This solvent is toxic, combustible, and polluting, despite its remarkable stability and oil extraction capabilities (Oliveira et al. 2012; Tir et al. 2012). Numerous replacements for hexane have now been investigated. Short-chain alcohols, primarily ethanol and isopropanol, are commonly discussed because of their lower side effects and enhanced operational safety (Tir et al. 2012). A high-quality protein fraction acceptable for human consumption can be found in defatted sesame seed cake (Sarkis et al. 2015). Because of its amino acid makeup, this protein fraction is regarded as an efficient additional source of vegetable proteins such as lentils (Escamilla-Silva et al. 2003). Furthermore, the optimal usage of a protein constituent is not solely determined by its nutritional properties but also by how varied processing circumstances affect its functional qualities (Khalid et al. 2003) because these attributes instantly affect the sensory features and, as a result, the final product quality.

8.3.1 Bioactive Compounds of Sesame Oilseed Cake

Sesame cakes or defatted seeds were shown to have diverse bioactive metabolites, highlighting the importance of agri-food waste as a source of bioactive metabolites. In this case, cutting-edge techniques like high-performance liquid chromatography combined with high-resolution mass spectrometry enabled the identification of bioactive metabolites in sesame oilseed cakes. Yeasmin et al. (2021) performed GCMS to analyse the secondary metabolites of defatted sesame cake. Depending on the relative area, they determined thirty bioactive compounds with the highest yield: 32.77% glycerin, 11.4% gamma-tocopherol (vitamin E), 10.85% hexadecanoic acid methyl ester, and 8.19% 13-docosenamide. Glycerin is antibacterial and antiviral in nature. Gamma tocopherol has greater antioxidant activity than alpha-tocopherol, representing approximately 90.5% of total tocopherols. Compared to other edible oils, sesame oil, which is abundant in tocopherols and lignans (especially sesamin and sesamolin), has greater oxidative stability (Hwang 2005). The methyl ester of hexadecenoic acid has antioxidant and antifungal effects (Pinto et al. 2017), and the presence of 13-docosonamide adds antioxidant activity (Kumaradevan et al. 2015). It was also shown that several metabolites, such as phenol, 2-methoxy-3-(2-propenyl)-, 3-allyl-6-methoxyphenol, 2-propyl-1-pentanol, and pentafluoropropionate, had antibacterial action at lower doses (Saravanan et al. 2014).

Nadeem (2022) performed LC-MS to visualize the effect of ultrasonication on the enhancement of metabolite concentration and identified the following compounds: Corylifolinin, samin, guaiacol, p-Coumaric, sesamolin, piperitol, ferulic acid, arachidic acid, linoleic acid, and eupafolin (Table 8.1a and 8.1b). Among identified compounds, Corylifolinin was the major constituent, following Arachidic acid, with a relative percentage of peak area of 25.9% and 19.2%, respectively. Tables 8.1a and b show the peak area percentage of all detected compounds in the sonicated extract sample of sesame oilseed cake compared to the control. In sonicated extract, for instance, the peak areas of detected metabolites in positive mode of liquid chromatography electrospray ionization mass spectrometry (LC-ESI-MS) were 25.9% (Corylifolinin), 11.4% (samin), 8.6% (guaiacol), 3.4% (p-Coumaric), 2.9% (sesamolin), and 1.7% (piperitol) at retention time of 1.49, 1.77, 2.23, 2.76, 3.25, and 16.44 min, respectively. In negative mode, the percentage peak areas of 19.2% (arachidic acid), 13.5% (ferulic acid), 4.5% (linoleic acid), and 5.6% (eupafolin), respectively, at the retention time of 18.48, 2.33, 30.66, and 38.97 min. The percentage areas of compounds were decreased in the non-sonicated extract compared to sonicated extract at a similar retention time and experimental set. Nadeem (2022) used the SEM technique to find out the structural modifications in the obtained sonicated sample. The images offered critical foundational findings for acquiring morphological changes in the target samples. The surface morphology was observed by attentively inspecting the SEM images. Figure 8.1 depicts SEM images of sesame oilseed cakes at various magnification ranges. These images show smooth surfaces with no pores or fissures and intact surface structures that are tightly connected. Sesame oilseed cake particle sizes decreased in sonicated extract compared to the non-sonicated one (Figure 8.1A–H). Vibrations caused by ultrasonic waves can form gaps that transfer energy to solid particles engrossed in the extraction. Furthermore, cavitation bubbles form nearer to the solid surface and collapse with greater amplitude, causing the cell wall to rupture and speeding up the transfer of required constituents stuck inside the solvent medium (Senrayan and Venkatachalam 2019).

TABLE 8.1A

Secondary Metabolites in the Extracted Sample of Sesame Seed Cake Identified by Liquid Chromatography Electrospray Ionization Mass Spectrometry (LC-ESI-MS) in Positive Mode

Rt (min)	Secondary Metabolites	Molecular Ion Mass (m/z)	Adduct Molecular Ion Mass (m/z)	Area% (SE sample)	Area% (NSE sample)
1.49	Corylifolinin	324.4	325 $[M+H]^+$	25.9	13.5
1.77	Samin	250.2	268 $[M+NH_4]^+$	11.4	PND
2.23	Guaiacol	124.1	166 $[M+ACN+H]^+$	8.6	PND
2.76	p-Coumaric	164.1	205 $[M+ACN]^+$	3.4	2.4
3.25	Sesamolin	370.4	388 $[M+ NH_4]^+$	2.9	1.5
16.44	Piperitol	356.4	374 $[M+NH_4]^+$	1.7	0.5

Note: ACN is acetonitrile, PND is a peak not determined, Rt is retention time, SE is sonicated extract, NSE is a non-sonicated extract, m/z is a mass-to-charge ratio, and the m/z value is considered the mass.

TABLE 8.1B

Secondary Metabolites in the Extracted Sample of Sesame Seed Cake Identified by Liquid Chromatography Electrospray Ionization Mass Spectrometry (LC-ESI-MS) in Negative Mode

Rt (min)	Secondary Metabolites	Molecular Ion Mass (m/z)	Adduct Molecular Ion Mass (m/z)	Area% (SE sample)	Area% (NSE sample)
2.33	Ferulic acid	194.1	191 $[M-3H]^-$	13.5	PND
18.48	Arachidic acid	312.5	311 $[M-H]^-$	19.2	12.2
30.66	Linoleic acid	280.4	279 $[M-H]^-$	4.5	2.3
38.97	Eupafolin	316.26	315 $[M-H]^-$	5.6	2.8

Note: ACN is acetonitrile, PND is a peak not determined, Rt is retention time, SE is sonicated extract, NSE is a non-sonicated extract, m/z is a mass-to-charge ratio, and the m/z value is considered the mass.

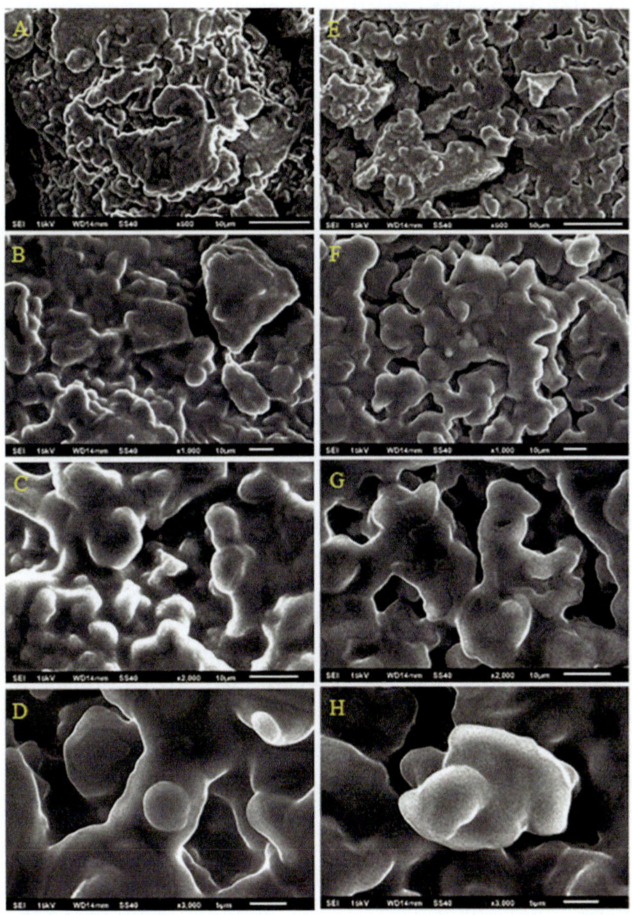

FIGURE 8.1 Scanning electron microscopy (SEM) images of sesame oilseed cake at ×500 (A, E), ×1000 (B, F), ×2000 (C, G), and ×3000 (D, H) magnification levels. (A–D) The surface morphology of a non-sonicated extract sample. (E–H) The surface morphology of a sonicated extract sample. The images of the sonicated extract sample show significant morphological alterations on the surface, which occurred via ultrasonication. The decrease in the size of the particle in the sonicated sample compared to the non-sonicated sample is also evident.

8.4 OILSEED CAKE COEXISTING WITH PHYTOTOXICITY

Organic amendments have improved soil nutrients, physical conditions, biological activity, and crop productivity. The enhancement in plant development attained by oil cakes was strongly related to a surge in soil nutritional condition, which served as fertilizer. The nutritional composition of the supplements and the vast amounts of these fertilizers given to the soil result in enhanced soil fertility and, therefore, plant development. This makes the plants more resistant to nematode assaults (Akhtar and Alam 1993). Phytotoxicity is thus a preventive issue in the overuse of oilseed cakes against phytonematodes. Except for sesame cake, significant phytotoxicity was associated with greater rates of oilseed cakes examined in a specific investigation. Based on previous research, it was possible to infer that oilseed cakes applied as organic amendments exhibited promising nematostatic action and served non-chemical approaches for managing PPNs. Oilseed cake–enriched soil has the potential to be a component of a comprehensive approach to nematicide substitute. The buildup of salts or ionic species (e.g., nitrites, nitrates, ammonia) or the establishment of phytopathogenic microbes may have a negative impact on root-knot nematodes (Rodriguez-Kabana et al. 1995). These organic amendment action methods are not always understood, and application methods are frequently empirical.

According to Bridge (1996) and Oka (2010), their mechanism of combating action against phytonematodes involves three important biological mechanisms:

- To increase the soil's ability to hold water and nutrients, they also help plants grow more vigorously and become more resistant to nematodes.
- They release certain compounds that could be nematicidal.
- They support nematode antagonists, nematode parasites and predators that depend on antimicrobial activity in the soil (e.g., nematophagous fungi, parasitic bacteria, and micro-arthropods).

PPNs may be suppressed by incorporating oilseed cake in crop soil (Radwan et al. 2004; Hafez and Sundararaj 1999; Tiyagi and Ajaz 2004; Sasanelli et al. 2003). Sesame callus extracts were also shown to include free and bound amino acids. Amino acids are not hazardous in and of themselves (Clark et al. 1959), but they can create toxic complexes with other compounds, such as complexes of amino acids and chlorogenic acids or phenolic glycosides, that are hazardous to microbes. Sesame meal releases methyl isothiocyanates which have the nematicidal effect that might help to control nematodes (Lear 1956). Oilseed cakes have recently emerged as an appealing source of bioactive chemicals, including phenolic and flavonoid compounds (Teh and Birch 2014; Ramachandran et al. 2007). Oilseed cake's secondary metabolites are trapped in complex and insoluble components such as lipoprotein bilayers, hull, lignin, cell walls, and vacuoles (Corrales et al. 2008). As a result, selecting an extraction procedure is crucial. Phenolics are a complex group of compounds such as flavonoids, isoflavones, tannins, cinnamic acid, caffeic acid, and phenylpropanoids. These compounds have shown nematicidal effects (Ntalli et al. 2009; Nguyen et al. 2013; Ohri and Pannu 2010). They also play a critical role in the hypersensitivity responses of plants to pathogens, as evidenced by their accumulation in feeding sites of *Meloidogyne* spp. (Oliveira et al. 2019). Karimi et al. (2022) studied the effects of sesame oil cake at 1.5%, 1%, and 0.5% (w/w) to control the *Meloidogyne javanica*, root-knot nematode within tomato roots in sterilized and non-sterilized soils under greenhouse conditions. The results showed that treatments of 1.5% sesame oil cake in non-sterilized soil reduced the final nematode population by 83%. The addition of oil cakes of sesame, fennel, and anise resulted in significant reductions in PPNs inhabiting eggplant (*Solanum melongena*), resulting in improved plant yield and growth contributing features (El-Sherif et al. 2010). According to Abid (1996), sesame oilseed and neem oil cakes dramatically reduced the development of galls on the root, resulting in increased plant growth. Sesame oilseed extract contains sesamin, sesamolin, and sesamol, found to be antagonists for nematodes (Radwan et al. 2009). Radwan et al. (2009)

acknowledged that the number of galls was dramatically reduced by oilseed cakes at 5 to 50 (g/kg soil) compared with control but had lower activity than carbofuran. Sesame-treated plants experienced the most significant reduction in root galling, followed by a cake made with soybeans, flax, cotton, and olives. The sesame cake performed similarly to flaxseed or soybean cake regarding gall reduction. The treatments might be grouped in descending order based on their reductions (%) for galls: 76.85% by carbofuran, 62.64% by sesame cake, 59.23% by soybean cake, 58.41% by flax cake, 50.91% by cotton cake, and 46.11% by olive. These data also demonstrated a significant reduction in galls on roots when the rate of the oilseed cake was raised.

Regarding the raised rates, *M. incognita* galling of tomato roots was considerably lowered at 50 g oil cake/kg soil compared to the control or other rates tested. Conversely, cotton, olive, or sesame cakes decreased galling by 91.54%, 86.38%, and 98.14%, respectively, when treated at the highest rate (50 g/kg). D'Addabbo et al. (2019) used four commercial biostimulants in a case study. The extract of sesame seed oil, seaweed, or neem cake was compared for efficacy on potted tomatoes against *M. incognita*, a root-knot nematode. All four biostimulants considerably reduced galls and nematode eggs on tomato roots. However, sesame and neem seed cakes were much more inhibitory than the others. Furthermore, practically all biostimulant treatments greatly improved tomato growth compared to the untreated control.

8.4.1 Integrated Management with Sesame Oilseed Cake

Sharma and Trivedi (2017) investigated the effect of sesame oilseed cake in combination with the fungus *Paecilomyces* and found it very efficient in reducing the disease incidence of root-knot nematode. Combining oilseed cakes of castor, mustard, sesame, neem, groundnut, and flax in soil with *P. lilacinus* improved plant development, decreased nematode population and development of galls on tomatoes, according to Khan and Saxena (1997). Nadeem (2022) performed a pot experiment using both sesame oilseed cake powder and a sonicated extract of the same before sowing fenugreek seeds and two biocontrol agents, *Bacillus subtilis* and *Pseudomonas putida*, to study their effect on *Meloidogyne javanica*, root-knot nematode. The results exhibit that the integrated treatment of sesame oilseed cake extract and *B. subtilis* are effective in enhancing plant growth and biochemical parameters and decreasing the pathological parameters of *M. javanica*. Some of the reports in which sesame oilseed cake was used in combination with other biocontrol agents are listed in Table 8.2.

8.4.2 Sesame Oilseed Cake's Antagonism Potential for PPNs on Pulse Crops

The pulses and PPNs are intricately related, and due to the parasitic nature of PPNs, enormous quantitative and qualitative losses have been reported (Khan 2008). According to Ali and Singh (2007), the most frequent nematodes that infect legumes appear to be the root-knot nematode, reniform

TABLE 8.2
Integrated Management with Sesame Oilseed Cake and Biocontrol Agents

Oilseed Cake + Biocontrol	Nematode	Plant	Reference
Sesame + *Paecilomyces lilacinus*	*M. incognita*	Tomato	Khan and Saxena (1997)
Sesame + *P. lilacinus*	*M. incognita*	Mung bean	Sharma and Trivedi (2017)
Sesame + *Glomus fasciculatum*	*Heterodera avenae*	Wheat	Sharma and Trivedi (2015)
Sesame + *G. mosseae*			
Sesame + *Bacillus subtilis*	*M. javanica*	Fenugreek	Nadeem (2022)
Sesame + Pseudomonas putida	*M. javanica*	Fenugreek	Nadeem (2022)

nematode, root lesion nematode, and cyst nematodes. PPNs are anticipated to result in global yield losses of 13.2% in *Cajanus cajan* (pigeon pea), 10.9% in *Phaseolus vulgaris* (field bean), and 13.7% in *Cicer arietinum* (chickpea; Sasser and Freckman 1987).

Sesame oilseed cake has also been used to control the PPNs. Numerous papers have discussed the use of oilseed cakes to control PPN-infecting pulse crops. Oilseed cakes such as neem, sesame, and mustard have been used. They are significantly effective in promoting plant development and suppressing the population of cyst nematodes on pigeon peas when well mixed with soil and left to decompose (Meena et al. 2009). Rizvi et al. (2012) demonstrated the role of oilseed cake and bioagents in green gram and lentil crop fields, found improved plant development indices such as plant weight, pod numbers, pollen fertility percentage, root nodulation, and chlorophyll content while decreasing the nematode population. They discovered that the combination inoculations outperformed the solo inoculations. The PPNs primarily targets the roots of pulse crops, interfering with their ability to form nodules, decreasing their capacity to fix nitrogen, and consequently reducing plant development. Numerous researchers have noted a decrease in nodules in plants with higher nematode infestations (Khan et al. 2005; Shafi et al. 2002). The majority of the cases under investigation revealed that oilseed cakes might help to boost the potential of leguminous crops to develop nodules on roots, promoting the growth of the plant and reducing nematode populations. Mishra and Gupta (1997) showed that mustard, neem, and oil cakes significantly boosted soybean plant growth (*Glycine max*). However, a reduction in root-knot infection was also observed. A multifold boost in plant growth metrics was also seen, as well as residual effects on the succeeding crop, chickpea. Tiyagi et al. (2001) showed similar results on lentil plant development metrics and PPN control.

8.5 CONCLUSION

PPNs cause massive losses to cultivated crops worldwide. PPNs have been controlled with nematicides and chemical soil fumigants in intensive agricultural production. However, due to the negative impact of synthetic nematicides, the processes behind managing PPNs require more investigation. Organic amendments have many potentials to advance sustainable agriculture effectively. Combining them with other cultural and biological agent control techniques can enhance their effectiveness in managing plant parasitic nematodes. Using oilseed cakes as soil organic amendments can effectively control the PPN based on mounting data. The nematicidal potential of sesame oilseed cake has been revealed. However, some recent studies reported that sesame oilseed cakes effectively managed the nematode population using an integrated approach. Sesame oilseed cakes also help increase soil nutrients, improve the soil's chemical and physical characteristics, and stimulate predators directly or indirectly. Further investigation requires knowing the mechanism of action against PPNs using sesame oilseed cake.

REFERENCES

Abad, P., J. Gouzy, J.M. Aury, P. Castagnone-Sereno, E.G. Danchin, E. Deleury, L. Perfus-Barbeoch, V. Anthouard, F. Artiguenave, V.C. Blok, and M.C. Caillaud. 2008. Genome sequence of the metazoan plant-parasitic nematode *Meloidogyne incognita*. *Nature Biotechnology* 26:909–915.

Abid, M. 1996. Studies on the control of root-knot nematodes (*Meloidogyne* spp.) with botanical toxicants. PhD Thesis, Department of Botany, University of Karachi, Karachi.

Afroz, M., S.N.K. Zihad, S.J. Uddin, R. Rouf, M.S. Rahman, M.T. Islam, I.N. Khan., E.S. Ali, S. Aziz, J.A. Shilpi, and L. Nahar. 2019. A systematic review on antioxidant and antiinflammatory activity of sesame (*Sesamum indicum* L.) oil and further confirmation of antiinflammatory activity by chemical profiling and molecular docking. *Phytotherapy Research* 33:2585–2608.

Akhtar, M. 1997. Current options in integrated management of plant-parasitic nematodes. *Integrated Pest Management Reviews* 2:187–197.

Akhtar, M., and M.M. Alam. 1990. Evaluation of nematicidal potential in some medicinal plants. *International Nematology Network Newsletter* 6:8–10.

Akhtar, M., and M.M. Alam. 1993. Utilization of waste materials in nematode control: A review. *Bioresource Technology* 45(1):1–7.

Akhtar, M., and A. Malik. 2000. Roles of organic soil amendments and soil organisms in the biological control of plant-parasitic nematodes: A review. *Bioresource Technology* 74:35–47.

Alam, M.M., M. Ahmad, and A.M. Khan. 1980. Effect of organic amendments on the growth and chemical composition of tomato, eggplant and chilli and their susceptibility to attack by *Meloidogyne incognita. Plant and Soil* 57:231–236.

Ali, S.S., and B. Singh. 2007. An overview of nematode constraints in pulses and their management. *Journal of Food Legumes* 20:1.

Bridge, J. 1996. Nematode management in sustainable and subsistence agriculture. *Annual Review of Phytopathology* 34:201e225.

Chavarria-Carvajal, J.A., and R. Rodriguez-Kabana. 1998. Changes in soil enzymatic activity and control of *Meloidogyne incognita* using four organic amendments. *Nematropica* 280:7–18.

Clark, R.S., J. Kuc, R.E. Henze, and F.W. Quackenbush. 1959. The nature and fungitoxicity of an amino acid addition product of chlorogenic acid. *Phytopathology* 49:594–597.

Corrales, M., S. Toepfl, P. Butz, D. Knorr, and B. Tauscher. 2008. Extraction of anthocyanins from grape by-products assisted by ultrasonics, high hydrostatic pressure or pulsed electric fields: a comparison. *Innovative Food Science and Emerging Technologies* 9:85–91.

D'Addabbo, T. 1995. The nematicidal effect of organic amendments: A review of the literature, 1982–1994. *Nematologia Mediterranea* 23:299–306.

D'Addabbo, T., S. Laquale, M. Perniola, and V. Candido. 2019. Biostimulants for plant growth promotion and sustainable management of phytoparasitic nematodes in vegetable crops. *Agronomy* 9:616.

El-Sherif, A.G., A.R Refaei, M.E. El-Nagar, and M.H. Salem. 2010. Impact of certain oilseed cakes or powder in comparison with oxamyl or urea on *Meloidogyne incognita* infecting eggplant. *Archives of Phytopathology and Plant Protection* 43:88–94.

Escamilla-Silva, E.M., S.H. Guzmán-Maldonado, A. Cano-Medinal, G. González-Alatorre. 2003. Simplified process for the production of sesame protein concentrate. Differential scanning calorimetry and nutritional, physicochemical and functional properties. *Journal of the Science of Food and Agriculture* 83:972–979.

FAOSTAT. 2020. Available online: www.fao.org/faostat/en/#data (accessed on 12 September 2022).

Faruk, M.I., M.A. Bari, M.S. Nahar, M.A. Rahman, and M.M. Hossain. 2001. Management of root knot nematode (*Meloidogyne*) of tomato with two organic amendments and a nematicide. *Bangladesh Journal of Plant Patholo*gy 17:27–30.

Hafez, S.L., and P. Sundararaj. 1999. Efficacy of seed crop meals for the management of Columbia root-knot nematode *Meloidogyne chitwoodi* on tomato under greenhouse conditions. *Nematropica* 29:171–177.

Hwang, L.S. 2005. Sesame oil. In: *Bailey's Industrial Oil and Fat Products*, ed. F. Shahidi, 537–576. Hoboken, NJ: John Wiley and Sons.

Karimi, S., Z. Azami-Sardooei, and F. Fekraf. 2022. Effects of the grapefruit wastes and sesame oil cake to control of the root-knot nematode *Meloidogyne javanica* on tomato plants. *Iranian Journal of Plant Protection Science* 52:157–172.

Khalid, E.K., E.E. Babiker, and A.H.E.L Tinay. 2003. Solubility and functional properties of sesame seed proteins as influenced by pH and/or salt concentration. *Food Chemistry* 82:361–366. doi:10.1016/S0308-8146(02)00555-1

Khan, M.R. 2008. Nematodes, an emerging threat to global forests: assessment and management. *Plant Pathology Journal* 11:9–113.

Khan, M.R., F.A. Mohiddin, S.M. Khan, and B. Khan. 2005. Effect of seed treatment with certain biopesticides on root-knot of chickpea. *Nematologia Mediterranea* 33:107–112.

Khan, T.A., and S.K. Saxena. 1997. Integrated management of root knot nematode *Meloidogyne javanica* infecting tomato using organic materials and *Paecilomyces lilacinus. Bioresource Technology* 61:247–250.

Kumaradevan, G., R. Damodaran, P. Mani, G. Dineshkumar, and T. Jayaseelan. 2015. Phytochemical screening and GC-MS analysis of bioactive components of ethanol leaves extract of *Clerodendrum phlomidis* (L.). *American Journal of Biological and Pharmaceutical Research* 2:142–48.

Lear, B. 1956. Results of laboratory experiments with Vapam for control of nematodes. *Plant Disease Reports* 40:847–852.

McKenry, M. 1994. Nematicides. In *Encyclopedia of Agriculture Science*, ed. C.J. Arntzen, vol. 3, 87–95. San Diego, CA: Acad. Inc.

Meena, P., S. Nehra, and P.C. Trivedi. 2009. Efficacy of decomposed organic cakes against *Heterodera cajani* infecting *Cajanus cajan. Asian Journal of Experimental Sci*ences 23:181–184.

Mishra, S.M., and P. Gupta. 1997. Effect of some oil cakes and saw dust application on root-knot (*Meloidogyne incognita*) infection and root nodulation of soybean. *Indian Phytopathology* 50:131–132.

Muller, R., and P.S. Gooch. 1982. Nematological reviews: Organic amendments in nematode control. An examination of the literature. *Nematropica* 12:319–326.

Nadeem, H. 2022. Studies on the potential of biopesticides for management of root-knot nematode. PhD thesis, Aligarh Muslim University, Aligarh.

Nguyen, D.M., D.J. Seo, K.Y. Kim, R.D. Park, D.H. Kim, Y.S. Han, T.H. Kim, and W.J. Jung. 2013. Nematicidal activity of 3, 4-dihydroxybenzoic acid purified from *Terminalia nigrovenulosa* bark against *Meloidogyne incognita*. *Microbial Pathogenesis* 59:52–59.

Nicol, J.M., R. Rivoal, N. Bolat, H. Aktas, H.J. Braun, M. Mergoum, A.F. Yildrim, A. Bagci, J.H. Eleckcioglu, and A. Yahyaoui. 2002. The frequency and diversity of the cyst and lesion nematode on wheat in the Turkish Central Anatolian Plateau. *Nematology* 4:272.

Noling, J., and J. Becker. 1994. The challenge of research and extension to define and implement alternatives to methyl bromide. *Journal of Nematology* 26:573.

Ntalli, N.G., U. Menkissoglu-Spiroudi, I.O. Giannakou, and D.A. Prophetou-Athanasiadou. 2009. Efficacy evaluation of a neem (*Azadirachta indica* A. Juss) formulation against root-knot nematodes *Meloidogyne incognita*. *Crop Protection* 28:489–494.

Ohri, P., and S.K. Pannu. 2010. Effect of phenolic compounds on nematodes-a review. *Journal of Applied and Natural Science* 2:344–350.

Oka, Y. 2010. Mechanisms of nematode suppression by organic soil amendments–a review. *Applied Soil Ecology* 44:101e115.

Oliveira, D.F., V.A. Costa, W.C. Terra, V.P. Campos, P.M. Paula, and S.J. Martins. 2019. Impact of phenolic compounds on *Meloidogyne incognita in vitro* and in tomato plants. *Experimental Parasitology* 199:17–23.

Oliveira, R., V. Oliveira, K.K. Aracava, and C.E.C. Rodrigues. 2012. Effects of the extraction conditions on the yield and composition of rice bran oil extracted with ethanol–a response surface approach. *Food and Bioproduct Processing* 90:22–31.

Pinto, M.E., S.G. Araújo, M.I. Morais, N.P. Sa, C.M. Lima, C.A. Rosa, E.P. Siqueira, S. Johann, and L.A. Lima. 2017. Antifungal and antioxidant activity of fatty acid methyl esters from vegetable oils. *Anais da Academia Brasileira de Ciências* 89:1671–1681.

Plaitho, Y., P. Rattanasena, P. Chaikham, and P. Prangthip. 2017. Biochemical and antioxidative properties of unprocessed and sterilized white and black sesame by-product from Northern Thailand. *Current Research in Nutrition and Food Science Journal* 5:196–205.

Potter, J.W., and T.H.A. Olthof. 1993. Nematode pests of vegetable crops. In: *Plant Parasitic Nematodes in Temperate Agriculture*, 171–207. Wallingford, UK: CAB International.

Radwan, M.A., M.M. Abu-Elamayem, K. ShMI, and E.K. Maadawy. 2004. Management of root knot nematode by integration of *Bacillus thuringiensis* with either organic amendments or carbofuran. *Pakistan Journal of Nemato*logy 22:135–142.

Radwan, M.A., E.K. El-Maadawy, S.I. Kassem, and M.M. Abu-Elamayem. 2009. Oil cakes soil amendment effects on *Meloidogyne incognita*, root-knot nematode infecting tomato. *Archives of Phytopathology and Plant Protection* 42:58–64.

Ramachandran, S., S.K. Singh, C. Larroche, C.R. Soccol, and A. Pandey. 2007. Oil cakes and their biotechnological applications–a review. *Bioresource Technology* 98:2000–2009.

Rizvi, R., I. Mahmood, S.A. Tiyagi, and Z. Khan. 2012. Conjoint effect of oilseed cakes and *Pseudomonas fluorescens* on the growth of chickpea in relation to the management of plant-parasitic nematodes. *Brazilian Archives of Biology and Technology* 55:801–808.

Rodriguez-Kabana, R., V. Estaun, J. Pinochet, and O. Marfa. 1995. Mixtures of olive pomace with different nitrogen sources for the control of *Meloidogyne* spp. on tomato. *Journal of Nematology* 27:575.

Saravanan P, G. Chandramohan, J. Mariajancyrani, and P. Shanmugasundaram. 2014. GC-MS analysis of phytochemical constituents in ethanolic bark extract of *Ficus religiosa* Linn. *International Journal of Pharmacy and Pharmaceutical Sciences* 6:457–460.

Sarkis, J.R., N. Boussetta, C. Blouet, I.C. Tessaro, L.D.F. Marczak, and E. Vorobiev. 2015. Effect of pulsed electric fields and high voltage electrical discharges on polyphenol and protein extraction from sesame cake. *Innovative Food Science and Emerging Technologies* 29:170–498.

Sasanelli, N., P. Greco, T. D'addabbo, M.I. Coiro, and F. Lamberti. 2003. The use of olive mill wastes for the control of root-knot nematodes. *Communications in Agricultural and Applied Biological Sciences* 68:135–138.

Sasser, J.N., and D.W. Freckman. 1987. A world perspective of nematology: The role of the society. In *Vistas on Nematology*, eds. J.A. Veech and D.W. Dickson, 7–20, Hyattsville, Maryland.

Senrayan, J., and S. Venkatachalam. 2019. A short extraction time of vegetable oil from *Carica papaya* L. seeds using continuous ultrasound acoustic cavitation: Analysis of fatty acid profile and thermal behavior. *Journal of Food Process Engineering* 42:e12950.

Shafi, A., I. Mahmood, and Z.A. Siddiqui. 2002. Integrated management of root-knot nematode *Meloidogyne incognita* on chickpea. *Thai Journal of Agricultural Science* 35:273–280.

Sharma, A. and P.C. Trivedi. 2015. Integrated management of *Heterodera avenae* infecting wheat using AM Fungi, oil Cakes and neem Products. *The Journal of Indian Botanical Society* 94:262–267.

Sharma, S., and P.C. Trivedi. 2017. Ecofriendly management of *M. incognita* with the combination of *Paecilomyces lilacinus* and organic amendments. *Journal of Phytological Research* 30: 1-5.

Suja, K.P., A. Jayalekshmy, and C. Arumughan. 2004. Free radical scavenging behavior of antioxidant compounds of sesame (*Sesamum indicum* L.) in DPPH system. *Journal of Agricultural and Food Chemistry* 52:912–915.

Sumbul, A., R. Rizvi, I. Mahmood, and R.A. Ansari. 2015. Oil-cake amendments: Useful tools for the management of phytonematodes. *Asian Journal of Plant Pathology* 9:91–111.

Tanda, A.S., and A.S. Atwal. 1988. Effect of sesame intercropping against the root-knot nematode (*Meloidogyne incognita*) in okra1. *Nematologica* 34:484–492.

Teh, S.S., and E.J. Birch. 2014. Effect of ultrasonic treatment on the polyphenol content and antioxidant capacity of extract from defatted hemp, flax and canola seed cakes. *Ultrasonics Sonochemistry* 21:346–353.

Tir, R., P.C. Dutta, and A.Y. Badjah-Hadj-Ahmed. 2012. Effect of the extraction solvent polarity on the sesame seeds oil composition. *European Journal of Lipid Science and Technology* 114:1427–1515.

Tiyagi, S.A., and S. Ajaz. 2004. Biological control of plant parasitic nematodes associated with chickpea using oil cakes and *Paecilomyces lilacinus. Indian Journal of Nematology* 34:44–48.

Tiyagi, S.A., A.V. Khan, and M.M. Alam. 2001. Role of oilseed cakes for the management of plant-parasitic nematodes and soil-inhabiting fungi on lentil and mungbean. *Archives of Phytopathology and Plant Protection* 33:453–472.

Wesemael W, N. Viaen, and M. Moens. 2011. Root-knot nematodes (*Meloidogyne* spp.) in Europe. *Nematology* 13:3–16.

Whitehead, A.G. 1998. *Plant Nematode Control*. Wallingford, UK: CAB International.

Yeasmin, S., T.A. Chowdhury, G.M. Rana, M.M. Rahman, and L. Ferdousi. 2021. Characteristics of sesame (*Sesamum indicum* l.) seed meal grown in the northern region of Bangladesh. *Biomedical Journal of Scientific and Technical Research* 38:29944–29949.

Youssef, M.M.A., and W.M. El-Nagdi. 2004. Cellular alteration of root-knot nematode, *Meloidogyne incognita* infected squash plant and intercropping sesame plant or sesame oil seed cake as control measures. *Egyptian Journal of Phytopathology* 32:77–85.

Youssef, M.M.A., and W.M. El-Nagdi. 2010. Effect of certain organic materials in controlling *Meloidogyne incognita* root-knot nematode infesting banana. *Archives of Phytopathology and Plant Protection* 43:660–665.

9 Oilseed Cake and Nematode Management in Legume Crops

Ikram Legrifi, Salah-Eddine Laasli, Rachid Lahlali and Fouad Mokrini

CONTENTS

9.1 INTRODUCTION

Plant-parasitic nematodes (PPNs) are among the principal biotic limitations to the qualitative and quantitative production of numerous economically important crops, including legumes, causing severe losses worldwide (Roopa and Gadag 2020). A total of 12.3% of agricultural losses are estimated due to PPNs, accounting for US$157 billion annually (Singh et al. 2015). PPNs are complicated organisms that have evolved adaptive ways to overcome natural obstacles over millions of years. Nematode control is difficult in this context. As a result, the most essential purpose is to minimize nematode populations and agricultural loss. Some of the most prevalent techniques for preventing and controlling PPNs are cultural practices, the use of resistant cultivars, and chemical nematicides (Ralmi et al. 2016; Tamilarasan and Manchikatla 2012).

For decades, nematicides and chemical soil fumigants have been used to regulate PPNs in intensive crop production systems. However, due to environmental and human health concerns, certain nematicides have been withdrawn from the market in recent years. In addition, nematicides may not always result in long-term disease control (Giannakou and Panopoulou 2019; Mokrini et al. 2018). The annual application of expensive and very harmful soil fumigants or non-fumigant nematicides often reduces crop losses. These chemicals offer serious health and environmental concerns, making their use unsustainable. Furthermore, the financial cost of pesticides is a significant obstacle for many farmers. Consequently, different PPN management strategies are necessary (Chaudhary and Kaul 2013).

The use of organic soil amendments against plant parasitic nematodes seems to be one of these approaches. It can control PPNs via enhancements in soil fertility and structure, changes in the resistance of plants, the production of nemato-toxic chemicals, and the activation of phytonematocidal microorganisms (Oka et al. 2007). Among the organic amendments, oilseed cakes have shown remarkable potential in nematode control measures (Goswami et al. 2006; Rizvi et al. 2012; Tiyagi et al. 2001). Oilseed cakes alter the physical and chemical features of the soil, making it unsuitable

DOI: 10.1201/9781003319252-9

for nematode activity and improving soil conditions for higher root development through enhancing nutrient uptake (Mohanty and Tayung 2021).

9.2 POTENTIAL OF OILSEED CAKES

9.2.1 Nematicidal Activity of Oilseed Cakes under In Vitro Study

Water-soluble fractions of oilseed cakes derived from castor, groundnut, mahua, and neem were harmful to nematodes such as *Helicotylenchus indicus*, *Rotylenchulus reniformis*, and *Tylenchorhynchus brassicae*. Also, they impeded *Meloidogyne incognita* larval hatching (Khan et al. 2013). Similarly, boiling extracts of mustard and cotton oilseed cake inhibited larval hatching of *M. incognita* up to 99.92% and 99.38% in water, respectively. *M. incognita* eggs were shown to be more sensitive to oilseed cakes (cotton, groundnut, karanj, kokam, linseed, mahua, sesamum, neem) than larvae (Lanjewar and Shukla 1986). Larvae of *M. incognita* were most effectively killed by neem cake extract (Gowda and Gowda 1997). Duong et al. (2014) reported that a 5% dilution of the solvent extracted from neem cake caused 100% larval mortality after 24 hours of exposure. The EC_{50} value of neem cake on J_2 larvae of *Meloidogyne* spp. was 0.51%. Mohanapriya et al. (2019) assessed the effects of three castor, peanut, and neem oilseed cakes on the hatching and mortality of *M. incognita* eggs. The research found that among the oilseed cakes, neem cake inhibited egg hatching at a concentration of 100% within 24 hours of the incubation period. In the mortality test, neem cake at a concentration of 100% demonstrated 96% death after 96 hours. *In vitro* tests revealed the nematicidal activity of colocynth, jojoba, moringa, and castor oilseed cakes at three doses (S, S/2, and S/4) against *M. incognita* J2 (Refaat et al. 2020). Colocyntha and jojoba oils were the most effective, followed by moringa oil. Castor oil was the least effective (Refaat et al. 2020). The nematicidal activity of castor and amaltas oilseed cakes against *M. incognita* was also investigated by Abbas et al. (2016) and found effective. All concentrations examined, namely 20%, 25%, 50%, and 100%, resulted in 100% larval mortality, as well as delayed and decreased egg hatch. However, higher concentrations of oilseed cake extract produced the greatest outcomes owing to the presence of more nemato-toxic compounds.

9.2.2 Nematicidal Potential of Oilseed Cakes in Soil

Previous research has shown that using oil cakes to control nematodes in the soil has been quite effective. The fields were naturally infected with *M. incognita* at 2–3 J_2/g soil and treated with two fungal bioagents, *Aspergillus niger* 1% wettable powder (WP) (3×10^7 cfu/g) and *Paecilomyces lilacinus* 1% W.P. (2×10^6 cfu/g), alone and in combinations with neem oilseed cake at 1.5 ha (Singh et al. 2011). Compared to farmer practices, the combined application of both bioagents at half dosage and neem oilseed cake was shown to be highly efficient in enhancing chickpea output and suppressing nematode growth. Reduced dosage treatment resulted in a two-fold increase in chickpea production, with a considerable decrease in the number of galls and egg masses of up to 71% and 83%, respectively. *M. incognita* was decreased by up to 61% in chickpea roots. Another study examined the nematicidal activity of crambe oil cake on *M. javanica* and *Pratylenchus brachyurus* (Tavares-Silva et al. 2015). *M. javanica* or *P. brachyurus* were injected into soybean seedlings before transplanting. Two months later, the aerial part was removed, the treatments (0, 0.5%, 1%, 2.5%, and 5% crambe cake per pot (% w/w) were used, and soybean was resown. Vegetative and nematological analyses were done after 60 days from the date of seed germination. In all experiments, 5% crambe cake treatment decreased seed germination. Higher dosages aided plant growth. Crambe cake efficiently controlled *M. javanica* in one trial when 1% was administered per pot containing 2 L of soil, and increasing crambe cake dosages increased *P. brachyurus* control in both tests. Research was conducted by Wani (2006) to study the effect of soil amendments using castor and neem cakes, both alone and in combination, on okra and lentils for the control of *M. incognita*. Overall, all treatments decreased root gall and increased plant growth and chlorophyll content. When the adjustments were coupled, the yield increased. The combination of nematode cakes and the other amendments provided the optimum nematode control and plant development. The biopesticidal potential of certain organic cakes, including cotton, castor, sesamum, mustard, and neem cake, was assessed at doses of 3g and 5g/pot for the control of *H. cajani* (Meena et al. 2009). These cakes were extensively mixed with soil to allow for decomposition. The treatments were shown to be very beneficial in promoting plant development and reducing nematode populations. In a greenhouse, the effectiveness of several organic amendments (linseed cake, mustard cake, neem cake) for controlling *H. cajani* on pigeon pea cv. Bahar was investigated (Singh 2004). All organic amendments highly reduced nematode cysts, males, egg sacs, and juveniles.

A study was conducted by Ashraf and Khan (2010) to assess the efficiency of oil cakes of castor, linseed, groundnut, mahua, and neem for the management of the root-knot nematode, *M. javanica*, under greenhouse circumstances. All treatments successfully controlled the nematode population. Using 1% mustard cake as a soil amendment significantly affected nematode-infected soybean roots (Parveen et al. 2019). Khan and Goswami (1995) demonstrated that the mortality of *M. incognita* juveniles increased with rising neem cake degradation duration up to 20 days and then decreased. In pot trials, Gutpa and Kumar (1997) discovered that groundnut and mustard cakes decreased the populations of *Tylenchorynchus* spp. and *Helicotylenchus* spp. Surprisingly, the decrease level increased with greater doses and longer treatment times. Neem, castor, mustard, and *Citrullus* oilseed cakes applied at a rate of 20 mg/pot (500 kg/ha) were effective in reducing the population of root-knot nematodes (Chaudhary and Kaul 2013).

9.2.3 Potential of Oilseed Cakes for the Management of PPNs in Leguminous Plants

Legumes are strongly linked with PPNs, resulting in quantitative and qualitative losses (Khan 2008). Root-knot nematode (*Meloidogyne incognita*), cyst nematodes (*Heterodera cajani*, *H. glycines*), reniform nematode (*Rotylenchulus reniformis*), and root lesion nematode (*Pratylenchus thornei*) are the most common worms that infest legumes crops (Ali and Singh 2007). PPNs are projected to cause 13.7% yield losses in chickpeas (*Cicer arietinum*), 13.2% in pigeon peas (*Cajanus cajan*), and 10.9% in field beans (*Phaseolus vulgaris*) globally (Sasser and Freckman 1987).

PPNs are more challenging to manage than other pathogens because they grow in the soil and potentially damage practically every component of the plant. Organic soil additives may be used effectively to control PPNs. Among numerous organic amendments, plant oil cakes have been discovered to be beneficial in controlling phytonematodes (Ashraf and Khan 2007; Goswami et al. 2006). The oilseed cake is a compacted combination of a meal left over after extracting vegetable oils from seeds such as cottonseed and groundnut, among others. These are good sources of protein and minerals and may be used as poultry and livestock feed. Oilseed cakes made from certain seeds, such as tung nuts and castor beans, are poisonous to various dangerous bacteria and are employed as fertilizers rather than feed. Food and vegetable farmers often use oil cakes as a source of plant nutrients and to prevent nematodes. Indeed, it may be spread straight to the soil as a powder or pellet, watered, and left to degrade (Sumbul et al. 2015). Their nematicidal action rises as the breakdown time increases. Plant seeds may also be treated with oilseed cake extracts before planting to protect the crop against worms. This tendency has lately piqued the attention of nematologists since it provides efficient management of the target nematode while avoiding contamination (Burman and Kaul 1995; El-Sherif et al. 2010, 2006; Youssef and El-Nagdi 2010).

Several studies have been undertaken on using oilseed cake to manage PPN-infested legume crops (Table 9.1). Castor oilseed seed cake considerably decreased the nematode population and egg masses in black grams (*Vigna mungo*) infected with 2000 recently hatched larvae of *M. incognita* (Kannan et al. 1987). Neem cake (at 2 t/haG1) reduced *R. reniformis* population by 86.2% while increasing the growth of green beans (*Phaseolus vulgaris*) (Padhi and Misra 1987). Sunflower and cotton oilseed cakes efficiently suppressed the population of *R. reniformis* on cowpea (*Vigna unguiculata*) and promoted vigorous plants, as reported by Yassin and Ismail (1994). Latif et al. (1999) and Umar and Simon (2008) discovered similar findings for nematode control on cowpea using neem cake. When applied 15 days before planting chickpea seeds, neem cake proved efficient for managing PPNs (Pandey and Singh 1990). On the roots of chickpea, castor, neem, flax, peanut, soybean, and sunflower oilseed cakes have been used to reduce *M. incognita* population and promote plant development (Tiyagi and Ajaz 2004). Likewise, oilseed cakes of karanj and neem seeds resulted in the suppression of *M. incognita* after application to chickpea roots (Yadav et al. 2005).

Similarly, Rizvi et al. (2012) reported that oilseed cakes of castor, groundnut, linseed, neem, sunflower, and soybean were found to be highly effective in reducing the population of soil nematodes of chickpeas. Oilseed cakes have also shown efficacy in managing cyst nematodes linked with pigeon peas (*Cajanus cajan*). Anver (2003) revealed that mustard, neem, duan, and castor seed cakes reduced nematode reproduction and, as a result, enhanced the woody stem density of pigeon peas. Singh (2004) investigated the effectiveness of two nematicides, phorate and carbofuran, as well as oilseed cakes of neem, mustard, and linseed in suppressing cyst nematode of *H. cajani* on pigeon pea. The highest plant growth was achieved in treatments with neem cake, followed by mustard cake, carbofuran, linseed, and phorate, but the maximum nematode population decrease was observed in treatments with both nematicides, followed by neem cake, mustard cake, and linseed cake. Oilseed cakes of castor, cotton, neem, mustard, and sesamum were considerably helpful in promoting plant development and suppressing cyst nematode populations on pigeon peas when thoroughly amended

TABLE 9.1
Management of Plant-Parasitic Nematodes in Different Legume Crops Using Oilseed Cakes

Legume Crops	Oilseed Cakes	Nematodes	References
Black gram (*Vigna mungo*)	Castor cake	*Meloidogyne incognita*	Kannan et al. 1987
Green bean (*Phaseolus vulgaris*)	Neem cake	*Rotylenchulus reniformis*	Padhi and Misra 1987
Cowpea (*Vigna unguiculata*)	Sunflower, cotton, and neem cake	PPNs including *R. reniformis*	Yassin and Ismail 1994; Latif et al. 1999; Umar and Simon 2008
Chickpea (*Cicer arietinum*)	Neem, castor, soybean, sunflower, karanj, groundnut, and linseed cake	PPNs including *M. incognita*	Pandey and Singh 1990; Tiyagi and Ajaz 2004; Yadav et al. 2005; Rizvi et al. 2012
Pigeon pea (*Cajanus cajan*)	Mustard, neem, duan, castor, linseed, cotton, and sesamum cake	PPNs including *Heterodera cajani* *and H. cajani*	Anver 2003; Singh 2004; Meena et al. 2009
Soybean (*Glycine max*)	Flax, castor mustard, mahua, and neem	*M. incognita*	Gupta and Mishra 2002
Mung bean (*Vigna radiate*)	Castor, duan, mustard, and neem	*M. incognita, R. reniformis, Tylenchorhynchus brassicae*, and *Helicotylenchus indicus*	Tiyagi and Alam 1995
Lentil (*Lens culinaris*)	Flax, castor, duan, peanut, and neem	*Meloidogyne incognita, R. reniformis, T. brassicae*, and *H. indicus*	Tiyagi et al. 2001

with soil and allowed to decompose (Meena et al. 2009). The ability of leguminous crops to fix nitrogen is linked to their capacity to create root nodules containing rhizobia. The PPN primarily attacks roots, interfering with their ability to produce nodules, limiting their nitrogen-fixing ability, and lowering plant development. Many researchers have documented a decrease in nodules in nematode-infested plants (Shafi et al. 2002).

In the majority of the cases studied, oilseed cake was beneficial in improving the root nodule formation capacity of leguminous crops, thereby promoting plant development and suppressing the nematode population. According to Gupta and Mishra (2002), flax, castor, mustard, mahua, and neem oilseed cake significantly improved soybean (*Glycine max*) growth. This effect could be attributed to the reduction of *M. incognita* infection, which will allow for increased nodulation by the root system, nodule weight, and/or release of nitrogen and other nutrients from the oilseed cake. Many researchers have also found a persistent impact of oilseed cake on PPN management. Application of castor, duan, mustard, and neem oilseed cakes to mung bean plants significantly reduced the population of PPNs (*M. incognita*, *R. reniformis*, *H. indicus*, *T. brassicae*; Tiyagi and Alam 1995). An increase in plant growth parameters and residual effects on the chickpea crop were also observed. Tiyagi et al. (2001) showed similar results on lentil plant development parameters and PPN control. The residual effects of several oilseed cakes (flax, castor, duan, peanut, and neem) were also observed on the mung bean crop.

9.3 ACTIVE CHEMICAL COMPONENTS OF OILSEED CAKES

Oilseed seed cakes are a solid residue discarded after the mechanical extraction of oilseeds containing lignin and cellulose in varying ratios (Gottipati and Mishra 2011; Sharma et al. 2013). They are of two types: edible and non-edible. Edible oilseed cakes have a high nutritional value, with a protein content ranging from 15% to 50% and are mainly used as animal feeds. The non-edible ones, such as *Azadirachta indica* (neem), *Jatropha curcas* (jatropha), *Madhuca indica* (mahua), and *Pongamia pinnata* (karanja), despite their high nutrient content, cannot be used as animal feed due to the presence of toxic components in them. Yet, the toxicity of these cakes can be used as biopesticides (Sharma et al. 2013).

Oilseed cake's composition and nutrient availability vary widely depending on seed or nut quality and the preprocessing steps. The removal of hulls, the fibrous outer covering enclosing the seed (dehulling); oil extraction method; and storage parameters are essential steps (Ramachandran et al. 2007; Mohanty et al. 2022). The mechanically pressed oilseed cakes contain more residual oil than the ones produced by the solvent extraction method (Nigam and Pandey 2009). The olive oilseed cake (OOC) is generated in the two-phase decanter process, known for a high rate of humidity (68%–71%, w/w) and the main constituents of OOC are polysaccharides (lignin 26%–30%, hemicellulose 7%–9% and cellulose 7%–9%, w/w), proteins 5%–7% (w/w), fatty acids 5%–8% (w/w), and polyphenols (oleuropein, hydroxytyrosol, tyrosol; Mirabella et al. 2014). These various chemical compositions make OOC a promising source of bioactive and nutritional compounds (Fainassi et al. 2021). Regarding physicochemical capacities, proteins and saponins are known to have surface-active properties due to the presence of hydrophilic and hydrophobic moieties in their chemical structures (Lam and Nickerson 2013). Polyphenols are also known to show surface-active properties (Souilem et al. 2014). However, they can also provide additional properties such as chemical stability and antioxidant activity (Di Mattia et al. 2010).

The cakes usually have high amounts of proteins. For example, the highest protein content is found in the oilseed cake of soybean and is an excellent source of amino acids. Approximately 15% of carbohydrates in soybeans are oligosaccharides like sucrose and raffinose. Although their nutritious quality, some oilseed cakes have their feedstock potential limited by toxic and antinutritional compounds. However, those substances can play important roles in controlling diseases in the soil. In addition, they release nutrients and act as organic fertilizers (Sharma et al. 2013). The soybean cake contains nutritional content, but anti-nutrients such as lectins, saponins, and phytates are reported in soybean cakes (Nigam and Pandey 2009). The rapeseed oilseed cake has a comparable amino acid balance as that of soybean cake. About 14.5% of carbohydrates in rapeseed cake are pectins. The anti-nutrients reported in rapeseed cakes are tannins, erucic acid, sinapine, and phytates (Mohanty et al. 2021). Cottonseed cake is reported with a good quantity of a toxic metabolite called gossypol (Nigam and Pandey 2009). *Jatropha* cake contains curcin, a highly toxic protein similar to ricin in castor, making it unsuitable for animal feed (Gübitz et al. 1999). The other major toxic compound in the meal is phorbol esters (a complex mixture of esters of tetracyclic diterpene). They were found to have tumour-promoting properties (Adolf et al. 1984). *Pongamia pinnata* cake, which is bitter and pungent, is used as a manure, fungicide, and insecticide. Although the cake is rich in proteins (30%), it is off-putting and toxic due to the presence of karanjin; pongamol; and an unusual amino acid, glabrin (Parmar et al. 1976; Sharma et al. 2013).

A study by Sharma et al. (2013) showed the extraction of active components existing in non-edible oilseed cake: azadirachtin from *Azadirachta indica* (neem), phorbol esters from *Jatropha* oilseed cake, a saponin from *M. indica* oilseed seed cake, and karanjin from *Pongamia pinnata* (karanja) oilseed cake. The aqueous, methanolic, and crude active components of oilseed cakes have been found to be effective against pests (Sharma et al. 2013). The same bioactive compounds have shown promising results against *M. incognita in vitro* (Nangki et al. 2022). Azadirachtin compounds found in neem oilseed cake are known to have a nematode toxic effect against *M. incognita*, *T. brassicae*, *R. reniformis*, and *M. javanica* in economic crops including wheat, mungbean, tomato, okra, tobacco, radish, and chilli (Akhtar 2000). A neem extract containing 10% azadirachtin has shown nematicidal properties against *M. javanica* juveniles in alkaline sandy soil (pH 8.5; Oka et al. 2007). Other nematicidal compounds, such as ammonia, could be involved and may have a direct effect, but in low concentrations, and it may occur during the oilseed cake's degradation (Abbasi et al. 2005).

9.4 MECHANISMS AND MODES OF ACTION

Plant chemicals probably evolved as a natural defence against diseases and parasites. Chemical constituents of the secondary metabolism play an important role in the defence mechanism of plants against pests, including nematodes (Akhtar 2000).

Oilseed cakes produce many elements during decomposition, directly or indirectly interfering with nematode activity. For instance, azadirachtin affects the target organism by inducing feeding deterrent effects on chemoreceptors, disrupting the endocrine system, influencing behaviour and reproduction through direct effects on tissues (Abbasi et al. 2005).

The higher nematicidal toxicity of crude phorbol esters is due to the stimulation of the cellular target protein kinase C (PKC) of *M. incognita* larvae. Activating PKC by phorbol esters leads to the phosphorylation of different proteins and the consequent organization of the cell cytoskeleton. However, a comparison of toxic bioactive constituents present in the non-edible oil seed cakes (NEOCs) and the nematicidal activity of aqueous extracts of oilseed cakes showed that the latter was found to be more effective. This might have been due to the combined synergistic effect of complex mixtures of thermostable compounds present in the cakes (Sharma et al. 2013).

The most prominent nematicidal compounds produced by the decomposition of additives in saturated soils are short-chain fatty acids, including acetic, butyric, formic, and propionic acids. These organic acids are generally produced by microorganisms fermenting carbon-rich organic matter under anaerobic conditions (Oka 2010). They are very toxic to PPNs *in vitro* (Taba et al. 2006). However, the nematicidal activity of organic amendments that release organic acids during decomposition in the soil is highly influenced by soil conditions, including oxygen concentrations, redox, pH, microbial activity, and soil temperature. Momma et al. (2006) reported higher nematicidal or fungicidal activity when soil was amended with carbon-rich materials under anaerobic conditions. Only non-ionized forms of the organic acids can cross the nematode's cell membrane and affect them: lowering the soil pH increases the percentage of non-ionized organic acids, thereby enhancing nematicidal activity and the direct adverse effect of low pH on nematodes (McElderry et al. 2005).

9.5 CONCLUSION

The excessive use of chemical pesticides for managing agricultural pests, notably PPNs, is becoming more undesirable owing to various negative repercussions. However, alternative control strategies may be applied in the integrated management of parasitic nematodes. Natural products, which may be safer than chemical substances, are one of these choices. Oil cakes have been discovered to offer the plant's most ideal and conveniently accessible by-products. The most successful and extensively used oilseed cake is neem cake, followed by a variety of others such as castor, mustard, and peanut. In the past few decades, research on oilseed cake amendments has followed the same investigative routes but with a greater emphasis on understanding the processes involved in nematicidal action. To summarize, no one technique can be used simply for the sustainable control of PPNs on economically significant crops, including legumes. But, combining several chemical and biological strategies along with oilseed cakes can effectively control the PPNs.

REFERENCES

Abbas, S. M., A. Siddiqui, M. W. Abbasi, M. J. Zaki, and S. Z. Akhtar. 2016. Phytochemical screeningand in vitro nematicidal activity of castor and amaltas seed extracts. *International Journal of Biological Research* 4:51–55.

Abbasi, P. A., E. Riga, K. L. Conn, and G. Lazarovits. 2005. Effect of neem cake soil amendment on reduction of damping-off severity and population densities of plant-parasitic nematodes and soilborne plant pathogens. *Canadian Journal of Plant Pathology* 27:38–45.

Adolf, W., H. J. Opferkuch, and E. Hecker. 1984. Irritant phorbol derivatives from four Jatropha species. *Phytochemistry* 23:129–132.

Akhtar, M. 2000. Nematicidal potential of the neem tree *Azadirachta indica* (A. Juss). *Integrated Pest Management Reviews* 5:57–66.

Ali, S. S., and B. Singh. 2007. An overview of nematode constraints in pulses and their management. *Journal of Food Legumes* 20:1.

Anver, S. 2003. Effect of different organic amendments with *Paecilomyces lilacinus* for the management of soil nematodes. *Archives of Phytopathology and Plant Protection* 36:103–109.

Ashraf, M. S., and T. A. Khan. 2007. Efficacy of *Gliocladium virens* and *Talaromyces flavus* with and without organic amendments against *Meloidogyne javanica* infecting eggplant. *Asian Journal of Plant Pathology* 1:18–21.

Ashraf, M., and T. Khan. 2010. Integrated approach for the management of *Meloidogyne javanica* on eggplant using oil cakes and biocontrol agents. *Archive of Phytopatholy and Plant Protection* 43:609–614.

Burman, U., and R. K. Kaul. 1995. Soil amendment induced changes in growth and physiology of nematode infested *Solanum melongena*. *Afroasian Journal of Nematology* 5:213–217.

Chaudhary, K. K., and R. K. Kaul. 2013. Efficacy of *Pasteuria penetrans* and various oil seed cakes in management of *Meloidogyne incognita* in chilli pepper (*Capsicum annuum* L.). *Journal of Agricultural Science and Technology* 15:617–626.

Di Mattia, C. D., G. Sacchetti, D. Mastrocola, D. K. Sarker, and P. Pittia. 2010. Surface properties of phenolic compounds and their influence on the dispersion degree and oxidative stability of olive oil O/W emulsions. *Food Hydrocolloids* 24:652–658.

Duong, D. H., X. Q. Ngo, D. G. Do, T. A. H. Le, V. T. Nguyen, and S. Nic. 2014. Effective control of neem (*Azadirachta indica* A. Juss) cake to plant parasitic nematodes and fungi in black pepper diseases in vitro. *Journal of Vietnamese Environment* 6:233–238.

El-Sherif, A. G., A. R. Refaei, M. E. El-Nagar, and H. A. A. El-Ghnam. 2006. Impact of certain organic amendments, *Bacillus thuringiensis* and oxamyl on *Tylenchulus semipenetrans* infecting lemon plant. *Journal of Plant Protection and Pathology* 31:6761–6770.

El-Sherif, A. G., A. R. Refaei, M. E. El-Nagar, and M. M. H. Salem. 2010. Impact of certain oil-seed cakes or powder in comparison with oxamyl or urea on *Meloidogyne incognita* infecting eggplant. *Archives of Phytopathology and Plant Protection* 43:88–94.

Fainassi, F., N. Taarji, F. Benkhalti, A. Hafidi, M. A. Neves, H. Isoda, and M. Nakajima. 2021. Emulsion formation and stabilizing properties of olive oil cake crude extracts. *Processes* 9:633.

Giannakou, I. O., and S. Panopoulou. 2019. The use of fluensulfone for the control of root-knot nematodes in greenhouse cultivated crops: Efficacy and phytotoxicity effects. *Cogent Food and Agriculture* 5:1–12.

Goswami, B. K., R. K. Pandey, K. S. Rathour, C. Bhattacharya, and L. Singh. 2006. Integrated application of some compatible biocontrol agents along with mustard oil seed cake and furadan on *Meloidogyne incognita* infecting tomato plants. *Journal of Zhejiang University Science B* 7:873–875.

Gottipati, R., and S. Mishra. 2011. A kinetic study on pyrolysis and combustion characteristics of oil cakes: Effect of cellulose and lignin content. *Journal of Fuel Chemistry and Technology* 39:265–270.

Gowda, P. D., and D. N. Gowda. 1997. In-vitro studies of oil-cake extracts on the survival of larvae of *Meloidogyne incognita* (Kofoid and White) Chitwood. *Plant Protection Bulletin (Faridabad)* 49:10–11.

Gübitz, G. M., M. Mittelbach, and M. Trabi. 1999. Exploitation of the tropical oil seed plant Jatropha curcas L. *Bioresource Technology* 67:73–82.

Gupta, P., and S. M. Mishra. 2002. Effect of some oil cakes and saw dust application on root-knot (*Meloidogyne incognita*) infection and root nodulation of soybean. *Indian Phytopathology* 50:131–132.

Gutpa, M. C., and S. Kumar. 1997. Efficacy of certain organic amendments and nematicides against *Tylenchorhynchus* spp. and *Helicotylenchus* spp. in soil. *Indian Journal of Nematology* 27:139–142.

Kannan, S., N. Natthuthurai, and J. Rathinavathy. 1987. *Alternatives to Synthetic Insecticides in Integrated Pest Management Systems.* Madurai: Centre for Research in Medical Entomology, pp. 223–231.

Khan, M. R. 2008. *Plant Nematodes: Methodology, Morphology, Systematics, Biology and Ecology* (1st ed.) Boca Raton, Florida, USA: CRC Press, p. 360.

Khan, M. R., and B. K. Goswami. 1995. Nematicidal efficacy of decomposed neem cake and fungal filtrates of *Paecilomyces lilacinus* and *Aspergillus flavus* against *Meloidogyne incognita. National Symposium Sustainable Agriculture in Sub-Humid Zone* 359–361.

Khan, M. W., A. M. Khan, and S. K. Saxena. 2013. Suppression of phytophagous nematodes and certain fungi in the rhizosphere of okra due to oil-cake amendments. *Acta Botanica Indica* 7(1):51–56.

Lam, R. S. H., and M. T. Nickerson. 2013. Food proteins: A review on their emulsifying properties using a structure–function approach. *Food Chemistry* 141:975–984.

Lanjewar, R. D., and V. N. Shukla. 1986. Vulnerability of larvae and eggs of *Meloidogyne incognita* to some oilcakes and fungicides. *Indian Journal of Nematology* 16:69–73.

Latif, Z. H., R. Ahmad, and M. I. Haq. 1999. Effect of seed treatment with neem cake, neem oil and Latex of Aak on the germination of cowpea and its vulnerability, root-knot nematode *Melodogyne incognita. Pakistan Journal of Phytopathology* 11:52–53.

McElderry, C. F., M. Browning, and J. A. Amador. 2005. Effect of short-chain fatty acids and soil atmosphere on *Tylenchorhynchus* spp. *Journal of Nematology* 37:71.

Meena, P., S. Nehra, and P. C. Trivedi. 2009. Efficacy of decomposed cakes against *Heterdera cajani* infecting *Cajanus cajan. Asian Journal of Experimental Science* 23:181–184.

Mirabella, N., V. Castellani, and S. Sala. 2014. Current options for the valorization of food manufacturing waste: A review. *Journal of Cleaner Production* 65:28–41.

Mohanapriya, R., G. Jothi, K. Poornima, N. Swarnakumari, and P. Renukadevi. 2019. Impact of humic acid and organic amendments on *Meloidogyne incognita* in vitro condition. *Annals of Plant Protection Sciences* 27:134–139.

Mohanty, A., P. R. Rout, B. Dubey, S. S. Meena, P. Pal, and M. Goel. 2022. A critical review on biogas production from edible and non-edible oil cakes. *Biomass Conversion and Biorefinery* 12:949–966.

Mohanty, S. S., and K. Tayung. 2021. Effect of different oil cakes against root knot nematode infecting tomato. *Journal of Emerging Technologies and Innovative Research* 8:910–918.

Mokrini, F., S. Janati, A. Houari, A. Essarioui, R. Bouharroud, and A. Mimouni. 2018. Management of plant parasitic nematodes by means of organic amendments–a review. *Revue Marocaine des Sciences Agronomiques et Vétérinaires* 6:337–344.

Momma, N., K. Yamamoto, P. Simandi, and M. Shishido. 2006. Role of organic acids in the mechanisms of biological soil disinfestation (BSD). *Journal of General Plant Pathology* 72:247–252.

Nangki, T., D. Byomakesh, and U. K. Anita. 2022. Effect of oilcakes on root-knot nematode (*Meloidogyne incognita*) infecting brinjal. *Indian Journal of Nematology* 52:21–27.

Nigam, P. S.-N., and A. Pandey. 2009. *Biotechnology for Agro-Industrial Residues Utilisation: Utilisation of Agro-residues.* Germany: Springer Science & Business Media.

Oka, Y. 2010. Mechanisms of nematode suppression by organic soil amendments–a review. *Applied Soil Ecology* 44:101–115.

Oka, Y., N. Shapira, and P. Fine. 2007. Control of root-knot nematodes in organic farming systems by organic amendments and soil solarization. *Crop Protection* 26:1556–1565.

Padhi, N. N., and R. P. Misra. 1987. Control of *Rotylenchulus reniformis* on French bean (*Phaseolus vulgaris* L.). *Indian Journal of Nematology* 17:130–131.

Pandey, G., and K. P. Singh. 1990. Effect of organic amendments on soil microflora and nematode fauna with special reference to *Meloidogyne incognita* in chick pea. *New Agriculturist* 1:65–70.

Parmar, B. S., K. L. Sahrawat, and S. K. Mukerjee. 1976. *Pongamia glabra*: Constituents & uses. *Journal of Scientific and Industrial Research* 35:608–611.

Parveen, G., R. Noreen, H. A. Shafique, V. Sultana, S. Ehteshamul-Haquel, and M. Athar. 2019. Role of rhizobia in suppressing the root diseases of soybean under soil amendment. *Planta Daninha* 37:1–8.

Ralmi, N. H. A. A., M. M. Khandaker, and N. Mat. 2016. Occurrence and control of root knot nematode in crops: A review. *Australian Journal of Crop Science* 10:1649–1654.

Ramachandran, S., S. K. Singh, C. Larroche, C. R. Soccol, and A. Pandey. 2007. Oil cakes and their biotechnological applications–a review. *Bioresource technology* 98:2000–2009.

Refaat, M. M., M. E. Mahrous, R. M. El-Ashry, and A. M. El-Marzoky. 2020. Nematicidal proprieties of commercial seed oils of certain medicinal plants on egg hatching and juvenile mortality of the root-knot nematode, *Meloidogyne incognita* in vitro. *Zagazig Journal of Agricultural Research* 47:487–497.

Rizvi, R., I. Mahmood, S. A. Tiyagi, and Z. Khan. 2012. Conjoint effect of oil-seed cakes and *Pseudomonas fluorescens* on the growth of chickpea in relation to the management of plant-parasitic nematodes. *Brazilian Archives of Biology and Technology* 55:801–808.

Roopa, K. P., and A. S. Gadag. 2020. Importance of biopesticides in the sustainable management of plant-parasitic nematodes. In *Management of Phytonematodes: Recent Advances and Future Challenges.* Singapore: Springer, pp. 205–227.

Sasser, J. N., and D. W. Freckman. 1987. A world perspective on nematology: The role of the society. *Vistas on Nematology* 7–14.

Shafi, A., I. Mahmood, and Z. A. Siddiqui. 2002. Integrated management of root-knot nematode *Meloidogyne incognita* on chickpea. *Thai Journal of Agricultural Science* 35:273–280.

Sharma, S., M. Verma, and A. Sharma. 2013. Utilization of non edible oil seed cakes as substrate for growth of *Paecilomyces lilacinus* and as biopesticide against termites. *Waste and Biomass Valorization* 4:325–330.

Singh, S., B. Bhagawati, and B. K. Goswami. 2011. Bio-management of root-knot disease of chick pea caused by *Meloidogyne incognita*. *Annals of Plant Protection Sciences* 19:159–163.

Singh, S., B. Singh, and A. P. Singh. 2015. Nematodes: A threat to sustainability of agriculture. *Procedia Environmental Sciences* 29:215–216.

Singh, V. K. 2004. Management of Heterodera cajani on pigeonpea with nematicides and organic amendments. *Indian Journal of Nematology* 34:213–214.

Souilem, S., I. Kobayashi, M. A. Neves, L. Jlaiel, H. Isoda, S. Sayadi, and M. Nakajima. 2014. Interfacial characteristics and microchannel emulsification of oleuropein-containing triglyceride oil–water systems. *Food Research International* 62:467–475.

Sumbul, A., R. Rizvi, I. Mahmood, and R. A. Ansari. 2015. Oil-cake amendments: Useful tools for the management of phytonematodes. *Asian Journal of Plant Pathology* 9:91–111.

Taba, S., M. Tomoyose, and Z. Moromizato. 2006. Control mechanism associated with rice bran-amended soil for root-knot nematode, *Meloidogyne incognita*. *International Journal of Nematology* 16:126.

Tamilarasan, S., and V. R. Manchikatla. 2012. Engineering crop plants for nematode resistance through host-derived RNA interference. *Cell & Developmental Biology* 2.

Tavares-Silva, C. A., C. R. Dias-Arieira, F. Rogério, H. H. Puerari, D. Mattei, T. R. B. da Silva, and O. Ferrarese-Filho. 2015. Control of *Meloidogyne javanica* and *Pratylenchus brachyurus* with crambe presscake. *Nematropica* 45:215–221.

Tiyagi, S. A., and S. Ajaz. 2004. Biological control of plant parasitic nematodes associated with chickpea using oil cakes and *Paecilomyces lilacinus*. *Indian Journal of Nematology* 34:44–48.

Tiyagi, S. A., and M. M. Alam. 1995. Efficacy of oil-seed cakes against plant-parasitic nematodes and soil-inhabiting fungi on mungbean and chickpea. *Bioresource Technology* 51:233–239.

Tiyagi, S. A., A. V. Khan, and M. M. Alam. 2001. Role of oil-seed cakes for the management of plant-parasitic nematodes and soil-inhabiting fungi on lentil and mungben. *Archives of Phytopathology and Plant Protection* 33:453–472.

Umar, I., and S. Y. Simon. 2008. The effect of oil seed cakes and ploughing against plant parasitic nematodes on cowpea (*Vigna unguiculata* (L.) Walp). *Agricultural Journal* 3:349–352.

Wani, A. H. 2006. Management of root-knoot nematode, *Meloidogyne incognita*, on okra and lentil by soil amendment with oil cakes and leaves of different plants. *Nematologia Mediterranea* 34:83–87

Yadav, Y. S., A. U. Siddiqui, and P. Aruna. 2005. Management of root-knot nematode *Meloidogyne incognita* infesting gram through oil cakes. *Journal of Phytological Research* 18:263–264.

Yassin, M. Y., and A. E. Ismail. 1994. Effect of some oilseed cakes as soil amendments and Aldicarb on cowpea plants infected with *Rotylenchulus reniformis* Lin. & Ol. in relation to soil type. *Anzeiger für Schädlingskunde, Pflanzenschutz, Umweltschutz* 67:176–178.

Youssef, M. M. A., and W. M. A. El-Nagdi. 2010. Effect of certain organic materials in controlling *Meloidogyne incognita* root-knot nematode infesting banana. *Archives of Phytopathology and Plant Protection* 43:660–665.

10 Oilseed Cakes and Their Biocarbon Products

A Sustainable Feedstock in Management of Nematodes in Fruit Crops

Kamini Devi, Shalini Dhiman, Jaspreet Kour, Tamanna Bhardwaj, Neerja Sharma, Kanika Khanna, Puja Ohri, Amrit Pal Singh and Renu Bhardwaj

CONTENTS

10.1 INTRODUCTION

Phytonematodes are the major limitation in the production of economically significant crops. These parasites attack the roots, rhizomes, bulbs, corms, and others, which hinder water movement. They also infect the plant's upper parts, inhibiting the metabolic pathways that affect the crop plant's yield. This makes the plant susceptible to other pathogens (fungi and bacteria). Therefore, to reduce the impact of nematodes, there has been increasing demand for the usage of by-products of plants and animals to manage crop productivity (Sumbul et al. 2015). There are different organic amendments, among which oil cakes have been produced that stimulate the nematode enemies, which in

DOI: 10.1201/9781003319252-10

turn attack the parasitic phytonematodes. This will reduce their number and enhance the supplementation of nutrients and other organic matter to the soil structure. Oil cakes are the remaining meal of the extracted vegetable oil from various seeds (groundnut, cotton, castor beans, etc.). The oil cakes from these seeds are used as biofertilizers to manage harmful microbes and parasitic nematodes. According to Nwankwo et al. (2016), adding organic matter to soil improves the management of *Meloidogyne* spp. and nutrient levels and decreases plants' dependency on artificial nematicides. Oil cakes were directly applied (pellets or powder form) to the soil, wetted, and allowed to decay to increase the nematicidal impact. The seeds were sometimes treated with oil cakes before sowing to control the effect of the parasitic nematodes on crops.

The oils were extracted by solvent extraction, hot or cold pressing, and supercritical fluid extraction from kernels. These processes include pretreatment and crushing after the refining of oil, which consists of cold (peanut, olive, sunflower, coconut oil) or hot pressing with respect to the kernel used for the production of oil cakes. The oilseeds were exposed to high temperatures (steam) for oil extraction via hot pressing to enhance the oil yield. Another process of oil extraction was the solvent extraction method, which included using appropriate solvents (cyclohexane, n-hexane, isopropyl ether, benzene, acetone, etc.) from crushed seeds. Using supercritical CO_2 to extract oil from broken seeds *via* the supercritical fluid extraction method led to a better and greater oil yield. These processes of oil extraction result in the formation of oil cakes (by-products) having different composition depending on the plant type. There were 150.6 million tons of remaining meal and oil cakes and 231.5 million tons of fats and oils generated globally in 2018 (OECD/FAO 2018; Siva-Sankari et al. 2022).

Oil cakes are used as feedstock for biocarbon production from edible and non-edible oil crops. The uncultivated and barren land was used to cultivate crops, including *Pongamia pinnata*, *Madhuca indica*, *Calophyllum inophyllum*, *Hevea brasiliensis*, *Jatropha curcas*, and others, for the production of non-edible oil, whereas cotton seed, sunflower, soybean, groundnut, and others are used to produce edible oil. Oilseed cakes were used as feedstock for the production of biocarbon, a carbonaceous material formed via thermo-chemical processes. The oil cake-derived biocarbon has broad application spectra and was used to enhance the remediation of toxic compounds, seed germination, ecological restoration, parasitic nematodes, and so on. The biocarbon production from oil cakes includes microwave-assisted carbonization, conventional carbonization, and hydrothermal carbonization. Moreover, after the production of biocarbon material, it was then activated by three methods: physical activation, microwave-enhanced chemical activation, and chemical activation to make activated carbonaceous material having a high porosity and a high surface area (Vivekanandhan et al. 2017; Elmaguanaa et al. 2018; Wang et al. 2018; Lee et al. 2018; Ao et al. 2018; Sankaranarayanan et al. 2019; Vivekanandhan 2020).

The applications of oil cake–derived biocarbon products include the remediation of pollutants (dyes, toxic organic compounds, toxic metal ions, etc.) and as effective catalysts in various reactions created for the construction of sustainable energy technologies like secondary batteries (fuel cells, batteries, etc.). The fruit crop was frequently threatened due to extensive susceptibility to parasitic phytonematodes (9–28% loss globally). It has been reported that applying oilseed cakes derived from cultivated crops efficiently reduced the nematode population in fruit crops (Yang et al. 2015; Bennett et al. 2016; Akinfalabi et al. 2017; Zhou et al. 2017).

10.2 NEMATODE LOSSES TO CROPS

Nematodes are usually microscopic and soil-borne pathogens. Phytonematodes are obligate parasites having a broad host range and damage the roots and aboveground parts of plants. It is estimated that crop losses due to nematodes are more than insect pests annually. Globally, about 12.3% (USD 173 billion) of crop yield is reduced due to the effect of these microscopic parasites (Kumar et al. 2020). Diseases caused by nematode parasites are hard to control because of their unseen nature; symptoms are not specific but similar to those caused by other biotic and abiotic stresses.

Nematodes cause infection not only independently, but they form disease complexes with other microorganisms, resulting in crop loss (Singh et al. 2015). Root-knot nematode *Meloidogyne* spp., having a wide range of hosts, is the most devastating pest among plant-parasitic nematodes that cause severe damage to crops. Tomato, capsicum, cucumber, and other common crops are infected by reniform nematodes and root-knot nematodes (*Meloidogyne incognita* and *M. javanica*). Today, a severe threat to fruit species has been amplified as new nematode species, such as *Meloidogyne enterolobii*, cause a threat to guava (Poornima et al. 2016). Various nematodes like root-knot, citrus, lesion, burrowing, and reniform nematodes put pressure on the growth and development of many fruits (citrus, guava, banana, pomegranate, etc.). It is estimated that about 20–40% of fruit crops are affected by plant nematodes worldwide (Kumar et al. 2020). Potato cyst nematodes (*Globodera pallida* and *G. rostochiensis*) are reported as economically important in temperate areas as they induce heavy loss of potato crops and are hence affirmed as quarantine pests all over the world. The presence of this parasitic nematode in some regions, such as Himachal Pradesh, Uttarakhand, and Jammu and Kashmir, restricts the trade and production of potatoes (Chandel et al. 2020). Some cereal cyst nematodes, such as *Heterodera* spp., reduced the yield of wheat, barley, and oats (Smiley et al. 2010).

10.3 DIFFERENT ORGANIC AMENDMENTS FOR NEMATODE MANAGEMENT

Organic amendments (organic substances derived from biomass) are used to improve the soil's physical and chemical properties. So, this agricultural practice is a good alternative for reducing nematode stress and enhancing plant tolerance by improving its nutrient status. Organic amendments control nematode infestation through the discharge of toxic substances during decomposition and stimulating antagonistic microbial activity. Different types of organic amendments like nematicidal plants, tea compost, protein-rich wastes, solid wastes, and manures have been assessed to overcome phytonematodes infection (Oka 2010; Ansari et al. 2019).

10.3.1 Plant Extracts

Using plant extracts is a cost-effective and eco-friendly method for managing nematode infection and improving soil structure and fertility (Feizi et al. 2014). Many plant extracts such as mustard plants (*Brassica carinata, B.rapa*), sunn hemp (*Crotalaria juncea*), eucalyptus, marigold (*Tagetesspp*,), castor bean (*Ricinus communis*), velvet bean (*Mucuna* spp.), and Rakshak gold (a neem-based product) have been used to suppress the root-knot nematode infection (El-Deriny et al. 2020). *Meloidogyne incognita*–infected tomato plants are treated with *Crotalaria virgulata* extract, resulting in reduced root gall formation (Jourand et al. 2004). The hatching action of root-knot nematodes is decreased by plant extracts of *Artemisia absinthium*, *Thymus vulgaris*, *Eucalyptus* spp., *Foeniculum vulgare*, and *Mentha piperita* (Ibrahim et al. 2006a). *Justicia adhatoda* and *Lantana camara* extracts restrict the entry of nematodes into roots (Ahmad et al. 2010, 2013).

10.3.2 Leaf Powder

M. incognita population is decreased in the roots of cucumber when the soil is treated with seed and leaf powder of 70.1% *Trigonella foenumgraecum* and 72.6% *Catharanthus roseus* (Mostafa et al. 2016). Under greenhouse conditions, the reproduction rate in *Rotylenchulus reniformis* infecting cantaloupe was lowered when the soil was supplemented with seed powder and dried peels of *T. foenum-graecum* and *Punica granatum*. The growth rate of *Meloidogyne* spp. infecting *Cucumis sativus* was reduced when the soil was treated with dried leaf powder of *Calendula officinalis*, *Gomphrena globose*, and *Nerium oleander* under in vivo conditions (Mostafa et al. 2016). The infection caused by reniform nematodes in cantaloupe was reduced using leaf powder of *N. oleander* and the dried seed of *B. juncea* (El-Deriny 2016).

10.3.3 Neem Formulations and Oilseed Cakes

It is reported that neem (*Azadirachta indica*) formulations are very effective against plant-parasitic nematode infection (El-Deriny et al. 2020). Neem (*Azadirachta indica*) has been used in various forms such as oil cakes, formulations, or whole plants (Oka 2010; Stirling 2018). The population of *M. graminicola* infecting rice plants is reduced by using neem (*Azadirachta indica*) extracts; therefore, this nematicidal activity boosts plant growth (Dongre and Sobita 2013).

Oilseed cakes are the residues collected after oil extraction from plant sources. Many oilseed cakes obtained from castor bean, neem, and mustard are effective against phytonematode (*Rotylenchulus reniformis*, *Meloidogyne incognita*, *Helicotylenchus indicus*, *Tylenchorhynchus brassicae*) infection. Applying anise, fennel, and sesame oilseed cakes also reduced the nematode population, resulting in increased plant growth and productivity (El-Sherif et al. 2010).

10.3.4 Animal Manure

Manure application in soil suppresses nematode infection and increases the microbial community in the rhizosphere. The population of citrus and potato cyst nematodes is decreased by using chicken manure, and yield is also improved (El-Deriny et al. 2020). The growth rate of *M. javanica* is reduced by applying cattle manure in potato fields, which then enhances the growth and productivity of the crop (El Hajji and Horrigne-Raouani 2012). The galling index is reduced in tomato roots when the soil is treated with compost made up of poultry, cattle, sheep, and horse manure, as the nematicidal activity of this compost decreases the nematode population (Kerkeni et al. 2007).

The extent of nematode elimination depends on numerous factors, such as the nematode species, the type of compost, and processing. The maximum suppression of *R. reniformis* was found with a lower carbon/nitrogen ratio in soil augmented with compost (Ismail et al. 2006).

10.4 EXTRACTION AND PRODUCTION OF OILSEED CAKE

Agricultural waste, especially oil cakes, is produced during different oilseed harvesting and processing phases. These oil cakes have the potential to generate a plethora of high-energy and valuable

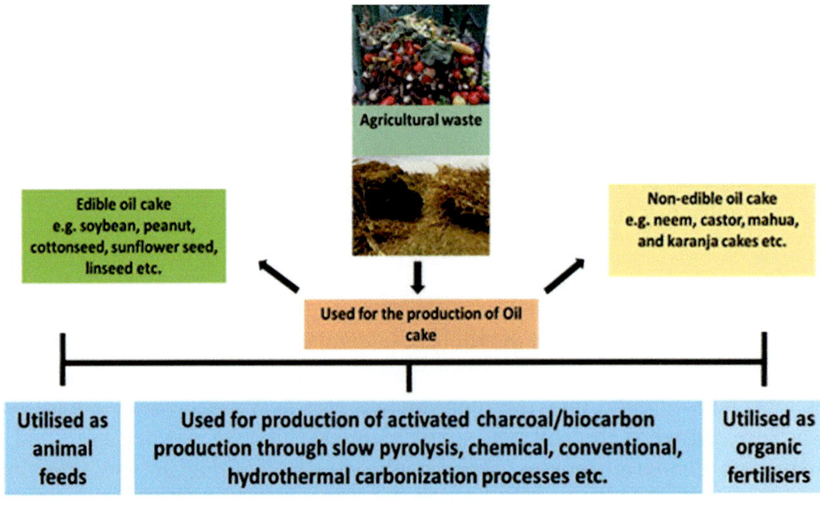

FIGURE 10.1 Utilization of different non-edible and edible oilseed cakes for the production of various activated biocarbon, organic fertilizers, and animal feeds.

products, including biodiesel, bioethanol, proteins, feed ingredients, enzymes, and a plethora of biochemicals like biogas, biochar, syngas, biocomposites, and biofertilizers. Oil cakes are a primary by-product of the extraction of oil from oilseeds. Historically, two oil extraction techniques have been used: mechanical and solvent extraction (Mohanty et al. 2021). In the first case, mechanical equipment like a screw press is used, whereas in the second, solvents such as hexane are used. The by-product acquired by pressing is referred to as oil cake, whereas the by-product produced during solvent extraction is referred to as oil meal (Sunil et al. 2015). There are two types of oil cakes: edible and non-edible (Figure 10.1). They both have a lot of nutrients in them. Cakes made from edible oil-bearing seeds have a high protein content, and these edible oil cakes are often utilized as animal feeds. However, non-edible oil cakes, which are rich in N, P, and K, such as neem, castor, mahua, and Karanja cakes, are utilized as organic fertilizers (Mohanty et al. 2021).

10.5 OILSEED CAKES ARE USED AS FEEDSTOCK FOR THE PRODUCTION OF BIOCARBON

The word *biocarbon* refers to a broad spectrum of carbonaceous compounds generated from renewable feedstocks through various thermochemical conversion processes. This carbonaceous material, also called biocarbon, includes carbon nanotubes, graphene sheets, graphene nanoparticles, activated carbon, biochar, and other particulate or bulk structures made of carbon (Siva-Sankari et al. 2022). As the need for biocarbon material grows, various feedstocks have been investigated as carbon sources. These biocarbon feedstocks include agro/forestry residues (such as husk, stalks, hull, tomato pomace, and rice straw), energy crops (such as miscanthus and switchgrass), short-term woody biomasses (such as poplars and willows), industrial coproducts (such as lignin, humin, distiller's dried grains with solubles, oil press cakes), and others.

Oil cake made from carbonaceous materials such as lignin (Vivekanandhan et al. 2015), humin (Tosi et al. 2020), distiller's dried grains with solubles (Reimer et al. 2020), and sugarcane bagasse (Adinaveen et al. 2013) is becoming a more important feedstock in industries than any other industrial co-product. While oil cakes have a limited number of conventional uses, increasing their production generates demand for their alternative high-volume/-value application. Additionally, there is a desire for sustainable feedstocks that may be utilized to synthesize diverse carbon compounds in place of fossil fuels successfully. Oil cakes have a wide range of chemical compositions, leading to carbon materials forming with a wide range of structural and morphological features. This allows them to be used in a wide range of applications. Because the biomass available for synthesizing carbon compounds is so varied, relying on a single biomass is difficult to accomplish. As a result, feedstock diversity is one of the most important aspects of long-term biocarbon production. In this regard, the efficient use of non-edible oil cakes for producing carbon materials solves supply chain difficulties. The oil cake conversion into biocarbon not only solves the issue of waste disposal but also provides a financial benefit via strong research and development efforts and increased returns on investment.

10.5.1 Oilseed Cake–Derived Biocarbon from Edible Oil

Edible oil cakes include those made from soybean, peanut, cottonseed, sunflower seed, linseed, and other seeds. Edible oil cakes are rich in protein (15–50%) and used as animal feed. These oil cakes reused for animal feed to a large extent due to the enormous volume that they produce and the rising demand for edible oil. Therefore, it is vital to identify other value-added uses for oil cakes to get more significant economic advantages. Moreover, edible oil cake–derived biocarbon production is one of the alternative value-added applications with greater economic benefit (as shown in Table 10.1.). Additionally, compared to commercial carbon, activated carbon exhibits a much higher specific surface area.

TABLE 10.1
Various Oilseed Cake–Derived Biocarbon Production

S. No.	Edible Oilseed Cake	Production of Biocarbon	Reference
1	Prickly pear seed cake	Chemically activated carbon formation from prickly pear seed cake with phosphoric acid	El Maguana et al. 2019
2	Olive cake	Activated carbon production using ZnCl2- and H3PO4-mediated chemical activation methods	Melouki et al. 2017
3	Mustard seed cake	Synthesis of carbon material at 715 + 5 degrees Celsius in the presence of nitrogen	Singh and Gautam 2017
4	Rapeseed and raspberry seed cakes	Physical activation method for the production of activated carbon	Smets et al. 2016
5	Mustard seed cake	Synthesized activated carbon from mustard oil cake in steam	Li and Khanal 2016
6	Olive cake	Potassium hydroxide (KOH)-activated activated carbons	Abdel-Ghani et al. 2016
7	Cottonseed cake	Production of activated carbon by treating oil cake with sulfuric acid at 160 degrees Celsius for 24 hours, followed by washing with distilled water and soaking in sodium bicarbonate to remove the free acid	Malathi et al. 2016
8	Linseed seed cake	H_3PO_4-based chemical activation process to turn linseed de-oiled cake into activated carbon, which they then used for the adsorptive removal of acridine orange and azure	Khan et al. 2016
9	Rapeseed cake	Na_2CO_3 biocarbon activation method	Uçar et al. 2015
10	Rapeseed cake	Hydrothermal technique for producing carbon nanoparticles	Purkayastha et al. 2014
11	Olive waste cake	H_3PO_4-based biocarbon activation method	Baccar et al. 2009; Hjaila et al. 2013
12	Safflower seed cake	$ZnCl_2$-based chemical activation technique for carbon activation	Angın et al. 2013

10.5.2 Oilseed Cake–Derived Biocarbon from Non-Edible Oil

Oil cakes mainly derived from neem, castor, sal, karanj, mahua, jatropha, and others constitute numerous harmful toxic compounds and are thus categorized as non-edible oil cakes. Converting them into diverse carbon compounds generates value-added possibilities, ultimately strengthening their relevant bioeconomy (Siva-Sankari and Vivekanandhan 2020). Many researchers have focused a great deal of their attention on non-edible jatropha oil cake as a potential raw material for producing carbonaceous materials (as shown in Table 10.2). Moreover, researchers also analyzed the characteristics of carbon compounds derived from various non-edible oil cakes. However, non-edible oil cakes, like edible ones, have not been studied extensively for biocarbon material manufacturing.

TABLE 10.2
Non-Edible Oilseed Cake–Derived Biocarbon Production

S. No.	Non-Edible Seed Cake	Production of Biocarbon	Reference
1	Castor seed cake	Used a slow pyrolysis approach to test the efficient conversion of castor seed cake into charcoal for methylene blue dye absorption	Silva et al. 2021
2	Castor seed cake	Produced activated charcoal from castor seed cake with H_3PO_4	Ferreira et al. 2020
3	*Jatropha* seed cake	Chemical activation procedure used for the production of activated carbon from jatropha oil cake using conventional and hydrothermal carbonization processes	Siva-Sankari and Vivekanandhan 2020
4	Jatropha and karanja seed cake	Made activated biochar from jatropha and karanja oilseed	Garg and Das 2018
5	Castor seed cake	Used K_2CO_3 to activate castor oil cake to make activated carbon	Sánchez-Cantú et al. 2018
6	*Pongamia pinnata* oil cake	Produced activated carbon from *Pongamia pinnata* char using KOH-mediated chemical activation with 1:2 and 1:4 impregnation ratios	Chaitra et al. 2017
7	De-oiled mahua seed cakes	Used a pyrolysis procedure (400–600 °C) to make charcoal from de-oiled mahua seed cakes	Mulimani and Navindgi 2017
8	*Jatropha*, *Pongammia pinnata*, and tung non-edible de-oiled seed cakes	Compared and generated biochar from *Jatropha*, *Pongammia pinnata*, and tung non-edible de-oiled seed cakes using hydrothermal conversion at 350, 400, and 450 °C and 20, 30, and 40 kg cm^{-2} and found that tung oil cake produces the most biochar as compared to the *Jatropha* and *Pongammia* de-oiled seed cakes	Kumar and Pant 2016
9	Castor seed cake	Used castor cake as a feedstock for biochar synthesis at different pyrolysis temperatures (200–600 °C). These researchers also discovered that greater the pyrolysis temperature causes lowered biochar output and increased carbon content	Kalinke et al. 2016
10	Castor seed cake	Used pyrolysis to manufacture char from castor seed cake and then subsequently used it as a feedstock for the production of activated carbon using H_3PO_4- and $ZnCl_2$-based chemical activation methods	Aldobouni et al. 2015
11	*Pongamia glabra* seed cake	Use of *Pongamia glabra* de-oiled cake as a novel feedstock for the manufacture of biochar and bio-oil	Chutia et al. 2014
12	*Jatropha* seed cake	Use of jatropha oil cake for the production of activated carbon using both chemical and thermal activation methods	Kurniawan and Ismadji 2011

10.6 ROLE OF OILSEED CAKE–DERIVED BIOCARBON PRODUCTS

The oil-derived biocarbon products are gaining much importance these days owing to their diversified applications. This section deals with multiple applications of biocarbon products in energy storage or conversion, environmental remediation, and catalysis. Various applications would positively fortify their commercial value in the coming times.

10.6.1 Energy Storage/Conversion

Multiple renewable resources were widely explored for manufacturing different types of electrochemical energy storage and conversion devices, such as fuel cells, batteries, and supercapacitors (Heidarinejad et al. 2020). Many other sources, including agro/forestry residues, energy crops, municipal organic solid wastes, and industrial co-products, were studied to synthesize activated carbon. Working on similar lines, Chaitra et al. (2017) worked on *Pongamia pinnata* seed oil cake to synthesize activated carbon *via* the KOH activation process. Furthermore, the obtained activated carbon material was examined as an electrode material for supercapacitor application. It displayed a specific capacitance of 177 F g^{-1} and 90% capacity retention even at a high current density of 50 A g^{-1}. Another study was conducted by Ferrero et al. (2015), who employed soybean oil cake for obtaining nitrogen-doped carbons. On testing the same with H_2SO_4 and Li_2- SO_4 electrolytes, they exhibited a very high specific surface area (2130 m^2 g^{-1}) and specific capacitance of 176 F g^{-1}. These biocarbon samples promised good results as their capacity retention was 90–95% up to 10,000 cycles, indicating good cycling stability under high cell voltages. Siva-Sankari and Vivekanandhan (2020) studied the application of activated biocarbon materials and derived pristine of jatropha oil cakes. They reported that a symmetric supercapacitor obtained using activated biocarbon through the hydrothermal carbonization (HTC) process showed better specific capacitance over the respective conventional carbonization process.

10.6.2 Remediation of Pollutants

Environmental remediation is one of the key applications of carbon material. It involves its role in the removal of oil products, metal ions, organic contaminants, and odour from aqueous mediums. The absorption activities were examined by potentiometric methods for efficient removal of Cu^{2+} and Eu^{3+} ions using olive cake carbon by Konstantinou et al. (2010). Baccar et al. (2009) reported that Tunisian olive waste cakes adsorbed heavy metal ions. They also stated that permanganate-treated activated biocarbon had enhanced metal ion uptake by three times. Govarthanan et al. (2016) worked on developing cost-effective mustard cake–based biocarbon material. They found that it removed metal ions such as As^{5+} and Pb^{2+} from industrial wastewater streams. In addition, oil cakes were also employed for dye adsorption based on factors such as pH, temperature, dye concentration, and time duration. Sánchez-Cantú et al. (2018) studied the role of castor oil cake–derived biochar in the adsorption of Pb^{2+}, Cd^{2+}, and Cu^{2+} ions. These oil cakes commendably removed toxic dyes from ecosystems. Ibrahim et al. (2006b) explained the removal of methylene blue (MB) by sunflower oil cake biochar. The dye performed better at pH 6 and 25°C temperatures. Another activated carbon employed for the efficient adsorption of MB is coffee-pressed cakes by Nunes et al. (2009). They found that activated carbon particle size and adsorption capacity have an inverse relation. On lowering the size of activated carbon particles, the adsorption capacity was enhanced. Baccar et al. (2010) explained the role of olive waste cake–derived activated carbon on the adsorption of commercial dyes. A wide range of oil cake residues can actively eliminate metal ions and dye from wastewater. Oilseed cakes are obtained from various plant sources such as linseed de-oiled cake (Khan et al. 2016), de-oiled mustard cake (Singh et al. 2017), safflower seed press cake (Alsalhiet al. 2018), sunflower oil cake (Angın et al. 2013), castor oil cake (Sánchez-Cantú et al. 2018), and others were used for fabricating activated carbon. These act in an efficient manner to expunge organic dyes and metal ions from the environment, making it clean and green.

10.6.3 Catalysis

Carbon materials are obtained from diverse renewable sources. They are gaining a lot of attention pertaining to their catalytic applications. They are highly cost effective and have superior performance over customary metal oxide–based catalysts (Bennett et al. 2016). Carbon-based solid acid catalysts were obtained post the sulfonization of biocarbon. The study conducted by Dawodu et al. (2014) explored the role of *Calophyllum inophyllum* seed cake-based carbon functionalized with an SO_3H group for efficient conversion of *C. inophyllum* seed oil into biodiesel. Furthermore, biocarbon catalysis showed a high conversion rate (99%) and excellent stability. Akinfalabi et al. (2017) described the effective transformation of palm fatty acid distillate into biodiesel using sulfonated activated carbon derived from palm seed cake. The addition of the $-SO_3H$ functional group enabled the catalyst to be reused, making it fit for reuse eight times. Relatedly, Konwar et al. (2014) evaluated oil cake waste from *M. ferrea* L. via sulfonated activated carbon solid acid catalysis. They performed phosphoric acid-based chemical activation followed by sulfonation. The final product obtained had better thermal stability (240 °C) and high acid exchange capacity. It effectively converted free fatty acids into relevant methyl esters. Mardhiah et al. (2017) employed *Jatropha curcas* seed cake to obtain a carbon-based solid acid catalyst. It was used in the transesterification process. It has a high conversion capacity and is highly cost effective. They found that 99.13% of free fatty acid conversion was attained at 7.5 wt. percentage catalyst loading. Additionally, *Jatropha curcas* seed cake–derived solid acid catalyst displayed excellent stability up to the fourth cycle. Figure 10.2 represents applications for oil cake–derived biocarbon products as potent catalysts, environment cleaners, and fabricating supercapacitors.

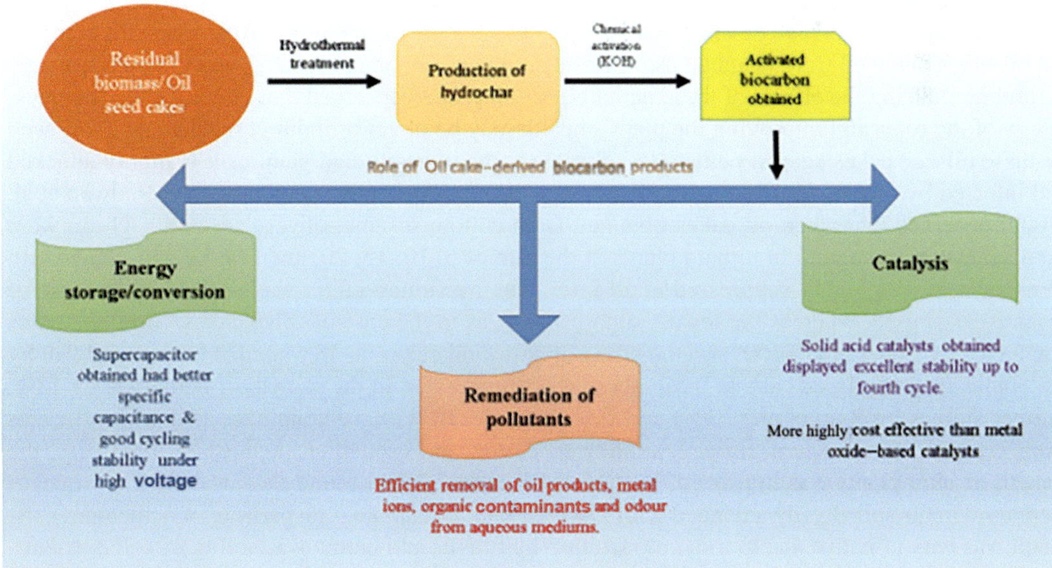

FIGURE 10.2 Applications of oilseed cake–derived biocarbon product.

10.7 ROLE OF OILSEED CAKES FOR NEMATODE MANAGEMENT IN FRUIT PLANTS

Oil cakes have been considered an effective method against plant nematodes, among various other forms of organic amendments (Goswami et al. 2007). These oil cakes are rich in nutrients and considered valuable. When these oil cakes are applied directly to the soil (pellets/powder) and allowed to decompose after watering, the nematicidal activity of oil cakes increases with increasing decomposition (Youssef and El-Nagdi 2010). Seed can be presoaked with these oil cakes to protect plants against nematodes (El-Sherif et al. 2010). Oil cakes can be beneficial to plants when applied in vitro or in vivo. It has been observed by Khan et al. (1974) that during in vitro conditions, oil cakes obtained from mahua, neem, castor, and groundnut are toxic to nematodes such as *R. reniformis*, *H. indicus*, and *T. brassicae*. They also inhibit the larval stage of *M. incognita*. Similarly it was observed by Lanjeswar and Shukla (1986) that when the boiled extracts of mustard and cotton were used, the hatching of larvae of *M. incognita* was suppressed up to 99.92% and 99.38% in water. Furthermore, it was observed that the oil cakes produced from neem are effective in inhibiting the growth of larvae of *M. incognita*, whereas the growth of *H. indicus* was suppressed from oil cake obtained from mustard (Deshmukh and Prasad 1969; Gowda and Gowda 1999). Studies conducted on the role of oil cakes in soil have shown that they are more effective in soil. It was observed that during in vitro experiments, mustard cakes were more helpful during the first week of growth of nematodes. In contrast, for 20 days, the effect of neem cakes was greater on the development of juveniles of *M. incognita*, and after 20 days, there was a decline in the impact of oil cakes (Goswami et al. 1993; Khan and Goswami 1995). Gutpa and Kumar (1997) found that during pot experiments conducted, groundnut and mustard cakes inhibited the growth of *Tylenchorynchus* sp. and *Helicotylenchus* sp. The oil cakes played an additive role in the development and sporulation of *P. lilacinus*.

Due to succulent and soft bodies and cultivation under moist conditions (which support nematode growth), vegetable crops are considered the most susceptible and affected crop by nematodes (Anwar et al. 2007). These crops are affected both qualitatively and quantitatively by nematodes, as they attack every part such as flowers, roots, leaves, the stem, bulbs, inflorescence, and corns (Webster 1972). *Globodera*, *Belonolaimus longicaudatus*, *Meloidgyne* spp., *R. similis*, and *Pratylenchus* spp. are some of the nematodes responsible for attacking crops, including fruit crops. In contrast, the oil cakes obtained from peanut, castor, linseed, mahua, coconut, mustard, and margosa are very helpful in reducing the effects of these nematodes. It has been observed that alterations in the physiology of the roots and improving the plant's health may be direct or indirect mechanisms followed by these oilseed cakes against nematodes. Tomato is the most-studied plant as it is highly attacked and infected by nematodes (Sikora and Fernandez 2005). Experiments conducted by Radwan et al. (2009) observed that when oil cakes obtained from cotton, sesame, olive, flax, and soybean were used against *M. incognita* in tomato plants at the rate of 5, 10, 15, 20, and 50g/kg soil, the growth of nematodes was highly suppressed at all rates. The maximum effect was noticed in the case of sesame cakes used, whereas the lowest suppression was in the case of olive cakes. Similar studies conducted by Lopes et al. (2009) on the effect of castor oil cakes on *M. javanica* have found that by the application of oilseed cake at 0.5%, there was suppression in the growth of nematode by 18%. Further studies by Ram et al. (2009) and Ganaie et al. (2011) have concluded that with oil cakes from neem, castor, mustard, linseed, sunflower, castor, and groundnut, against *M. incognita*, the growth of okra plants was improved. Kumar and Khanna (2006) found that when neem oil cakes were used in the soil already amended with *Trichoderma harzianum* (egg parasite of nematodes), the result was very effective due to root colonization by fungus and better availability of soil nutrients. Similar results were observed by Bharadwaj and Sharma (2006), where the growth and development of tomato plants were enhanced by the combined application of fungus and neem oil cakes.

Cereals are considered the most important food source, and maize, rice, and wheat are among the top crops in this category. Due to the high susceptibility of cereal crops to nematodes, there is a high loss to crop yield and thus the economy. Among the most common root-knot nematodes that affect cereal crops are *M. graminicola*, *H. avenae*, *Ditylenchus angustus*, *Anguina tritici*, and *T. vulgaris* (Khan and Jairajpuri 2010). Prasad et al. (2005) observed that due to the application of neem, castor, and *Simarouba glauca* at the rate of 2.5 and 5 g/kg against *M. graminicola*, the growth of rice was improved.

Fruit crops act as economically and commercially important crops because of their consumption and nutritional value. Unfortunately, the cultivation and growth of these crops are declining due to their high susceptibility to nematodes (Khan 2008). Among various nematodes, *Rotylenchulus reniformis*, *Radopholus similis*, and *Tylenchulus semipenetrans* are among the top ones that attack the fruit crops, and application of soybean, cotton, neem, cottonseed, and sesame oil cakes is considered very effective methods against these nematodes in *Citrus reticula* seedlings (Mohammad et al. 1980). Harish and Gowda (2001) conducted experiments and found that due to the combined application of neem cake, *T. viridie* and carbofuran (500g/plant + 10g/plant + 40g/plant) reduced the population of *Radopholus similis* and improved the growth of plant and fruit in banana. Similar observations were made by Sundraraju and Kiruthika (2009) and Srivastava (2002) against *M. incognita* in banana and papaya by applying neem cake and *P. lilacinus*. Pot experiments conducted by Yang et al. (2015) on the nematicidal effect of *Camellia* seed cake in bananas against *Meloidogyne javanica* found that the cake causes suppression in the growth of nematodes and promotes the growth and development of the plant.

Studies by Azmi et al. (2000) have concluded that due to the application of neem cakes in berseem, the population of *Tylenchorhynchus*, *Meloidogyne*, and *Pratylenchus* spp. was suppressed very effectively. Similarly, other studies by Pandey (2000) and Boia and Neog (2006) have found that neem cakes are very helpful in suppressing the growth of *M. incognita* that are affecting the growth and development of *Mentha arvensis* and tea plants, respectively.

10.8 CONCLUSION

Oilseed cake's utilization to manage harmful nematodes and microorganisms has been found to be beneficial for plant growth and yield. The oil cakes derived from different seeds reduced the parasitic phytonematodes population by disintegrating their eggs. However, the neem oilseed cake was used extensively in managing nematode populations, followed by mustard, castor, and ground oil cakes, among others. The oilseed cake–derived biocarbon can potentially remove and adsorb the toxic metal ions and other pollutants from the contaminated sources. Thus, there is a need for an integrated method for the sustainable execution of oilseed cakes and biochemical agents for managing nematodes in crops. Increasing demand for biocarbon production from oilseed cakes using renewable resources leads to an effective step toward managing nematodes. The actual mechanism in the action of nematodes has yet to be interrogated.

REFERENCES

Abdel-Ghani, Nour T., Ghadir A. El-Chaghaby, Mohamed H. ElGammal, and El-Shaimaa A. Rawash. "Optimizing the preparation conditions of activated carbons from olive cake using KOH activation." *New Carbon Materials* 31, no. 5 (2016): 492–500.

Adinaveen, T. L. John Kennedy, J. Judith Vijaya, and G. Sekaran. "Studies on structural, morphological, electrical and electrochemical properties of activated carbon prepared from sugarcane bagasse." *Journal of industrial and Engineering Chemistry* 19, no. 5 (2013): 1470–1476.

Ahmad, F., M. A. Rather, and M. A. Siddiqui. Nematicidal activity of leaf extracts from *Lantana camara* L. against *Meloidogyne incognita* (kofoid and white) chitwood and its use to manage roots infection of *Solanum melongena* L. *Brazilian Archives of Biology and Technology* 53 (2010): 543–548.

Ahmad, F., M. A. Siddiqui, and Olubukola O. Babalola. "Characterization of nematicidal activity of plant residues and their application with moisture approach against *Meloidogyne incognita* in tomato." *African Journal of Agricultural Research* 8, no. 1 (2013): 93–101.

Akinfalabi, Shehu-Ibrahim, Umer Rashid, Robiah Yunus, and Yun Hin Taufiq-Yap. "Synthesis of biodiesel from palm fatty acid distillate using sulfonated palm seed cake catalyst." *Renewable Energy* 111 (2017): 611–619.

Aldobouni, I. A. A. B. Fadhil, and I. K. Saied. "Conversion of de-oiled castor seed cake into bio-oil and carbon adsorbents." *Energy Sources, Part A: Recovery, Utilization, and Environmental Effects* 37, no. 24 (2015): 2617–2624.

Alsalhi, Mohamad Saleh, Sandhanasamy Devanesan, Akram Ahmed Alfuraydi, and Mysoon MF Al-Ansari. "Synthesis of silver nanoparticles using sesame oil cake." U.S. Patent Application 15/599,429, filed November 22, 2018.

Angın, D., E. Altintig, and T. Ennil. "Bioresource technology influence of process parameters on the surface and chemical properties of activated carbon obtained from biochar by chemical activation." *Bioresource Technology* 148 (2013): 542–549.

Ansari, Rizwan Ali, Aisha Sumbul, Rose Rizvi, and Irshad Mahmood. "Organic soil amendments: Potential tool for soil and plant health management." In: *Plant Health Under Biotic Stress*, pp. 1–35. Springer, Singapore, 2019.

Anwar, Safdar A., A. Zia, M. Hussain, and M. Kamran. "Host suitability of selected plants to *Meloidogyne incognita* in the Punjab, Pakistan." *International Journal of Nematology* 17, no. 2 (2007): 144.

Ao, Wenya, Jie Fu, Xiao Mao, Qinhao Kang, Chunmei Ran, Yang Liu, Hedong Zhang, Hedong Gao, Zuopeng Li, Jing Liu, Xiaoying Dai, Jianjun. "Microwave assisted preparation of activated carbon from biomass: A review." *Renewable and Sustainable Energy Reviews* 92 (2018): 958–979.

Azmi, Mujib I., K. C. Pandey, and R. B. Bhaskar. "Effect of some treatments on the nematode fauna in an IPM experiment." *Indian Journal of Nematology* 30, no. 1 (2000): 105.

Baccar, R., P. Blánquez, J. Bouzid, M. Feki, and M. Sarrà. "Equilibrium, thermodynamic and kinetic studies on adsorption of commercial dye by activated carbon derived from olive-waste cakes." *Chemical Engineering Journal* 165, no. 2 (2010): 457–464.

Baccar, R., J. Bouzid, M. Feki, and A. Montiel. "Preparation of activated carbon from Tunisian olive-waste cakes and its application for adsorption of heavy metal ions." *Journal of Hazardous Materials* 162, no. 2–3 (2009): 1522–1529.

Bennett, James A., Karen Wilson, and Adam F. Lee. "Catalytic applications of waste derived materials." *Journal of Materials Chemistry A* 4, no. 10 (2016): 3617–3637.

Bharadwaj, Anuja, and Satyawati Sharma. "Biocontrol of *Meloidogyne incognita* in *Lycopersicon esculentum* with AM Fungi and Oil Cakes." *Plant Pathology Journal* no. 2 (2006): 166–172.

Boia, B. C., and P. P. Neog. "Effect of oil cakes for management of *Meloidogyne incognita* in tea." *Annals of Plant Protection Sciences* 14, no. 2 (2006): 522–523.

Chaitra, K. Reddy Narendra, Krishna Venkatesh, N. Nagaraju, and Nagaraju Kathyayini. ""Green" carbon with hierarchical three-dimensional porous structure derived from-*Pongamia pinnata* seed oil extract cake and NiCo2O4-Ni (OH) 2/Multiwall carbon nanotubes nanocomposite as electrode materials for high performance asymmetric supercapacitor." *Journal of Power Sources* 356 (2017): 212–222.

Chandel, Y. S., Surendra Singh Bhadu, Rajan Salalia, Sharmishtha Thakur, Satya Kumar, Vishal Singh Somvanshi, Abhishek Mukherjee, and R. K. Walia. "Prevalence and spread of potato cyst nematodes, *Globodera* spp. in northern hilly areas of India." *Current Science* 118, no. 12 (2020): 1946–1952.

Chutia, Rahul Singh, Rupam Kataki, and Thallada Bhaskar. "Characterization of liquid and solid product from pyrolysis of *Pongamia glabra* deoiled cake." *Bioresource Technology* 165 (2014): 336–342.

Das Purkayastha, Manashi, Ajay Kumar Manhar, Manabendra Mandal, and Charu Lata Mahanta. "Industrial waste-derived nanoparticles and microspheres can be potent antimicrobial and functional ingredients." *Journal of Applied Chemistry* 2014 (2014).

Dawodu, Folasegun A., Olubunmi Ayodele, Jiayu Xin, Suojiang Zhang, and Dongxia Yan. "Effective conversion of non-edible oil with high free fatty acid into biodiesel by sulphonated carbon catalyst." *Applied Energy* 114 (2014): 819–826.

Deshmukh, M. G., and S. K. Prasad. "Effect of water soluble extracts of oil-cakes on the population of *Hoplolaimus indicus* Sher, 1963." *Indian Journal of Entomology* 31 (1969): 273–276.

Dongre, Mukesh, and S. Sobita. "Efficacy of certain botanical extracts in the management of *Meloidogyne graminicola* of rice." *International Journal of Agricultural Science and Research* 3 (2013): 91–98.

El-Deriny, Marwa M. "Integrated control of certain plant parasitic nematodes infecting cucurbitaceae plants." PhD diss. PhD Thesis. Egypt: Faculty of Agriculture, Mansoura University Egypt, 2016.

El-Deriny, Marwa M., Dina S. S. Ibrahim, and Fatma A. M. Mostafa. "Organic additives and their role in the phytoparasitic nematodes management." In: *Management of Phytonematodes: Recent Advances and Future Challenges*, pp. 73–93. Springer, Singapore, 2020.

El Hajji, Lobna, and N. Horrigne-Raouani. "Efficiency of solarization and incorporation of cattle manure in management of *Meloidogyne javanica* (Treub.) Chitwood and *Globodera pallida* (Stone) Behrens associated with potato." *Pest Technology* 6 (2012): 75–78.

El Maguana, Y., N. Elhadiri, M. Bouchdoug, M. Benchanaa, and A. Jaouad. "Optimization of preparation conditions of activated carbon from walnut cake using response surface methodology." *Moroccan Journal of Chemistry* 6, no. 1 (2018): 92–105.

El Maguana, Y., N. Elhadiri, M. Bouchdoug, M. Benchanaa, and A. Jaouad. "Activated carbon from prickly pear seed cake: Optimization of preparation conditions using experimental design and its application in dye removal." *International Journal of Chemical Engineering* 2019 (2019).

El-Sherif, A. G., A. R. Refaei, M. E. El-Nagar, and M. M. Hagar Salem. "Impact of certain oilseed cakes or powder in comparison with oxamyl or urea on *Meloidogyne incognita* infecting eggplant." *Archives of Phytopathology and Plant Protection* 43, no. 1 (2010): 88–94.

Feizi, A., E. Mahdikhani-Moghadam, M. Azizi, and H. Roohani. "Inhibitory effect of *Allium cepa* var. aggregatum, *Salvia officinalis* and *Kelussiaodor atissima* essence on the root-knot nematode (*Meloidogyne javanica*) and extraction of active ingredients." *Journal of Plant Protection* 28 (2014): 220–225.

Ferreira, Lucélio Mendes, Rafael Rodolfo de Melo, Alexandre Santos Pimenta, Tatiane Kelly Barbosa de Azevedo, and Cleber Brito de Souza. "Adsorption performance of activated charcoal from castor seed cake prepared by chemical activation with phosphoric acid." *Biomass Conversion and Biorefinery* (2020): 1–12.

Ferrero, G. A., A. B. Fuertes, and M. Sevilla. "From soybean residue to advanced supercapacitors." *Scientific Reports* 5, no. 1 (2015): 1–13.

Ganaie, Mucksood A. Azhar A. Rather, and Mansoor A. Siddiqui. "Pathogenicity of root knot nematode *Meloidogyne incognita* on okra and its management through botanicals." *Archives of Phytopathology and Plant Protection* 44, no. 17 (2011): 1683–1688.

Garg, Sonal, and Piyali Das. "High-grade activated carbon from pyrolytic biochar of Jatropha and Karanja oil seed cakes–Indian biodiesel industry wastes." *Biomass Conversion and Biorefinery* 8, no. 3 (2018): 545–561.

Goswami, B.K., R.K. Pandey, J. Goswami, and D.D. Tewari. "Management of disease complex caused by root knot nematode and root wilt fungus on pigeonpea through soil organically enriched with vesicular arbuscular Mycorrhiza, karanj (Pongamia pinnata) oilseed cake and farmyard manure." *Journal of Environmental Science and Health - Part B Pesticides, Food Contaminants, and Agricultural Wastes* 42 (2007): 899–904.

Goswami, B. K., U. Rao, and M. Manton. "Efficacy of decomposed products of some oil-cakes and ABCD collected at different time intervals on hatching and mortality of *Meloidogyne incognita*." *Current Nematology* 4 (1993): 215–217.

Govarthanan, Muthusamy, Min Cho, Jung-Hee Park, Jum-Suk Jang, Young-Joo Yi, Seralathan Kamala-Kannan, and Byung-Taek Oh. "Cottonseed oilcake extract mediated green synthesis of silver nanoparticles and its antibacterial and cytotoxic activity." *Journal of Nanomaterials* 2016 (2016).

Gowda, P. D., and D. N. Gowda. "In-vitro studies of oil-cake extracts on the survival of larvae of *Meloidogyne incognita* (kofoid and white) chitwood." *Plant Protection Bulletin* 49 (1999): 10–11.

Gutpa, M. C., and S. Kumar. "Efficacy of certain organic amendments and nematicides against *Tylenchorhynchus* spp. and *Helicotylenchus* spp. in soil." *Indian Journal of Nematology* 27 (1997): 139–142.

Harish, M., and D. N. Gowda. Management of the burrowing nematode, *Radopholus similis* (Cobb, 1893) Thorne, 1949 infesting banana. *Indian Journal of Nematology* 31 (2001): 23–25.

Heidarinejad, Zoha, Mohammad Hadi Dehghani, Mohsen Heidari, Gholamali Javedan, Imran Ali, and Mika Sillanpää. "Methods for preparation and activation of activated carbon: A review." *Environmental Chemistry Letters* 18, no. 2 (2020): 393–415.

Hjaila, Kefah, Rym Baccar, Montserrat Sarrà, C. M. Gasol, and Paqui Blánquez. "Environmental impact associated with activated carbon preparation from olive-waste cake via life cycle assessment." *Journal of Environmental Management* 130 (2013): 242–247.

Ibrahim, S. C., M. A. K. M. Hanafiah, and M. Z. A. Yahya. "Removal of cadmium from aqueous solutions by adsorption onto sugarcane bagasse." *American-Eurasian Journal of Agricultural & Environmental Sciences* 1, no. 3 (2006b): 179–184.

Ibrahim, Said K., Abdallah F. Traboulsi, and Samih El-Haj. "Effect of essential oils and plant extracts on hatching, migration and mortality of *Meloidogyne incognita*." *Phytopathologia Mediterranea* 45, no. 3 (2006a): 238–246.

Ismail, A. E., A. E. Rawia, and W. M. A. El-Nagdi. "Effect of different composts, biofertilizers and olive pomace as soil amendments on *Rotylenchulus reniformis*, growth and chemical analysis of jasmine Egypt." *Journal of Applied Science Research* 2, no. 11 (2006): 909–916.

Jourand, Philippe, Sylvie Rapior, Mireille Fargette, and Thierry Mateille. "Nematostatic effects of a leaf extract from *Crotalaria virgulata* subsp. grantiana on *Meloidogyne incognita* and its use to protect tomato roots." *Nematology* 6, no. 1 (2004): 79–84.

Kalinke, Cristiane, Antonio Sálvio Mangrich, Luiz H. Marcolino-Junior, and Márcio F. Bergamini. "Biochar prepared from castor oil cake at different temperatures: A voltammetric study applied for $Pb2^{+}$, Cd2+ and $Cu2^{+}$ ions preconcentration." *Journal of Hazardous Materials* 318 (2016): 526–532.

Kerkeni, A., N. Horrigue-Raouani, and M. Ben Khedher. "Suppressive effect of five compost extracts against the root-knot nematode *Meloidogyne incognita*." *Nematologia mediterranea* 35 (2007): 15–21.

Khan, M.R. *Plant Nematode Methodology, Systematics, Biology and Ecology*. Science Publishers, Enfield, New Hampshire, 2008, p. 360.

Khan, M. R., and B. K. Goswami. "Nematicidal efficacy of decomposed neem cake and fungal filtrates of *Paecilomyces lilacinus* and *Aspergillus flavus* against *Meloidogyne incognita*." Proceedings of the National Symposium Sustainable Agriculture in Sub-Humid Zone, March 3–5, 1995, Visva-Bharati, Sriniketan, India, pp. 359–361.

Khan, M. R., and M. S. Jairajpuri. Nematode Infestation in Food Crops-National Scenario. In: *Nematode Infestations Part I: Food Crops*, Khan, M.R. and M.S. Jairajpuri (Eds.), National Academy of Sciences, India, 2010, pp. 1–16.

Khan, M. W., A. M. Khan, and S. K. Saxena. "Rhizosphere fungi and nematodes of egg plant as influenced by oil cake amendments." *Indian Phytopathology* 27 (1974): 480–484.

Khan, T. A., E. A. Khan, Shahjahan. "Adsorptive uptake of basic dyes from aqueous solution by novel brown linseed deoiled cake activated carbon: Equilibrium isotherms and dynamics." *Journal of Environmental Chemical Engineering* 4 (2016): 3084–3095. https://doi.org/10.1016/j.jece.2016.06.009

Konstantinou, Maria, and Ioannis Pashalidis. "Competitive sorption of Cu (II) and Eu (III) ions on olive-cake carbon in aqueous solutions-a potentiometric study." *Adsorption* 16, no. 3 (2010): 167–171.

Konwar, Lakhya Jyoti, Rupali Das, Ashim Jyoti Thakur, Eero Salminen, Päivi Mäki-Arvela, Narendra Kumar, Jyri-Pekka Mikkola, and Dhanapati Deka. "Biodiesel production from acid oils using sulfonated carbon catalyst derived from oil-cake waste." *Journal of Molecular Catalysis A: Chemical* 388 (2014): 167–176.

Kumar, Dinesh, and K. K. Pant. "Insitu upgradation of biocrude vapor generated from non-edible oil cake's hydrothermal conversion over aluminated mesoporous catalysts." *Renewable Energy* 95 (2016): 43–52.

Kumar, S., and A. S. Khanna. Role of *Trichoderma harzianum* and neem cake separately and in combination against root-knot nematode on tomato. *Indian Journal of Nematology* 36 (2006): 247–249.

Kumar, Vinod, Matiyar Rahaman Khan, and R. K. Walia. "Crop loss estimations due to plant-parasitic nematodes in major crops in India." *National Academy Science Letters* 43, no. 5 (2020): 409–412.

Kurniawan, Alfin, and Suryadi Ismadji. "Potential utilization of *Jatropha curcas* L. press-cake residue as new precursor for activated carbon preparation: Application in methylene blue removal from aqueous solution." *Journal of the Taiwan Institute of Chemical Engineers* 42, no. 5 (2011): 826–836.

Lanjeswar, R. D., and V. N. Shukla. "Vulnerability of larvae and eggs of *Meloidogyne incognita* to some oil-cakes and fungicides." *Indian Journal of Nematology* 16 (1986): 69–73.

Lee, Jongkeun, Kwanyong Lee, Donghwan Sohn, Young Mo Kim, and Ki Young Park. "Hydrothermal carbonization of lipid extracted algae for hydrochar production and feasibility of using hydrochar as a solid fuel." *Energy* 153 (2018): 913–920.

Li, Yebo, and S. K. Khanal. *Bioenergy: Principles and Applications*. John Wiley & Sons, Hoboken, New Jersey, 2016.

Lopes, E. A., S. Ferraz, O. D. Dhingra, A. Ferreira, and L. G. Freitas. "Soil amendment with castor bean cake and jack bean seed powder to control *Meloidogyne javanica* on tomato roots". *Nematologia Brasileira* 33, no. 1 (2009): 106–109.

Malathi, S., N. Krishnaveni, and R. Sudha. "Adsorptive removal of lead (II) from an aqueous solution by chemically modified cottonseed cake." *Research on Chemical Intermediates* 42, no. 3 (2016): 2285–2302.

Mardhiah, H. Haziratul, Hwai Chyuan Ong, H. H. Masjuki, Steven Lim, and Yean Ling Pang. "Investigation of carbon-based solid acid catalyst from Jatropha curcas biomass in biodiesel production." *Energy Conversion and Management* 144 (2017): 10–17.

Melouki, Redouane, Philip L. Llewellyn, Sana Tazibet, and Youcef Boucheffa. "Hydrogen adsorption on activated carbons prepared from olive waste: Effect of activation conditions on uptakes and adsorption energies." *Journal of Porous Materials* 24, no. 1 (2017): 1–11.

Mohammad, H. J., Husain Si, and A. J. Al Zarari. "Effect of oil cakes amended soil on the growth of *Citrus reticulata* L. and the control of Citrus nematode in Iraq." *Phytopathology Mediterranea* 19 (1980): 153–154.

Mohanty, Anee, Prangya Ranjan Rout, Bipro Dubey, Sumer Singh Meena, Parimal Pal, and Mukesh Goel. "A critical review on biogas production from edible and non-edible oil cakes." *Biomass Conversion and Biorefinery* (2021): 1–18.

Mostafa, Fatma, A. Refaei, A. Khalil, and Marwa El-Deriny. "Potential use of botanicals rich in alkaloids for controlling *Meloidogyne incognita* and *Rotylenchulus reniformis* infecting cucurbits." *Egyptian Journal of Agronematology* 15, no. 1 (2016): 29–43.

Mulimani, H. V., and M. C. Navindgi. "Production of solid fuel biochar from de-oiled seed cake by pyrolysis." *Journal of Mechanical and Civil Engineering* 14 (2017): 57–61.

Nunes, Anne A., Adriana S. Franca, and Leandro S. Oliveira. "Activated carbons from waste biomass: An alternative use for biodiesel production solid residues." *Bioresource Technology* 100, no. 5 (2009): 1786–1792.

Nwankwo, E. N., D. C. Onuseleogu, C. U. Ogbonna, and A. O. E. Okorocha. "Effect of neem leaf extracts (*Azadirachta indica*) and synthetic pesticide (Carbofuran) on the root-knot nematode (*Meloidogynes* pp.) of cowpea (*Vigna unguiculata* L. Walp)." *International Journal of Entomology and Research* 3, no. 1 (2016): 1–6.

OECD/FAO "OECD-FAO Agricultural Outlook", OECD Agriculture statistics (database), 2018. http://dx.doi.org/10.1787/agr-outl-data-en.

Oka, Y. "Mechanisms of nematode suppression by organic soil amendments–a review." *Applied Soil Ecology* 44, no. 2 (2010): 101–115.

Pandey, R. "Additive effect of three orgabic materials and nematicides on the reproduction of *Meloidogyne Incognita* and yield of *Mentha arvensis*." *Nematropica* (2000): 155–160.

Poornima, K., P. Suresh, P. Kalaiarasan, S. Subramanian, and K. Ramaraju. "Root knot nematode, *Meloidogyne enterolobii* in guava (*Psidium guajava* L.) a new record from India." *Madras Agricultural Journal* 103, no. 10–12 (2016): 359–365.

Prasad, J. S., K. S. Varaprasad, Y. R. Rao, E. Srinivasa Rao, and M. Sankar. "Comparative efficacy of some oil seed cakes and extracts against root-knot nematode (*Meloidogyne graminicola*) infection in rice." *Nematologia Mediterranea* 33, no. 2 (2005): 191–194.

Radwan, M. A., E. K. El-Maadawy, S. I. Kassem, and M. M. Abu-Elamayem. "Oil cakes soil amendment effects on *Meloidogyne incognita*, root-knot nematode infecting tomato." *Archives of Phytopathology and Plant Protection* 42, no. 1 (2009): 58–64. DOI: 10.1080/03235400600940830

Ram, L., A. U. Siddqui, and A. Parihar. "Management of root-knot nematode, *Meloidogyne incognita* infecting okra using oil cakes." *Indian Journal of Nematology* 39, no. 1 (2009): 125–127.

Reimer, Christoff, Michael R. Snowdon, Singaravelu Vivekanandhan, Xiangyou You, Manjusri Misra, Stefano Gregori, Deborah F. Mielewski, and Amar K. Mohanty. "Synthesis and characterization of novel nitrogen doped biocarbons from distillers dried grains with solubles (DDGS) for supercapacitor applications." *Bioresource Technology Reports* 9 (2020): 100375.

Sánchez-Cantú, M., V. J. Janeiro-Coronel, J. A. Galicia-Aguilar, and J. D. Santamaría-Juárez. "Effect of the activation temperature over activated carbon production from castor cake and its evaluation as dye adsorbent." *International Journal of Environmental Science and Technology* 15, no. 7 (2018): 1521–1530.

Sankaranarayanan, Sivashunmugam, Perumalsamy Vishnukumar, Muruganandham Hariram, Singaravelu Vivekanandhan, Carolina Camus, Alejandro H. Buschmann, and Rodrigo Navia. "Hydrothermal synthesis, characterization and seed germination effects of green-emitting graphene oxide-carbon dot composite using brown macroalgal bio-oil as precursor." *Journal of Chemical Technology & Biotechnology* 94, no. 10 (2019): 3269–3275.

Sikora, R. A., and E. Fernandez. "Nematodes parasites of vegetables." In: *Plant Parasitic Nematodes in Subtropical and Tropical Agriculture*, Luc, M. R.A. Sikora and J. Bridge (Eds.), pp. 319–392. CABI, Wallingford, 2005.

Silva, R.V.S., Gonçalves, A.D., Vinhal, J.O., Cassella, R.J., Santos, R.C., Sasso, M.A.D., Peixoto, B.S., Borba-Santos, L.P., Rozental, S., Azevedo, D.A., Romeiro, G.A. Bioproducts from the pyrolysis of castor seed cake: basic dye adsorption capacity of biochar and antifungal activity of the aqueous phase." *Journal of Environmental Chemical Engineering* 9, no. 1 (2021): 104825.

Singh, Kaman, and Manisha Gautam. "Development of inexpensive biosorbents from de-oiled mustard cake for effective removal of As (V) and Pb (II) ions from their aqueous solutions." *Journal of Environmental Chemical Engineering* 5, no. 5 (2017): 4728–4741.

Singh, Satyandra, Bijendra Singh, and A. P. Singh. "Nematodes: A threat to sustainability of agriculture." *Procedia Environmental Sciences* 29 (2015): 215–216.

Siva-Sankari, M., and S. Vivekanandhan. "*Jatropha* oil cake based activated carbon for symmetric supercapacitor application: A comparative study on conventional and hydrothermal carbonization processes." *Chemistry Select* 5, no. 4 (2020): 1375–1384.

Siva-Sankari, M., S. Vivekanandhan, M. Misra, and A. K. Mohanty. "Oil cakes as sustainable agro-industrial feedstock for biocarbon materials." *ChemBioEng Reviews* 9, no. 1 (2022): 21–41.

Smets, Koen, Mats De Jong, Iwona Lupul, Grazyna Gryglewicz, Sonja Schreurs, Robert Carleer, and Jan Yperman. "Rapeseed and raspberry seed cakes as inexpensive raw materials in the production of activated carbon by physical activation: Effect of activation conditions on textural and phenol adsorption characteristics." *Materials* 9, no. 7 (2016): 565.

Smiley, R. W., and Yan, G. P. "Cereal Cyst Nematodes: Biology and management in Pacific Northwest wheat, barley and oat crops". PNW Extension Bulletin 620, (2010) Oregon State University, Corvallis.

Srivastava, S. S. "Efficacy of organic amendments in the management of root knot nematode (*Meloidogyne incognita*) infecting papaya." *Indian Journal of Nematology* 32, no. 2 (2002): 183–184.

Stirling, Graham R. "Biological control of plant-parasitic nematodes." In *Diseases of nematodes*, pp. 103–150. CRC Press, New York, 2018.

Sumbul, Aisha, Rose Rizvi, Irshad Mahmood, and R. Ali Ansari. "Oil-cake amendments: Useful tools for the management of phytonematodes." *Asian Journal of Plant Pathology* 9, no. 3 (2015): 91–111.

Sundraraju, P., and P. Kiruthika. "Effect of bio-control agent, *Paecilomyces lilacinus* along with neemcake and botanicals for the management of *Meloidogyne incognita* on banana." *Indian Journal of Nematology* 39, no. 2 (2009): 201–206.

Sunil, L. Prakruthi Appaiah, P. K. Prasanth Kumar, and A. G. Gopala Krishna. "Preparation of food supplements from oilseed cakes." *Journal of Food Science and Technology* 52, no. 5 (2015): 2998–3005.

Tosi, Pierluigi, Gerard P. M. van Klink, Charlotte Hurel, Claire Lomenech, Alain Celzard, Vanessa Fierro, Clara Delgado-Sanchez, and Alice Mija. "Investigating the properties of humins foams, the porous carbonaceous materials derived from biorefinery by-products." *Applied Materials Today* 20 (2020): 100622.

Uçar, Suat, Murat Erdem, Turgay Tay, and Selhan Karagöz. "Removal of lead (II) and nickel (II) ions from aqueous solution using activated carbon prepared from rapeseed oil cake by Na2CO3 activation." *Clean Technologies and Environmental Policy* 17, no. 3 (2015): 747–756.

Vivekanandhan, S. "Biochar as Sustainable Reinforcement for Polymer Composites". In: *Encyclopedia of Renewable and Sustainable Materials*, S. Hashmi and I.A. Choudhury (Eds.), pp. 10–22. Elsevier: Amsterdam, Netherlands, 2020. 10.1016/B978-0-12-803581-8.11290-1.

Vivekanandhan, Singaravelu, Manjusri Misra, and Amar Kumar Mohanty. "Microscopic, structural, and electrical characterization of the carbonaceous materials synthesized from various lignin feedstocks." *Journal of Applied Polymer Science* 132, no. 15 (2015).

Vivekanandhan, Singaravelu, Makoto Schreiber, Sankaralingam Muthuramkumar, Manjusri Misra, and Amar Kumar Mohanty. "Carbon nanotubes from renewable feedstocks: A move toward sustainable nanofabrication." *Journal of Applied Polymer Science* 134, no. 4 (2017).

Wang, Tengfei, Yunbo Zhai, Yun Zhu, Caiting Li, and Guangming Zeng. "A review of the hydrothermal carbonization of biomass waste for hydrochar formation: Process conditions, fundamentals, and physicochemical properties." *Renewable and Sustainable Energy Reviews* 90 (2018): 223–247.

Webster, J. M. "Nematodes and biological control." *Economic Nematology* (1972): 469–496.

Yang, Xiujuan, Xuan Wang, Kang Wang, Lanxi Su, Hongmei Li, Rong Li, and Qirong Shen. "The nematicidal effect of Camellia seed cake on root-knot nematode *Meloidogyne javanica* of banana." *PloS one* 10, no. 4 (2015): e0119700.

Youssef, M. M. A., and W. M. A. El-Nagdi. "Effect of certain organic materials in controlling *Meloidogyne incognita* root-knot nematode infesting banana". *Archives of Phytopathology and Plant Protection* 43 (2010): 660–665.

Zhou, Meng, Joshua Gomez, Binsong Li, Ying-Bing Jiang, and Shuguang Deng. "Oil tea shell derived porous carbon with an extremely large specific surface area and modification with MnO_2 for high-performance supercapacitor electrodes." *Applied Materials Today* 7 (2017): 47–54.

11 Nematode Management in Vegetable Crops Using Oilseed Cakes

B.S. Chandrawat, Raju Dhayal, Kavita Choudhary, Sarthak Ranjan and Hemraj Gurjar

CONTENTS

11.1 INTRODUCTION

A vegetable is a plant in which fruits, leaves, seeds, bulbs, stems, roots, tubers, or flower parts are used for food. Vegetables play a major role in human food. Vegetables provide dietary fibre and are major sources of essential vitamins A, C, and E; elements; and minerals. Vegetables in our daily diet help reduce the incidence of various diseases like cardiovascular disease, stroke, cancer, and other chronic ailments, among others (Büchner et al. 2010; Terry 2011).

Various factors are responsible for the low yield of vegetables. Plant-parasitic nematodes (PPNs) are known as the most dangerous and important pest of vegetable crops, which cause huge yields and economic losses (Collange et al. 2011). Vegetable plants are susceptible to nematodes, and soil's moist condition favours nematode infestation. PPNs mainly feed on underground parts like bulbs, roots, rhizomes, and corms and reduce uptake and transport of water. Some nematodes also cause infection on aboveground parts of plants such as flowers, leaves, buds, and so on. Among PPNs, *Meloidogyne* spp., *R. reniformis*, *Pratylenchus* spp., *T. brassicae*, *Globodera* spp., *B. longicaudatus*, *R. similis*, *Trichodorus* spp., and *Paratrichodorus* spp. cause infection in vegetable crops

DOI: 10.1201/9781003319252-11

(Figure 11.1). However, root-knot nematodes (RKNs; *Meloidogyne* spp.) are the main restraint for producing vegetables worldwide (Sasser 1980). Various *Meloidogyne* species cause physiological and morphological changes in the roots. RKNs induce gall formation, affecting nutrient and water uptake within the host (Figure 11.2) (Anwar and Van Gundy 1993).

The worldwide annual loss of total production in vegetable and fruit (horticultural) crops because of PPN has been 8.8–14.6% and 100–157 billion US dollars (Sasser and Freckman 1987). In India, PPNs cause severe damage to several vegetables, such as 34% loss in carrots, 26% loss in potatoes, 23% loss in tomatoes, 22% loss in bottle gourds, 21% loss in brinjal, 19.5% loss in okra, 15% loss in chilli, 13.5% loss in bitter gourds, 12% loss in cucumbers, and 10% loss in capsicum (Kumar et al. 2020). Sasser and Freckman (1987) recorded a 12.2% loss worldwide in chilli crops because of plant parasitic nematodes. Jain et al. (2007) recorded a 12.85% national loss in chilli due to *Meloidogyne* spp. and 210 million rupees in monetary terms. PPN management is much more complex than other soil-borne disease pathogens because nematodes cause an infection on plant roots, which are underground parts of plants. Often, nematodes damage plants, which become stressed from the deficiency of plant nutrients and thus are complicated to accommodate. Several management strategies, such as resistant cultivars, sanitation, crop rotation with non-host crops, crop residual destruction, and chemical nematicides, have been known to manage the population of nematodes below the ETL (damaging/economic threshold level). Among all practices, chemical nematicides are the most effective option in controlling PPNs. Currently, uses of chemical nematicides have shown to be prompt action and widely used. The chemical nematicides cause residue problems in agricultural practices, creating an ecological disturbance in nature. Because of negative impacts of chemical nematicides on the environment and health, farmers are strongly attracted to organic farming, which is a priority area for growing safe and healthy food and long-term sustainability in the soil. Organic materials are the best option to source micro- and macro-nutrients for plants, increase crop yield, and save the environment. Adding organic matter to soil increases available nutrients (i.e., carbon, potassium, calcium, and magnesium), changes nitrogen levels in soil, and helps manage plant diseases. It also improves soil fertility, structure, and nematode-toxic compounds release. These released compounds are antagonists to nematodes.

11.2 OILSEED CAKES FOR THE MANAGEMENT OF PPNS

Organic material is added to soil to boost biological and physical properties. This practice enhances soil's nutritional status, moisture conservation, and management of nematodes and other pests. Oilseed cake is obtained after oil extraction from plant seeds. These are two categories: (1) edible oil cakes and (2) non-edible oil cakes. Oilseed cakes are used in a common practice to manage plant diseases. Several oilseed cakes are reported to effectively manage the PPNs in crops (Goswami et al. 2007). Oilseed cakes are often applied within the soil in powder or pellet form fifteen to twenty days before sowing, then watered and left for decomposition. The nematicidal effect of oilseed cake will increase with the decomposition time. A root dip treatment of various seedlings has been used in oilseed extracts before transplanting and sowing, providing effective management against the PPN and saving the environment (Youssef and El-Nagdi 2010).

11.2.1 In Vitro Test of Oilseed Cake Extract against PPN

To test the nematicidal properties of oilseed cakes, extracts of various oilseed cakes have been tested against PPNs. Oilseed cake extracts of neem, castor, mustard, mahua, groundnut, and karanj have been used against the nematodes, namely *Hoplolaimus indicus*, *Rotylenchulus reniformis*, and *Tylenchorynchus brassicae* (Sumbul et al. 2015). Khan et al. (1974a) reported that oilseed cake extract effectively kills the second-stage juveniles (J2) of *Meloidogyne incognita*. The hatching of J2 was also inhibited by oilseed cake extract. The larval hatching was suppressed up to 99.92% and 99.38% when boiled cotton and mustard oilseed cake extracts, respectively, were used. Mustard cake extract proved effective in managing *H. indicus* (Deshmukh and Prasad 1969), and neem oilseed cake extract was shown to be effective against *M. incognita* (Gowda and Setty 1978; Gowda and Gowda 1999). Neem and karanj oilseed cake extracts were the foremost sturdy treatments that decreased egg hatching and killed the larvae of *M. incognita* (Rao et al. 1986). Javed et al. (2014) showed that 10% of extracts of neem cake caused 85% immobility and 28% mortality of J2 of *Meloidogyne javanica*. *M. javanica* mortality was also reported in oilseed cake extract of neem, karanj, mustard, castor, and mahua. These extracts also inhibited larval hatching (Choudhary 2021).

11.2.2 Effect of Soil-Amended Oilseed Cakes against PPNs

Various research results have shown that soil-amended oilseed cake application successfully managed PPNs. Goswami et al. (1993) showed that decayed neem, karanj, mustard, mahua, and groundnut oilseed cakes increased RKN mortality in rhizospheric soil. Khan and Goswami (1995) found that by increasing the neem cake decaying period to up to twenty days, the mortality of J2 of *Meloidogyne* spp. also increased. Gupta and Kumar (1997) reported that soil-amended groundnut and mustard oilseed cake reduced the population of *Tylenchorynchus* spp. and *Helicotylenchus* spp. A reduction in nematode population increased at the higher dose of oilseed cake and longer treatment periods. Sitaramaiah (1979) assessed the effect of temperature on the decaying of margosa cake and the survival of *Tylenchorynchus elegans*. He found that *T. elegans* population was significantly reduced at 20°C and above within the amended soil. Nagesh et al. (2001) showed that soil-amended neem oilseed cake supplemented with inorganic fertilizers (N_2, P_2O_5, and K_2O) was shown to have a potential impact against nematodes. The earlier combined treatment enhanced growth of the mycelial and sporulation of *P. lilacinus*, which acted as an antagonist to nematodes.

11.2.3 Mode of Action of Oilseed Cakes

Oilseed cakes contain nematicidal compounds that are released during decomposition in soil (Table 11.1). Various compounds, such as azadirachtin, salannin, nimbin, isoleucine, mithionine, karanjin, tannin, saponin, and amino acids, have been reported (Mcsorley 2011). These chemical compounds accelerate the development of plant roots and the overall growth and protect plants from nematodes.

TABLE 11.1
List of Some Important Nematicidal Compounds of Oilseed Cakes

Oilseed Cake	Chemical Compound
Neem cake	Azadirachtin, salannin, and nimbin
Castor cake	Isoleucine and mithionine
Karanj cake	Karanjin, tannin, and trypsin
Mahua cake	Saponin
Jatropha cake	Jatropin, vitexin, and isovitexin
Mustard cake	Gluco-sinolates and methionine

11.3 IMPACT OF OILSEED CAKES ON NEMATODE-INFESTED VEGETABLE CROPS

Several pieces of literature have been available regarding managing PPNs using oilseed cake in different vegetables. Singh and Sitaramaiah (1973) reported the impact of oilseed cakes, including peanut, margosa, linseed, castor, mahua, coconut, and mustard oilseed cake, against nematode infection in vegetables. Other researchers have also reported the effects of castor, mahua, neem, groundnut, and mustard oilseed cakes alone or in combinations for managing nematodes on cauliflower, tomato, okra, chilli, eggplant, and cabbage (Alam et al. 1980; Alam 1991).

11.3.1 Tomato

The depreciation in PPN population and the betterment in growth of the plant by incorporating soil-amended oilseed cakes have been found (Hameed 1968; Javed et al. 2007; Lopes et al. 2009; Abbasi et al. 2005; Roldi et al. 2013; Sahu et al. 2018; Parveen et al. 2019). Alam et al. (1980) found fewer juveniles in tomato-grown seedling's roots in amended soil neem oil cake. The application of a 4% dose of neem seed cake was reported toxic for nematodes, but it increased plant growth. Singh et al. (1980) managed the nematode population of *H. indicus*, *T. brassicae*, *Hoplolaimus indicus*, and *M. incognita* using soil-amended castor, mustard, neem, mahua, and groundnut oilseed cake. The application of karanj and neem cakes at 400 kg/ha was reported to be effective in reducing the population of RKN in tomatoes (Darekar et al. 1990). Goswami and Meshram (1991) reported neem oilseed cake was effective against *Pratylenchus penetrans* and *M. hapla*. Olive, cotton, flax, soybean, and sesame oilseed cake was used @ 5, 10, 15, 20, or 50 g per kg soil against *M. incognita* infesting tomato plants. All the oilseed cake suppressed the nematode population in the soil and root galling (Radwan et al. 2009). The application of castor and mustard oilseed cake extracts as root dip significantly reduced nematode infection and improved tomato and eggplant plant growth (Abid and Maqbool 1991). Akhtar (2000) explained that azadirachtin in neem cakes when released causes a significant reduction in the nematode population. Applying a 1% neem oilseed cake in soil reduced the populations of *Pratylenchus penetrans* and *M. hapla* up to 67–90% in tomatoes (Abbasi et al. 2005). Chandrawat et al. (2020) applied mustard, castor, karanj, mahua, and neem oilseed cake at 2.5 and 5.0 q/ha. They found increased enzymatic activity, including peroxidase (PO), polyphenol oxidase (PPO), phenylalanine ammonia lyase (PAL), and Superoxide dismutase (SOD), improved tomato plant growth and reduced *M. incognita* nematode population. Mohanty and Tayung (2021) reported managing *M. incognita* using neem, mahua, groundnut, and mustard cake. Meena et al. (2021) tested karanj, castor, mahua, mustard, and neem cake and vermicompost efficacy against *M. incognita* and *Fusarium oxysporum* disease complex in tomatoes and reported a decrease in disease incidences.

11.3.2 Okra

PPNs associated with okra are reported to be managed using oilseed cakes. Groundnut, castor, and neem oilseed cakes application have suppressed *T. brassicae* and *M. incognita* in okra (Khan et al. 1979). Patel et al. (1985) used mustard cake for the management of RKNs on okra. Alam (1991) found mustard, mahua, groundnut, castor, and neem oilseed cakes effective against PPNs. These oilseed cakes decreased RKNs and reniform nematode population and improved okra plant growth, including capacity of water absorption (Anver and Alam 1996). The integration of castor, groundnut, neem, mustard, and duan oil cakes with deep ploughing was highly influential in decreasing the population of *H. indicus*, *R. reniformis*, *T. brassicae*, *T. filiformis*, and *M. incognita* in okra (Siddiqui and Alam 1987). An application of castor, mustard, and neem cakes reduces the RKN population and improves okra growth, per Ram et al. (2009). Ganaie et al. (2011) also found a similar effect using castor, groundnut, linseed, and sunflower cake application. Baheti et al. (2019) used neem cake at 6 q/ha and found that treatment was most effective for managing *M. incognita* in okra.

Castor and neem oilseed cakes have also been used to keep nematode population density below the threshold (Wani 2006).

11.3.3 Eggplant

The management of RKN and enhancement in plant growth have been reported in eggplant by using soil-amended groundnut, castor, and neem oilseed cake (Khan et al. 1974b). Tagi et al. (2022) tested polang and mahua oilseed cake against RKNs at 0.5, 1.0, and 1.5 g/kg soil. They also compared the previously given treatment with neem oilseed cake at 1.0 g/kg soil. The results revealed that neem cake at 1.0 and mahua cake at 1.5 g/kg soil exhibited significantly better plant growth parameters and reduced root galls.

11.3.4 Chilli

The management of PPN on chilli was reported using sesame, duan, groundnut, and mustard oilseed cakes (Trivedi et al. 1978). The seed cakes including mahua, mustard, castor, neem, and groundnut, significantly reduced the nematode density in soil (Alam et al. 1980; Alam 1991). Chandrawat (2017) observed the effect of different oilseed cakes, namely castor, mahua, karanj, mustard, and neem oilseed cakes. Among them, mustard and neem oilseed cakes at 5.0 q/ha were shown to be the best to enhance chilli growth and reduce reproduction in nematodes. Among enzymes, the PO was found to be highest, followed by SOD, while PAL and PPO were in lower quantities in chilli roots.

11.3.5 Cluster Bean

Baheti et al. (2018) tested jatropha, castor, and neem oilseed cake powder in cluster beans at 1, 2, and 4 q/ha against RKN *M. incognita*. Their study showed that maximum plant growth and minimum nematode population were obtained when neem cake was applied at 4 q/ha.

11.3.6 Other Vegetables

Various researchers have worked on oilseed cake application for PPN management. Sharma and Raj (1987) reported the management of nematodes on potatoes using neem. A similar approach was taken by Chakraborti (2000) and Alam (1991) for nematode management in pointed gourds,

TABLE 11.2
Effect of Various Oilseed Cakes for the Management of Plant-Parasitic Nematodes on Different Vegetables

S. No.	Crop	Nematode	Oil Cake	Reference
1.	Chilli (*Capsicum annum*)	*Meloidogyne incognita*	Castor, mahua, karanj, and mustard	Chandrawat (2017)
1.			Neem + *T. viride*, Gingelly + *T. viride*, Garlic + *T. viride*, Mahua + *T. viride*, *T. viride*	Nirosha et al. (2018)
2.	Eggplant (*Solanum melongena*)	*Meloidogyne incognita*	*P. lilacinus* + mustard cake, neem cake extract + spores of *P. lilacinus*, neem leaf suspension + spores of *P. lilacinus* Castor cake extract + *T. harzianum*	Zareena and Das (2014)
1.		*M. javanica*	Castor, mahua, karanj, mustard, and neem cake	Choudhary (2021)

TABLE 11.2
(Continued)

S. No.	Crop	Nematode	Oil Cake	Reference
3.	Okra	*M. incognita*	Castor, mustard, and neem	Ram et al. (2009)
1.	(*Abelmoschus esculentus*)	*M. incognita*	Linseed, groundnut, sunflower, and castor	Ganaie et al. (2011)
1.		*Meloidogyne* spp.	Sesame seed extract	Kankam et al. (2015)
1.		*Meloidogyne incognita*	Neem, mustard and karanj oil cake	Baheti et al. (2019)
4.	Tomato (*Solanum lycopersicum*)	*Meloidogyne incognita*	Groundnut, karanj, jatropha, mustard, polango, mahua, neem, and kusum oil cake	Sahu et al. (2018)
1.		Root diseases	Mustard and cotton cake	Parveen et al. (2019)
1.		*M. incognita*	Castor, mahua, karanj, and mustard	Chandrawat et al. (2020)
1.		*M. incognita* and *F. oxysporum* f. sp. *lycopersici*	Karanj cake, mahua cake, mustard cake, castor cake, neem cake, and vermicompost	Meena et al. (2021)
1.		*Meloidogyne incognita*	Neem cake, mahua cake, groundnut cake, and mustard cake	Mohanty and Tayung (2021)
5.	Coriander (*Coriandrum sativum* L.)	*Hoplolaimus columbus* and *M. javanica*	Castor and mustard oilseed cake	Khan et al. (2021)
6.	Lettuce	*Meloidogyne incognita*	Crambe cake, bokashi, cottonseed composted, shredded woodchips composted and whey protein	Dias-Arieira et al. (2015)
7.	Onion	*Meloidogyne incognita*	poultry refuse, mustard oil cake, rice bran, and sawdust	Faruk (2019)
8.	Watermelon	*M. incognita*	Poultry refuse and mustard oilcake	Faruk et al. (2012)
9.	Pumpkin	*Meloidogyne incognita*	*Trichoderma harzianum* and mustard oil cake	Lalhruaitluanga and Hazarika (2020)
10.	Carrot	*Meloidogyne* spp.	Poultry manure and neem leaves	Agyarko et al. (2006)
11.	Cucumber	*M. javanica* and *Fusarium oxysporum*	Neem, castor, karanj, mahua, and mustard oil cakes	Monika (2021)

cabbage, and cauliflower using oilseed cakes, including mustard, mahua, neem, groundnut, and castor. The sesame cake has also been applied for nematode management on squash (Youssef and El-Nagdi 2010). The details regarding oilseed cake application against nematodes are given in Table 11.2.

11.4 COMPARATIVE ASPECTS OF OILSEED CAKES AND CHEMICAL NEMATICIDES

Studies have proved the efficacy of oilseed cakes in comparison to inorganic chemicals for PPN management in vegetables. Results revealed that observed data were not uniform. The chemical nematicides were found more effective than oilseed cake in a few cases, while in some other studies, oilseed cakes were reported superior over chemical nematicides. Aldicarb response was superseded by neem oil cake according to Bhattacharya and Goswami (1987a, 1987b). Similarly, carbofuran response was superseded by mustard and karanj oil cake (Goswami and Meshram 1991) against

RKNs infesting tomato. Neem oilseed cake at 80q/ha was recorded as highly effective as compared to carbofuran @ 2 kg a.i./ha in tomato against RKNs (Jain and Gupta 1997). Aldicarb produced effective results in potatoes (Sharma and Raj 1987) and carbofuran in okra (Abuzar and Haseeb 2006) compared to neem oil cake. Goswami and Meshram (1991) showed effective results of karanj and mustard oil cakes compared carbofuran in reference to better plant growth and increased tomato root capacity in inhibiting *M. incognita* juvenile penetration. Hossam et al. (2012) tested the effect of neem and poultry, composts, organic amendments, inorganic fertilizers, and chemical nematicide (nemacur 10 G) against *M. incognita* on tomatoes. Results proved that all treatments reduced nematode reproduction, fecundity, the number of galls, and increased antioxidant substances (Figure 11.3).

11.5 ADDITIVE ASPECT OF OILSEED CAKES AND BIOAGENTS

A combination of oilseed cake, namely flax, neem, groundnut, castor, mustard, and sesame, with *Paecilomyces lilacinus* in soil resulted in increased plant growth as well as decreased nematode population and root galling on tomatoes (Khan and Saxena 1997). Among all, the *P. lilacinus* and groundnut cake combination was found best. The neem cake and *Pasteuria penetrans* combination has also shown similar results (Javed et al. 2008). Neem cake integration with *P. penetrans* + *P. lilacinus* showed a maximum increase in tomato plant growth and the number of seedlings per bed infected with *M. incognita* (Reddy et al. 1997). A combined application of neem cake and *Trichoderma harzianum* was highly effective against *M. incognita*. Neem cake and vesicular-arbuscular mycorrhiza (VAM) integration improved eggplant growth, increased fruit yield, and reduced the population of *M. incognita* (Borah and Phukan 2004). Castor, mustard, and neem oil cakes combined with Arbuscular mycorrhizal fungi (AMF) resulted in reduced root galling as well as nematode multiplication and increased plant growth as well as fruit yield (Bhardwaj and Sharma 2006). The improvement in plant growth was found due to better soil nutrition availability through neem oil cake and excess colonization by fungus on roots (Kumar and Khanna 2006). The application of mustard cake and furadon with *Pupurescillium lilacinum* and *T. viridae* increased plant growth and reduced the nematode population in soil as well as in roots (Tripathi and Singh 2006). Per results obtained by researchers, it is concluded that oilseed cake increases the parasitism of bioagents against plant parasitic nematodes and reduces disease incidence. *P. lilacinus* integration with groundnut oil cake showed a maximum increase in the growth of brinjal plants and decreased the *M. javanica* population, followed by *P. lilacinus* combined with linseed, mahua, neem, and castor oilseed cake and *C. oxysporum* and neem cake integration, followed by mahua, groundnut, castor, and linseed cake (Ashraf and Khan 2010). Chaudhary and Kaul (2013) tested *Pasteuria penetrans* combination with neem, castor, *Citrullus*, and mustard oil cakes in chilli against RKNs. Among them, a combination of *P. penetrans* and castor showed a maximum reduction in root-knot galling (84.75%) and final population (85.74%) and better plant growth parameters as compared to the control and other treatments. Best control of root-knot disease, increased yield, and plant growth has also been accomplished when bioagents, namely *P. lilacinus*; *T. harzianum*; and botanicals such as neem, mustard, neem leaf, and castor cake extract, were used (Zareena and Das 2014). A *Glomus fasciculatum* and neem cake combination has been found better against *M. incognita* (Rizvi et al. 2015). Nirosha et al. (2018) have also tested a combination of oilseed cake and bioagents (neem + *T. viride*, gingelly + *T. viride*, garlic + *T. viride*, mahua + *T. viride, T. viride*) against *M. incognita* in chilli.

11.6 ADVANTAGES

1. The decomposed oilseed cake may be directly toxic to nematodes by producing nematotoxic compounds.
2. Organic amendments can produce secondary metabolites, which have allelopathic effects and can induce resistance in plants against the penetration of PPNs (Parihar et al. 2015).

3. The oilseed cakes may increase microbial activity in the plant root zone, reducing the soil's nematode population.
4. Applying oilseed cakes may increase organic material to the soil, resulting in increased pores and decreased size of pores. This condition may help reduce nematode activity within the soil.

11.7 CHALLENGES

1. A high dose means a large amount of oilseed cake is required per unit area. This makes it more challenging to ensure the availability of oilseed cake in large-scale farming.
2. Excess labour requirements mean the cost of farming will also be increased.
3. The lower availability of oilseed cakes in the local market and financial issues in developing countries are barriers.
4. Environmental perturbations and an excess amount of oilseed cakes can enhance the soil's micronutrients, which ultimately causes surface and groundwater pollution with P_2O_5 and N_2.

11.8 CONCLUSION

Oilseed cakes provide a suitable substrate for better plant growth and microorganisms in various ways. Among oilseed cakes, neem oilcake is widely used and found most effective against PPNs, followed by castor, mustard, and karanj oilseed cake. Research on oil cake application in the last few years has increased, but the focus must be on the actual mechanism, the rate of application, and the time of application for nematode management. For the eco-friendly management of PPNs, the focus is also on the integrated effect of oilseed cake and bioagents and various possible methods.

FIGURE 11.1 Common disease symptoms caused by nematodes on vegetable crops. (A) Wilting disease in okra; (B, C) wilting disease in tomato.

FIGURE 11.2 Disease symptoms caused by root-knot nematode (RKN) on vegetables. (A) Yellowing leaves in eggplant; (B) root galls in tomato; and (C) root galls in cucumber plant.

FIGURE 11.3 The plant growth improvement in root dip-treated (A) tomato and (B) eggplant seedlings in oilseed cake extract.

REFERENCES

Abbasi, P.A., E. Riga, K.L. Conn, and G. Lazarovits. 2005. Effect of neem cake soil amendment on reduction of damping-off severity and population densities of plant-parasitic nematodes and soilborne plant pathogens. *Canadian Journal of Plant Pathology* 27:38–45.

Abid, M., and M.A. Maqbool. 1991. Effect of bare-root dip treatment in oilseed cake and neem leaf extract on the root-knot development and growth of tomato and eggplant. *Pakistan Journal of Nematology* 9:13–16.

Abuzar, S., and A. Haseeb. 2006. Efficacy of carbofuran, bavistin, neem, *Trichoderma harzianum* and *Aspergillus niger* against *Meloidogyne incognita* and *Fusarium oxysporum* f. sp. *vasinfectum* disease complex on okra. *Indian Journal Nematology* 36:282–284.

Agyarko, K., P.K. Kwakye, M. Bonsu, B.A. Osei, and K.A. Frimpong. 2006. The effect of organic soil amendments on root-knot nematodes, soil nutrients and growth of carrot. *Journal of Agronomy* 5(4):641–646.

Akhtar, M. 2000. Nematicidal potential of the neem tree *Azadirachta indica* (A. Juss). *Integrated Pest Management Reviews* 5:57–66.

Alam, M.M. 1991. Control of plant-parasitic nematodes with oilseed cakes on some vegetables in field. *Pakistan Journal of Nematology* 9:21–30.

Alam, M.M., M. Ahmad, and A.M. Khan. 1980. Effect of organic amendments on the growth and chemical composition of tomato, eggplant and chilli and their susceptibility to attack by *Meloidogyne incognita*. *Plant and Soil* 57:231–236.

Anver, S., and M.M. Alam. 1996. Effect of some oilseed cake against *Meloidogyne incognita* and *Rotylenchulus reniformis* infecting okra. *Annals of Plant Protection Sciences* 4:176–178.

Anwar, S.A., and S.D. Van Gundy. 1993. Effect of *Meloidogyne incognita* on root and shoot growth parameters of susceptible and resistant varieties of tomato. *Afro-Asian Journal of Nematology* 3:152–160.

Ashraf, M.S., and T.A. Khan. 2010. Integrated approach for the management of *Meloidogyne javanica* on eggplant using oil cakes and biocontrol agents. *Archives of Phytopathology and Plant Protection* 43:609–614.

Baheti, B.L., S.S. Bhati, and Hanuman Singh. 2019. Efficacy of different oilseed cake as soil amendment for the management of root-knot nematode, *Meloidogyne incognita* infecting okra (*Abelmoschus esculentus* L.). *International Journal of Current Microbiology and Applied Sciences* 8(12):799–808.

Baheti B.L., P.K. Yadav, C.P. Nama, and S.K. Khandelwal. 2018. Environment friendly management of root-knot nematode, *Meloidogyne incognita* infecting cluster bean (*Cyamopsis tetragonoloba* L.). *Indian Journal of Nematology* 48(1):56–61.

Bhardwaj, A., and S. Sharma. 2006. Biocontrol of *Meloidogyne incognita* in *Lycopersicon esculentum* with AM fungi and oil cakes. *The Journal of Plant Pathology* 5:166–172.

Bhattacharya, D., and B.K. Goswami. 1987a. A study on the comparative efficacy of neem and ground nut oil cakes against root-knot nematode, *Meloidogyne incognita* as influenced by microorganisms on sterilized and unsterilized soil. *Indian Journal of Nematology* 17:81–83.

Bhattacharya, D., and B.K. Goswami. 1987b. Comparative efficacy of neem and groundnut oilseed cake with aldicarb against *Meloidogyne incognita* in tomato. *Revue de Nematologie* 10:467–470.

Borah, A., and P.N. Phukan. 2004. Comparative efficacy of *Glomus fasciculatum* with neem cake and carbofuran for the management of *Meloidogyne incognita* on brinjal. *Indian Journal of Nematology* 34:129–132.

Büchner, F.L., H.B. Bueno-de-Mesquita, M.M. Ros, K. Overvad, C.C. Dahm, L. Hansen, A. Tjønneland, F. Clavel-Chapelon, M.C. Boutron-Ruault, M. Touillaud, R. Kaaks, S. Rohrmann, H. Boeing, U. Nöthlings, A. Trichopoulou, D. Zylis, V. Dilis, D. Palli, S. Sieri, P. Vineis, R. Tumino, S. Panico, P.H. Peeters, C.H. van Gils, E. Lund, I.T. Gram, T. Braaten, M.J. Sánchez, A. Agudo, N. Larrañaga, E. Ardanaz, C. Navarro, M.V. Argüelles, J. Manjer, E. Wirfält, G. Hallmans, T. Rasmuson, T.J. Key, K.T. Khaw, N. Wareham, N. Slimani, A.C. Vergnaud, W.W. Xun, L.A. Kiemeney, and E. Riboli. 2010. Variety in fruit and vegetable consumption and the risk of lung cancer in the European prospective investigation into cancer and nutrition. *Cancer Epidemiology, Biomarkers & Prevention* 19(9):2278–2286. doi: 10.1158/1055-9965.EPI-10-0489.

Chakraborti, S. 2000. Management approaches for root knot nematode in pointed gourd. *Indian Journal of Nematology* 30:136–140.

Chandrawat, B.S. 2017. Assessment of defence enzymes in determining resistance in tomato and chilli against root-knot nematode, *Meloidogyne incognita*. Ph.D. thesis, RCA, MPUAT, Udaipur.

Chandrawat, B.S., A.U. Siddiqui, S.S. Bhati, and Vinod Saharan. 2020. Response of defence related enzymes in tomato treated with oilseed cake against root-knot nematode, *Meloidogyne incognita*. *International Journal of Current Microbiology and Applied Sciences* 9(11):1100–1111.

Choudhary, K. 2021. Sustainable management of root-knot nematode (*Meloidogyne* spp.) on brinjal. M.Sc. thesis, SKNAU, Jobner.

Chaudhary, K.K., and R.K. Kaul. 2013. Efficacy of *Pasteuria penetrans* and various oil seed cakes in management of *Meloidogyne incognita* in chilli pepper (*Capsicum annuum* L.) *Journal of Agricultural Science and Technology* 15:617–626.

Collange, B., M. Navarrete, G. Peyer, T. Mateille, and M. Tchamitchian. 2011. Root-knot nematode (*Meloidogyne*) management in vegetable crop production: the challenge of an agronomic system analysis. *Crop Protection* 30:1251–1262.

Darekar, K.S., N.L. Mhase, and S.S. Shelke. 1990. Effects of placement of non-edible oilseed cakes on the control of root-knot nematodes on tomato. *International Nematology Network Newsletter* 7:5–7.

Deshmukh, M.G., and S.K. Prasad. 1969. Effect of water-soluble extracts of oilseed cake on the population of *Hoplolaimus indicus* Sher, 1963. *Indian Journal of Entom*ology 31:273–276.

Dias-Arieira, C.R., D. Mattei, H.H. Puerari, and R.C.F. Ribeiro. 2015. Use of organic amendments in the management of root-knot nematode in lettuce. *Horticultura Brasileira* 33:488–492.

Faruk, Iqbal. 2019. Efficacy of organic soil amendments and a nematicide for management of root-knot nematode *Meloidogyne* spp. of onion. *American Journal of Bioscience and Bioengineering* 7(4):57–64.

Faruk, M. I., M. M. H. Mustafa, and T. K. Dey. 2012. Efficacy of two organic amendments and a nematicide for management of root-knot nematode *(Meloidogyne incognita)* of watermelon. *African Journal of Agricultural Research* 7(25):3695–3700.

Ganaie, M.A., A.A. Rather, and M.A. Siddiqui. 2011. Pathogenicity of root knot nematode *Meloidogyne incognita* on okra and its management through botanicals. *Archives of Phytopathology and Plant Protection* 44:1683–1688.

Goswami, B.K., and N.J. Meshram. 1991. Studies on comparative efficacy of mustard and karanj oil seed cakes with a nematicide, carbofuran against root-knot nematode, *Meloidogyne incognita* in tomato. *Indian Journal of Nematology* 21:66–70.

Goswami, B.K., R.K. Pandey, J. Goswami, and D.D. Tewari. 2007. Management of disease complex caused by root knot nematode and root wilt fungus on pigeon pea through soil organically enriched with vesicular arbuscular mycorrhiza, karanj (*Pongamia pinnata*) oilseed cake and farmyard manure. *The Journal of Environmental Science and Health, Part B, Pesticides, Food Contaminants, and Agricultural Wastes* 42:899–904.

Goswami, B.K., U. Rao, and M. Manton. 1993. Efficacy of decomposed products of some oilseed cake and ABCD collected at different time intervals on hatching and mortality of *Meloidogyne incognita. Current Nematology* 4:215–217.

Gowda, D.N., and K.G.H. Setty. 1978. Comparative efficacy of neem cake and methomyl on the growth of tomato and root-knot development. *Current Research* 7:118–120.

Gowda, P.D., and D.N. Gowda. 1999. *In-vitro* studies of oilcake extracts on the survival of larvae of *Meloidogyne incognita* (Kofoid and White) Chitwood. *Plant Protection Bulletin* 49:10–11.

Gupta, M.C., and S. Kumar. 1997. Efficacy of certain organic amendments and nematicides against *Tylenchorhynchus* spp. and *Helicotylenchus* spp. in soil. *Indian Journal of Nematology* 27:139–142.

Hameed, S.F. 1968. Relative efficacy of different oil cakes on the incidences of root-knot nematodes. *Madras Journal of Agriculture Sciences* 55:168–171.

Hossam, S.E., A.A. Farahat, A.A. Alsayed, and N.A. Mahfoud. 2012. Response of antioxidant substances and enzymes activities as a defense mechanism against root-knot nematode infection. *Notulae Botanicae Horti Agrobotanici Cluj-Napoca* 40:132–142.

Jain, R.K., and D.C. Gupta. 1997. Efficacy of neem (*Azadirachta indica*) cake as nursery bed treatment in the management of root knot nematode (*Meloidogyne javanica*) infecting tomato. *Indian Journal of Nematology* 27:249–251.

Jain, R.K., K.N. Mathur, and R.V. Singh. 2007. Estimation of losses due to plant parasitic nematodes on different crops in India. *Indian Journal of Nematology* 37:219.

Javed, N., S. El-Hassan, S. Gowen, B. Pemproke, and M. Inam-ul-Haq. 2008. The potential of combining *Pasteuria penetrans* and neem (*Azadirachta indica*) formulations as a management system for root-knot nematodes on tomato. *European Journal of Plant Pathology* 120:53–60.

Javed, N., S.R. Gowen, S.A. El-Hassan, and M. Inam-ul-Haq. 2014. Efficacy of neem (*Azadirachta indica*) formulations on biology of root-knot nematodes (*Meloidogyne* javanica) on tomato. *Microbial Pathogenesis* 75:16–20.

Javed, N., S.R. Gowen, M. Inam-ul-Haq, and S.A. Anwar. 2007. Protective and curative effect of neem (*Azadirachta indica*) formulations on the development of root-knot nematode *Meloidogyne javanica* in roots of tomato plants. *Crop Protection* 26:530–534.

Kankam, F., E.N. Sowley, and M. Alhassan. 2015. Management of root-knot nematode (*Meloidogyne* spp.) on okra (*Abelmoschus esculuntus* (L.) *Moench*) with aqueous sesame seed extract. *International Journal of Agronomy and Agricultural Research* 6(4): 24–31.

Khan, A., S.S. Shaukat, K.A. Khanzada, S. Waheed, and J. Akhtar. 2021. Nematicidal potential of oilcake against plant nematodes associated with *Coriandrum sativum* L. in Balochistan, Pakistan. *Bangladesh Journal of Scientific and Industrial Research* 56(1):61–68.

Khan, M.R., and B.K. Goswami. 1995. Nematicidal efficacy of decomposed neem cake and fungal filtrates of *Paecilomyces lilacinus* and *Aspergillus flavus* against *Meloidogyne incognita.* In Proceedings of the National Symposium Sustainable Agriculture in Sub-Humid Zone, March 3–5, Visva-Bharati, Sriniketan, India, 359–361.

Khan, M.W., A.M. Khan, and S.K. Saxena. 1974a. Rhizosphere fungi and nematodes of eggplant as influenced by oil cake amendments. *Indian Phytopathology* 27:480–484.

Khan, M.W., M.M. Alam, A.M. Khan, and S.K. Saxena. 1974b. Effect of water soluble fractions of oil cake and bitter principles of neem on some fungi and nematodes. *Acta Botanica Indica* 2:120–128.

Khan, M.W., A.M. Khan, and S.K. Saxena. 1979. Suppression of phytophagous nematodes and certain fungi in the rhizosphere of okra due to oilcake amendments. *Acta botanica Indica* 7:51–56.

Khan, T.A., and S.K. Saxena. 1997. Integrated management of root knot nematode *Meloidogyne javanica* infecting tomato using organic materials and *Paecilomyces lilacinus. Bioresource Technology* 61:247–250.

Kumar, S., and A.S. Khanna. 2006. Role of *Trichoderma harzianum* and neem cake separately and in combination against root-knot nematode on tomato. *Indian Journal of Nematology* 36:247–249.

Kumar, Vinod, M.R. Khan, and R.K. Walia. 2020. Crop loss estimations due to plant-parasitic nematodes in major crops in India. *National Academy Science Letters* 43(5):409–412.

Lalhruaitluanga, P.C., and K. Hazarika. 2020. Bio-management of *Meloidogyne incognita* in pumpkin. *International Journal of Current Microbiology and Applied Sciences* 9(11):762–770.

Lopes, E.A., S. Ferraz, O.D. Dhingra, P.A. Ferreira, and L.G. Freitas. 2009. Soil amendment with castor bean oilcake and jack bean seed powder to control *Meloidogyne javanica* on tomato roots. *Nematologyogia Brasileira* 33:106–109.

McSorley, R. 2011. Overview of organic amendments for management of plant-parasitic nematodes, with case studies from Florida. *Journal of Nematology* 43:69–81.

Meena, Pinki, B.S. Chandrawat, R.R. Ahir, B. Gugarwal, S. Meena, N. Nitharwal, and V. Sharma. 2021. Management of tomato wilt complex caused by *Meloidogyne incognita* and *Fusarium oxysporum* f. sp. *lycopersici* through organic amendments. *Journal of Entomology and Zoology Studies* 9(1):1691–1693.

Mohanty, S.S., and K. Tayung. 2021. Effect of different oil cakes against root knot nematode infecting tomato. *Journal of Emerging Technologies and Innovative Research* 8(6):910–918.

Monika. 2021. Interaction of *Fusarium oxysporum* and *Meloidogyne* spp. in cucumber wilt and its management. M.Sc. thesis, SKNAU, Jobner.

Nagesh, M., P.P. Reddy, and N. Raman. 2001. Influence of oil cakes in combination with inorganic fertilizers on growth and sporulation of *Paecilomyces lilacinus* and its antagonism on *Meloidogyne incognita* infecting tomato. *Nematologia Mediterranea* 29:23–27.

Nirosha, A., S. Rajeshkanna, and G. Mikunthan. 2018. Effect of oil cakes and garlic aqueous based formulations of *Trichoderma viride* on management of *Meloidogyne incognita* in chilli. *Open Agriculture* 3:214–219.

Parihar, K., B. Rehman, M.A. Ganai, M. Asif, and M.A. Siddiqui. 2015. Role of oil cakes and *Pochonia chlamydosporia* for the management of *Meloidogyne javanica* attacking *Solanum melongena* L. *Journal of Plant Pathology & Microbiology* S1:004.

Parveen, G., N. Mukhtar, Kaleemullah, and M.K. Sheikh. 2019. Enhancement of growth and suppressing the root diseases of tomato plant by using organic amendment. *Pure and Applied Biology* 8(2):1092–1099.

Patel, H.R., N.A. Thakar, and C.C. Patel. 1985. Azolla and mustard cake against nematicides for root-knot management in okra. *Sour Madras Agriculture Journal* 72:593–594.

Radwan, M.A., E.K. El-Maadawy, S.I. Kassem, and M.M. Abu-Elamayem. 2009. Oil cakes soil amendment effects on *Meloidogyne incognita*, root-knot nematode infecting tomato. *Archives of Phytopathology and Plant Protection* 42:58–64.

Ram, L., A.U. Siddiqui, and A. Parihar. 2009. Management of root-knot nematode, *Meloidogyne incognita* infecting okra using oil cakes. *Indian Journal of Nematology* 39:125–127.

Rao, C. V., A. Many, and P.K. Rao. 1986. Effect of plant products on egg hatch and larval mortality of *Meloidogyne incognita. Proceedings of the Indian Academy of Sciences Animal Sciences* 95(4):397–401.

Reddy, P., M. Nagesh, and V. Devappa. 1997. Effect of integration of *Pasteuria penetrans*, *Paecilomyces lilacinus* and neem cake for the management of root-knot nematodes infecting tomato. *Pest Management in Horticultural Ecosystems* 3:100–104.

Rizvi, R., G.S. Safiuddin, R.A. Ansari, S.A. Tiyagi, and I. Mahmood. 2015. Sustainable management of root-knot disease of tomato by neem cake and *Glomus fasciculatum. Cogent Food & Agriculture* 1:1.

Roldi, M., C.R. Dias-Arieira, J.J. Severino, S.D.M. Santana, T.S. Dadazio, P.M. Marini, and D. Mattei. 2013. Use of organic amendments to control *Meloidogyne incognita* on tomatoes. *Nematropica* 43:49–55.

Sahu, S., M.K. Patra, and B. Dash. 2018. Management of root knot nematode (*Meloidogyne incognita*) in tomato (cv. Pusa ruby) using different oil cakes. *International Journal of Current Microbiology and Applied Sciences* 7(3):2527–2532.

Sasser, J.N. 1980. Root-knot nematode: a global menace to crop production. *Plant Disease* 64(1):36–41.

Sasser, J.N., and D.W. Freckman. 1987. A World Perspective on Nematology: The Role of the Society. In: *Vistas on Nematology*, Veech, J.A. and D.W. Dickson (Eds.). Society of Nematologyogists, Hyattsville, Maryland, 7–14.

Sharma, R.K., and D. Raj. 1987. Effect of nematicides and organic amendments on root-knot nematodes infecting potato. *International Nematology Network Newsletter* 4:8–10.

Siddiqui, M.A., and M.M. Alam. 1987. Utilization of marigold plant wastes for the control of plant parasitic nematodes. *Biological Wastes* 21:221–229.

Singh, I., S.K. Sharma, and P.K. Sakhuja. 1980. Effect of oil cakes extract on the hatching of root-knot nematode, *Meloidogyne incognita. Indian Journal of Mycology and Plant Pathology* 9:115–116.

Singh, R.S., and K. Sitaramaiah. 1973. Effect of organic amendments, green manuring and inorganic fertilizers on root knot of vegetable crop. Research Bulletin, Experiment Station and College of Agriculture, GBPUAT, Pant Nagar 6:289.

Sitaramaiah, K. 1979. Effect of temperature on decomposition of margosa cake and on survival of *Tylenchorhynchus elegans. Indian Journal of Nematology* 7:167–169.

Sumbul, A., R. Rizvi, I. Mahmood, and R.A. Ansari. 2015. Oil-cake amendments: useful tools for the management of phytonematodes. *Asian Journal of Plant Pathology* 9(3):91–111.

Tagi, N., B. Dash, and A.K. Ujjwal. 2022. Effect of oilseed cake on root-knot nematode (*Meloidogyne incognita*) infecting brinjal. *Indian Journal of Nematology* 52(1):21–27.

Terry, L. 2011. *Health-Promoting Properties of Fruits and Vegetables.* CABI, Nosworthy Wa, Wallingfor, Oxfordshire.

Tripathi, P.K., and C.S. Singh. 2006. Effect of some compatible biocontrol agents along with mustard cake and furadon on *Meloidogyne incognita* infecting tomato plant. *Indian Journal of Nematology* 36:293–295.

Trivedi, P.C., A. Bhatnagar, and B. Tiagi. 1978. Control of root-knot nematodes on *Capsicum annum* by application of oilcake. *Indian Phytopathology* 31:75–76.

Wani, A.H. 2006. Management of root-knot nematode, *Meloidogyne incognita*, on okra and lentil by soil amendment with oil cakes and leaves of different plants. *Nematologia Mediterranea* 34:83–87.

Youssef, M.M.A., and W.M.A. El-Nagdi. 2010. Effect of certain organic materials in controlling *Meloidogyne incognita* root-knot nematode infesting banana. *Archives of Phytopathology and Plant Protection* 43:660–665.

Zareena, S.K., and V.V. Das. 2014. Root-knot disease and its management in brinjal. *Global Journal of Bioscience and Biotechnology* 3(1):126–127.

12 Oilseed Cakes and Biocontrol Agents

Ticking on the Zero Hours of Root-Knot Nematode Infections

Tamanna Bhardwaj, Indu Sharma, Dhriti Kapoor, Arun Dev Singh, Shalini Dhiman, Priyanka Sharma, Kamini Devi, Neerja Sharma, Tunisha Verma, Amrit Pal Singh, Kanika Khanna, Puja Ohri and Renu Bhardwaj

CONTENTS

12.1 INTRODUCTION

Root-knot nematodes (RKNs) are highly pervasive pests. They severely harm economically important crops, including fruits, ornamentals and vegetables. *Meloidogyne arenaria*, *M. incognita*, *M. hapla* and *M. javanica* are the most prevalent phytophagous RKNs. The genus *Meloidogyne* poses a threat to many plant species (over 2000). Annually, they result in over USD 80 billion in losses, reducing total production by 12% and 25%, respectively (Mitiku et al. 2018). To control diseases caused by RKNs to improve plant health, chemicals are used in enormous amounts. Nevertheless, the use of chemicals is restricted due to stringent EU legislation (Migunova et al. 2021). This encouraged researchers to explore green nematode control strategies that can replace chemical-based nematicides. Biocontrol agents (BCAs) are emerging as a crucial tools for integrated pest management (Hildalgo-Diaz and Kerry 2008). Microbial-based pesticides and growth regulators are potential candidates for developing eco-friendly nematode control measures (Singh et al. 2017). Both nematophagous fungi and bacteria serve as potential candidates as biological control agents against RKNs. Some fungal genera, *Paecilomyces*, *Trichoderma*, *Arthrobotrys* and *Pochonia*, work well as *Meloidogyne* sp. inhibitors (Al-Hazmi et al. 2019; Kassam et al. 2021). Bacterial species belonging to the genera *Pseudomonas*, *Azotobacter* and *Bacillus* are closely associated with plant

DOI: 10.1201/9781003319252-12

roots and display antagonistic activity against RKNs (Verma et al. 2019). They produce antimicrobial metabolites, which have broad-spectrum action. The chemical nature of these compounds varies with respect to different genera of bacteria. For instance, *Bacillus* spp. has been studied to generate antibiotic metabolites like surfactin, fengycin and iturin, whereas *Pseudomonas* produces compounds like pyrrolnitrin, phenazine and 2, 4-diacetylphloroglucinol (DAPG) (Kenawy et al. 2019; Dimkić et al. 2022). Among the BCAs, Cobb et al. (1917) were the first to propose the use of mononchids to curb the population of plant parasitic nematodes due to their predatory nature upon them. Several scientists (Cassidy 1931; Banage 1963; Esser 1987) backed using predatory nematodes as BCAs for managing RKNs. Predatory nematodes mainly belong to the Aphelenchida, Mononchida, Diplogasterida and Dorylaimida orders. They are classified based on their feeding apparatus and food preference (Bilgrami 1997). The application of organic amendments (OAs) to soil aids in bettering soil properties and soil microbiomes. OAs alter soil nematode fauna by increasing the beneficial nematodes predating on RKNs. OAs also serve as an efficient environment-friendly alternative to harmful soil fumigants and nematicides. Among various OAs, oil cakes from multiple plants have proved to be remarkably potent against different nematodes (Ashraf and Khan 2007). Due to their nematocidal properties, many nematologists have gained interest in oilseed cakes (OSCs). Many reports stated the role of oil cakes, including olive cake, mustard cake and neem cakes, in overturning the destructive effects of parasitic-plant nematodes (D'Errico and Di Maio 1980). This chapter focuses on the current threat of RKN infection on crops, BCAs and their mechanisms and the action of predatory nematodes and oilseed cakes in managing RKNs.

12.2 RKNS AND THEIR THREAT TO CROPS

RKN are obligate, sedentary rhizospheric parasites belonging to the *Meloidogyne* genus having destructive ability in the agricultural field. RKNs attack the vascular system of roots, causing gall formation and nutrient deficiency (Figure 12.1). The damaged water channel in the host plant leads

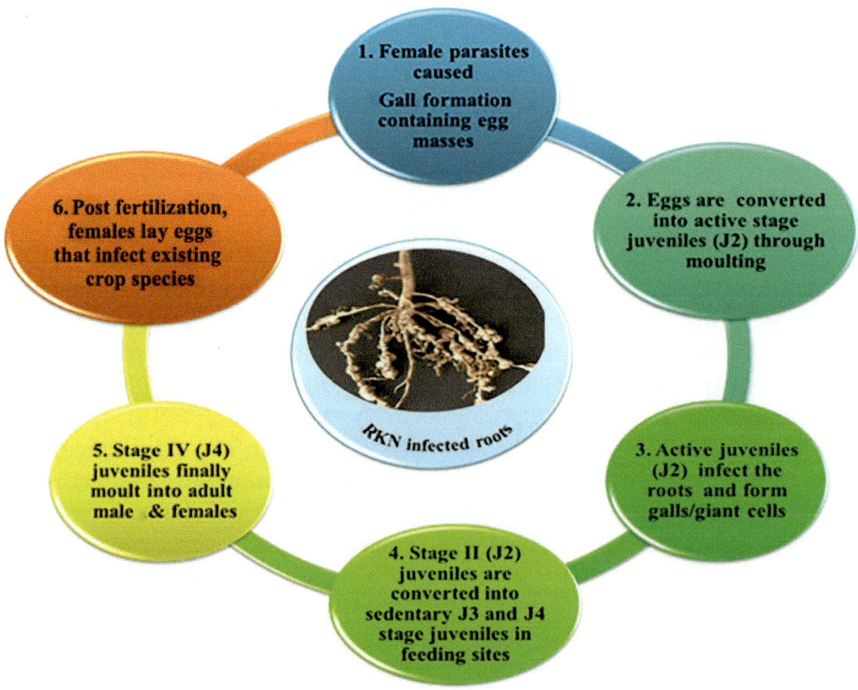

FIGURE 12.1 Root-knot nematode infection and their threat to crop species.

to reduced crop productivity. They are primarily dispersed in tropical and subtropical regions. For instance, *Meloidogyne incognita* Kofoid and White, *M. arenaria* Neal, *M. javanica* Treub and *M. hapla* Chitwood were reported as the main species destroying crops and vegetables, resulting in large economic losses (Sikora and Fernandez 2005). The first three species are found in tropical and subtropical regions, while *M. hapla* infects vegetable species grown in temperate zones and thus affects the quality and quantity of vegetables (Seid et al. 2015). *Meloidogyne* spp. adversely affects over 2000 agricultural crops like tomatoes, eggplants, sugarcane, melons, tea, coffee, carrots and others (Postnikova et al. 2015).

These nematodes have stylets for sucking nutrients from roots. Female parasites release some secretary proteins that induce the formation of multinucleated cells. This process results in the release of proteins that, on consumption by nematodes through feeding tubes, lead to the formation of root galls. The symptoms generated after the induction of RKN infection are chlorosis, root galling (1–10mm in diameter), permanent wilting, leaf shedding and abnormal root development, resulting in massive crop destruction (Arshad et al. 2021; Philbrick et al. 2020). Because of its great reproductive potential and broad host specificity, *Meloidogyne* sp. infection is tough to control (Philbrick et al. 2020). The nematode infection decreases the chlorophyll content and damages the photosystems, causing wilting of the plant body and therefore lowering production (El-Sagheer 2019; Guerrero et al. 2021).

12.3 OSCS FOR RKN MANAGEMENT

Efficient pest management programmes are required to combat the spread of diseases to achieve high yields and biomass. Many studies have explored the idea of using integrated pest management programmes to alleviate the negative impact caused by pests (Allen and Bath 1980). Nematode management has always been a huge problem for researchers as they make diagnosing the disease difficult due to their widespread distribution, extensive host range and association with other pests and pathogens (Akhtar et al. 1997). Current management strategies in developed countries rely mainly on nematicides, the development of resistant cultivars and cultural practices. However, due to the detrimental effects of chemical nematicides on human and environmental health, many nematicide products have been withdrawn from the market in search of more effective control strategies (Duncan et al. 1991). Farmers are under immense pressure to fulfil the increasing demand for food, ensuring long-term sustainability with a diminishing use of pesticides (Renčo et al. 2013).

Considerable progress has been made in using various agro-industrial wastes, including manure, OSCs and compost, as soil supplements for controlling phytonematodes. Among different forms of organic amendments, the OSCs of numerous plants have been demonstrated to be competent against nematodes (Goswami et al. 2007; Ashraf and Khan 2007). OSCs are remnants of the oil industry produced by extracting vegetable oils from seeds. These de-oiled cakes are classified into two types: edible plants such as mustard (*Brassica juncea*) and groundnut (*Arachis hypogaea*) and non-edible such as neem (*Azadirachta indica*) and castor (*Ricinus communis*; (Tiyagi et al. 2011; Rizvi et al. 2012; Sharma et al. 2013). These OSCs are used directly as organic additions as powder or pellets to improve soil fertility. This concept has sparked considerable interest among nematologists to deliver efficient nematode control. Soil amendment with oil cakes has achieved much success due to the nematocidal properties of oil cake extract. Water-soluble fractions of oil cakes extracted from castor, mustard, neem and duan when applied to pigeon peas proved to be beneficial in suppressing nematode proliferation, consequently increasing the bulk density in the stem (Anver 2003). It was reported by Goswami et al. (1993) that decomposed products of groundnut, mahua, mustard, karanj and neem OSCs increase the mortality of *M. incognita*. Similar results were observed in the case of cowpea (*Vigna unguiculata*), which showed vigorous plant growth and successfully controlled the population of *Rotylenchulus reniformis* when treated with cotton and sunflower OSCs (Yassin and Ismail 1994). Among different plant parasitic nematodes, *Meloidogyne* species is the most common destructive plant-parasitic nematode (PPN) that is suppressed by modifying the soil with composted oil cakes (D'Errico and Di Maio 1980), including mustard cake (Hameed 1968; Nadeem et al. 2021a), olive cake, jatropha and neem cakes (Albalkhi et al. 2004). Across Asian countries, neem and neem products (seed powders, seed extracts, oil, sawdust) are used and have excellent nematocidal properties. Several other plant metabolites, such as terpenoids, also have nematocidal properties (Akhtar and Mahmood 1994). A study on French beans (*Phaseolus vulgaris*) discovered that neem oil cake effectively suppressed the population of *Rotylenchulus reniformis*. Also Latif et al. (1999) found similar results for protective action of neem cake–amended soil on cowpeas against RKNs.

Neem and groundnut cakes have been found to be more active in unsterilised soil than in sterilised soil, resulting in improved plant growth response and a lower nematode population. Numerous reports exist on using oil cakes in vegetables for nematode control. Sitaramaiah (1979) stated that oilseed cakes of margosa, peanut, castor, mustard, linseed, mahua and coconut effectively reduced *Meloidogyne javanica* infection on vegetable crops to varying degrees. Research on oil cake amendments has followed the same investigation pathways in the last few decades. Still, a more significant concern is required for understanding the mechanisms related to nematicidal action and the effect of integrating biocontrol agents and oilseed cake.

12.3.1 Using OSCs as OAs for RKN Management

OAs follow the suppression mechanisms against PPNs by applying compounds like crop residues, animal manures, plant by-products and peat-like products. These amendments may be biotic and abiotic or a combination of both factors and contribute to enhancing the soil organic matter content or can also create micro-habitats that are less conducive to harmful plant nematodes, including RKNs. Also, OAs release volatile organic compounds (VOCs) that are toxic to nematode growth. Indirect mechanism leads to a surge in the number of antagonistic organisms within the soil ecosystem, which either compete for nutrient uptake or increase the parasitic populations, limiting the growth and reproduction of pathogenic microorganisms. OA decomposition also leads to the generation of multiple VOCs, which either act as nemato-static or nemato-toxic (Rosskopf et al. 2020). OAs can be developed as commercial nematicides like methyl isothiocyanates (MITCs). MITCs are produced from the Brassicaceae family members and derived from species like *Brassica napus*, *Brassica juncea* and other plants, such as *Sinapis alba*. These commercially developed compounds act against the parasitic nematodes when applied to *in vitro* and *in vivo* soil conditions (Lazzeri et al. 1993; Zasada et al. 2009). It was stated by Rich and Rahi (1995) that nematotoxic compounds derived from *Crotalaria juncea* plant have allelopathic compounds like pyrrolizidine, alkaloids and monocrotaline. These compounds acted as antagonists and reduced nematode populations.

Many of the OAs are able to enhance the soil microflora. This soil microflora acts against nematode populations. It was reported by Mian et al. (1982) that soil microflora improved when soils were applied with 1% chitin. These soil microflorae also mediated the secretions of chitinolytic enzymes. Also, there are reports showing that the OAs are composed of certain chemical compounds capable of the induction of resistance responses in plants. Qtu et al. (1997) showed that early chitinase isozyme activity is observed in nematode-resistant soybeans. A list of different OAs, their organic sources and their biochemical compositions is provided in a tabulated form (Table 12.1).

TABLE 12.1
Multiple Organic Amendments (OAs) and Their Associated Organic Sources

S.No.	OAs	Biochemical Composition	Applicable on Plant Pathogen/Nematode	References
1.	Brassica green manure (*B. juncea*, *B. napus* etc.)	Labile C, nutrients, and glucosinolates	*Phytophthora erythroseptica*, *Rhizoctonia solani* and *Sclerotinia sclerotiorum*	Larkin and Griffin 2007
2.	Sudan grass, pea, broccoli	Labile C, cyanogenic glycosides and nutrients	*Verticillium dahlia*	Ochiai et al. 2007; Yadav et al. 2018
3.	Liquid swine manure	Potentially high content of NH_4, N and volatile fatty acids (VFAs)	*Verticillium* spp.	Conn et al. 2007
4.	Dairy manure solids	VFAs	Root rots of snap bean and sweet corn, cucumber damping-off	Lee et al. 2014
5.	Liquid hog manure	VFAs	*Pratylenchus* spp.	Bao et al. 2013
6.	Pelletised poultry manure	N rich	*Fusarium* spp.	Borrego-Benjumea et al. 2014
7.	Chicken manure	N rich (2–5%)	*Verticillium* spp.	Reddy 2017
8.	Bone meal, soy meal, meat and blood	N content (7.3–7.7%)	*Verticillium* spp.	Lazarovits 2010; Chaves et al. 2014
9.	Broccoli residue	Labile C	*V. dahliae*	Strauss and Kluepfel 2015
10.	Brassica seed meal	Nutrients and glucosinolates	*Pratylenchuspenetrans*, *Rhizoctonia* spp. and *Meloidogyne incognita*	Zasada et al. 2009; Mazzola et al. 2007
11.	Brassica pellets and flour	Nutrients and glucosinolates	*Fusariumoxysporum*	Gilardi et al. 2016
12.	Animals, plants and their by-product-based compost	Organic matter and useful microorganisms	*R. solani*, *P. ultimum*, *Phytophthora* spp., *V. dahlia* and *F. oxysporum*	Bonanomi et al. 2007; Pugliese et al. 2015
13.	Only plant waste–based compost	Organic matter and beneficial microorganisms	*F. oxysporum*	Yogev et al. 2006
14.	Suppressive peat	Rich in manure and beneficial microorganisms	*Pythium sylvaticum*	Rosskopf et al. 2020

12.4 RISE OF BCAS AGAINST RKNS

Numerous treatments and approaches have been developed worldwide to control, restrain and alleviate RKNs. Many researchers developed safe, eco-friendly, and effective pesticides for diminishing the effect of RKNs. BCAs were used to suppress the proliferation of pathogens and other pests, leading to eliminating plant disease (Table 12.2). BCAs might be achieved in several ways, like

TABLE 12.2
Fungal/Bacterial Strains Used as Biological Control Agents against Root-Knot Nematodes

S. No.	Bioagents	Nematode Species	Reference
1.	*Pseudomonas* sp. (*P. putida, P. stutzeri, P. aeruginosa, P. fluorescens*)	*M. incognita*	Avelar 2021
2.	*Arbuscularmycorrhizal* fungi (AMF)	*M. incognita*	Herrera-Parra et al. 2021
3.	*Serratiamarcescens*	*M. incognita* and *M. javanica*	El-Ashry et al. 2020; Zhao et al. 2021
4.	*Bacillus* sp. (*B. coagulans, B. cereus, B. licheniformis, B. megaterium, B. firmus*)	*M. arenaria, M. incognita, M. graminicola, M. javanica,*	Subedi et al. 2020; Ahmad et al. 2021
5.	*Pasteuria penetrans*	*Meloidogyne* sp.	Subedi et al. 2020
6.	*Bacillus cereus*	*M. javanica, M. incognita*	Jiang et al. 2020
7.	*Corynebacterium paurometabolous*	*M. incognita*	Subedi et al. 2020
8.	*Bacillus subtilis*	*M. javanica*	Mazzuchelli et al. 2020
9.	*Bacillus pumilus*	*M.arenaria*	Forghani and Abolfazl 2020
10.	*Rhizobium etli*	*M. javanica*	Xia et al. 2019
11.	*Bacillus licheniformis*	*M. incognita*	El-Nagdi et al. 2019
12.	*Bacillus megaterium*	*M. graminicola, M. incognita*	Mostafa et al. 2018
13.	*Aspergillus niger*	*M. arenaria, M. incognita, M. javanica*	Al-Ani et al. 2018
14.	*Bacillus pumilus*	*M. incognita, Heterodera glycine*	Lastochkina et al. 2017
15.	*Trichodermaviride*	*M. graminicola*	Narasimhamurthy et al. 2017
16.	*Pseudomonas aeuroginosa*	*M. javanica*	Fatima and Anjum 2017
17.	*Verticillium chlamydosporium*	*M. incognita*	Giné et al. 2016
18.	*Paecilomyces lilacinus*	*M. arenaria M. incognita M. javanica*	Najafi et al. 2016
19.	*Pochonia chlamydosporia*	*M. incognita*	Parihar et al. 2015

absorbing similar nutrients from the same rhizosphere of pathogens, and are used to control the pests partially, significantly or entirely. It has been studied that many genera of fungi like *Paecilomyces*, *Verticillium*, *Trichoderma* and *Pochonia* were used against RKNs (Kiriga et al. 2018). *Pochonia chlamydosporia* was one of the fungi species isolated and tested as a BCA against RKNs (Peiris et al. 2020; Tazi et al. 2021). *Pochonia chlamydosporia* parasitised nematode eggs by preventing them from hatching within 24 hours. It has been reported that *Purpureocillium lilacinum* parasitised the sedentary stage (eggs piled up in a gelatinous bed) of RKNs (Juyal et al. 2018; Raghavan et al. 2018; Kiriga et al. 2018). Due to this potential, *Purpureocillium lilacinum*was utilised on tomato plants against *M. incognita* and *M. javanica* (Dahlin et al. 2019).

Pseudomonas fluorescens, *Bacillus subtilis* and *Bacillus sphaericus* were used as competitor rhizobacteria that controlled the effects of RKN (Ahmad et al. 2021; Migunova et al. 2021). Other fungal genera, including *Aureobacterium rhizobium*, *Alcaligenes*, *Paeni bacillus*, *Agrobacterium*, *Telluria*, *Chryseo bacterium*, *Klebsiella*, *Phyllo bacillus*, *Corynebacterium*, *Xanthomonas* and *Enterobacter*, were used for suppressing phytopathogens. *Pasteuria penetrans* (endospores), *Bacillus firmus* (commercial rhizobacteria), *Purpureocillium* and *Trichoderma* genera were used as the most potent BCAs against RKN (Engelbrecht 2019; Abd-Elgawad et al. 2020; Baron et al. 2020) action. BCAs were microbial contenders that diminished phytopathogens and other plant diseases in various ways. Over the preceding years, BCAs were used as a substitute for chemical pesticides and fertilisers to manage RKN (Table 12.2). There was an enhancement in the characteristics of BCA via modification of their genes that allowed them to persist in a limited environment (Odelade et al. 2019; Poveda et al. 2020). Once they are established in the rhizospheric zone, they compete and multiply with growing roots. The management of phytopathogens has been done by using isolated fungi or bacterial species from the endosphere, phyllosphere or rhizosphere. The microbes like viruses, mites, collembolans, turbellarians, protozoans, predatory nematodes, tardigrades and enchytraeids have been employed as BCAs to control RKNs. These microbial genera were utilised as an antagonist to hinder the invasion and establishment by producing extracellular enzymes for detoxification of the RKN in a plant (Guerrero et al. 2021; Zehra et al. 2021).

Microbial biological control agents (MBCAs) like bacteriophages and mycoviruses have also been utilised to manage plant diseases. It has been reported that 101 MBCAs, which were helpful in controlling the RKN infection. (Ji et al. 2020). They directly interacted with other phytopathogens via hyper-parasitism or antibiosis mechanisms. The MBCAs directly produce antimicrobial complexes (potent disease control compounds) that inhibit the growth of pathogens and reduce their number (Palyzová et al. 2019; Los et al. 2020; Dann et al. 2021). One or more pathways to inhibit the pathogen's growth cycle, including the production of pathogen-suppressing compounds (active compounds, enzymes, or antibiotics), that is, antibiosis, and the struggle for space and food is referred to as competitive shift (Singh et al. 2020). This biological management system appeared to be cost-effective and more manageable to farmers contributing durability and viability of crop plants.

12.4.1 Fungi as BCAs

The fungus has a unique ability to fight against PPNs, particularly RKNs. When nematodes have fungal infections, their egg membranes and cuticular skins play a significant role. Nematode-invading fungi feed on nematode cuticular skin proteins like chitin, fibre and collagen, which are the building blocks of the cuticle (Dackman et al. 2021). The first step of predation involves penetration. This penetration happens before the fungus moves inside the host's/prey's body through digestion, allowing it to meet its nutritional needs (Hajieghrari et al. 2008). The manner of inhibition of PPNs by various fungal strains, that is, nematophagous fungal strains is elaborated in Table 12.3.

12.4.1.1 Mechanism of Antagonistic Fungi

Fungal biological control agents kill various arthropods or pests *via* two primary mechanisms: (1) mycoparasitism and (2) entomo-pathogenicity. Mycoparasitism is the relationship between a fungal parasite and its host. In contrast to mycoparasitism, entomo-pathogenicity is used by entomopathogenic fungi that are pathogen-killing parasitic fungi (Boakye et al. 2022). Many well-known fungus genera are also known as well-characterised killers of RKNs, *Meloidogyne* spp. These are *Trichoderma*, *Fusarium*, *Pochonia*, *Purpureocillium*, *Arthrobotrys* and *Paecilomyces* (Kassam et al. 2021). *Trichoderma* spp. is reported to infect root surfaces of various plants, and this has been connected to its capacity to compete with nematodes (Mukhtar et al. 2021) and mitigate root-knot disease (mainly

TABLE 12.3
Illustrations of Various Fungal Strains That Act as a Biocontrol Agent against Root-Knot Nematodes

S.No.	Nematode Species	Host	Fungi	Mode of Action	Reference
1	*Meloidogyne incognita*	*Capsicum annuum*	*Rhizoglomus fasciculatum* (*AMF*) and *Paecilomyces lilacinus* (*NF*)	Enhanced yield (fruit output) with controlled infection of *M. incognita*	Giri et al. 2022
2	*Me. javanica*	Eggplant	*Glomus mosseae*, *Gigaspora gigantea*	Increased plant resilience to infection of RKN	Sharma et al. 2021
3	*Meloidogyne* spp.	*Coffea arabica*	*AMF* consortia	Better plant growth and protection against RKN	Vallejos-Torres et al. 2021
4	*M. incognita*	*Lycopersicum esculentum*	*Purpureocillium lilacinum*	Excellent efficacy to reduce *Meloidogyne incognita*	Xu et al. 2021
5	*Meloidogyne incognita*	*Lycopersicum esculentum*	*Arthrobotrys oligospora*	Curbed growth of *M. incognita* by trapping mechanism and secreting poisonous compounds	Soliman et al. 2021
6	*Meloidogyne javanica*	Green gram	*Trichoderma spp.*	Causes reduction in eggs and galls production and improved plant-growth attributes	Mukhtar et al. 2021
7	*Meloidogyne incognita*	Tomato	*Pochonia chlamydosporia*	Reduces infection by attacking RKN *Meloidogyne incognita* eggs and females	Ghahremani et al. 2019
8	*Pratylenchus* spp., *Heterodera* spp., *Tylenchorhynchus* spp., *Paratylenchus* spp., *Trichodorus* spp., *Helicotylenchus* spp., *Rotylenchus* spp.	Wheat plant	*Clonostachys rosea* strain IK72	Decreased parasitic nematodes and increases wheat shoot length and weight	Iqbal et al. 2018
9	*Meloidogyne javanica*	Pineapple	*Trichoderma* spp. and *Purpureocillium lilacinum*	Influenced nematode multiplication and host response, directly and indirectly, indicating their potential for control of *M. javanica*on pineapples	Kiriga et al. 2018
10	*Meloidogyne enterolobii*	Tomato and banana	*Pochonia chlamydosporia* and *Purpureocillium lilacinum*	Attacks *Meloidogyneenterolobii* eggs and females	Silva et al. 2017

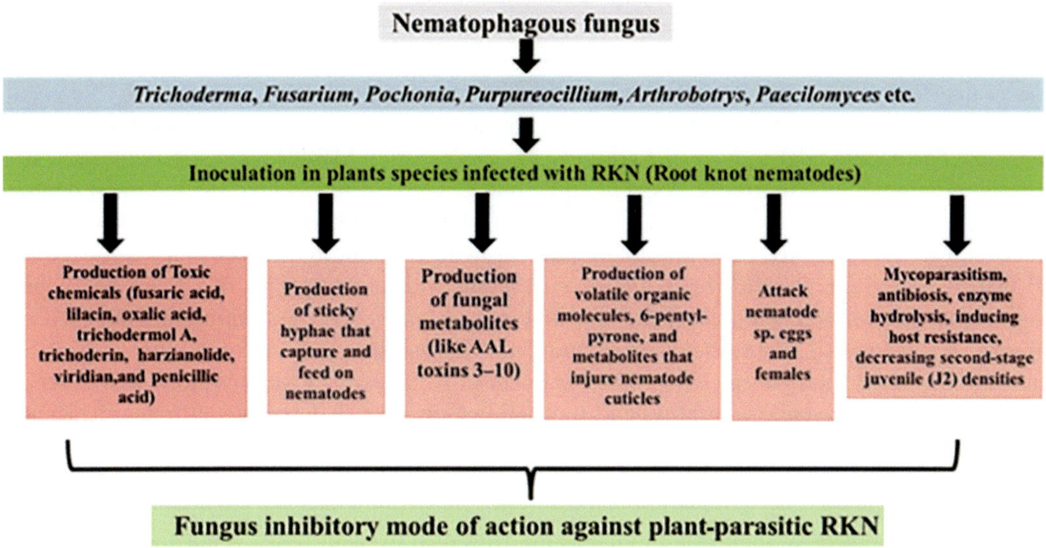

FIGURE 12.2 Action of biological control agents against RKN.

through the production of toxic chemicals, mycoparasitism, antibiosis, enzyme hydrolysis, inducing host resistance, decreasing J2 densities and egg production (also shown in Figure 12.2; Kiriga et al. 2018; Deb et al. 2020). *Trichoderma* spp. are also reported to generate volatile organic molecules, 6-pentyl-pyrone, and metabolites that injure nematode cuticles (Basit et al. 2021). In addition to direct antagonism, *Trichoderma* sp. can fight *Meloidogyne* by making fungal metabolites (like AAL toxins 3–10), thus making the plant more resistant (Al-Ani et al. 2018; Abdelrahman et al. 2019). Thus, *Trichoderma* spp. promotes plant growth by preventing nematode penetration. Other fungal members such as *Pochonia chlamydospores* and *Purpureocillium lilacinum* have been shown to attack *Meloidogyne* sp. eggs and females (Silva et al. 2017), while nematophagous fungus *Arthrobotrys* spp. generates sticky hyphae that capture and feed on nematodes (Cao et al. 2018; Soliman et al. 2021). Many fungi produce chemical compounds that play a significant role in plant-pathogen biocontrol actions (De et al. 2019).

Fungi have been reported to have a high reproduction rate, high specificity to their target and a short generation time; these factors improve the potential for fungi to be used as BCAs against RKNs (Boakye et al. 2022). Also, most fungal species have defensive mechanisms that allow them to protect plants successfully against illnesses caused by plant pathogenic fungi (Figure 12.2). Regardless of how fungi seize or parasitise their hosts, the relationship between host–pathogen and fungus or prey requires penetration of the body's outermost layer. Following this invasion, the pathogen colonises the body's internal tissues through digestion, allowing it to meet its nutritional requirements (Branine et al. 2019; Tawidian et al. 2019). They are saprophytes, yet these fungi can trap nematodes in the soil in larval or free adult stages. Hydrolytic enzymes found in the fungal cuticle, like lipases and chitinase, might be able to kill insects and pests (Litwin et al. 2020). Hydrolysis of the cuticle by extracellular proteases has been demonstrated to create traps (Liang et al. 2019; Shalaby et al. 2020). These enzymes also puncture the vitelline layer of the nematode eggshell, enabling mycelium to enter, multiply, and, thus, impair the early stages of embryonic development. When there is insufficient food in the nematode egg, many fungal hyphae penetrate the eggshell and then spread to other eggs in the egg mass (Islam et al. 2021). Thus, fungi help in preventing nematode infection in plants.

12.4.2 Bacteria as BCAs

RKNs are one of the most harmful PPNs, which cause significant damage to host plants. Using bacteria as an effective BCA has been the long-term practice for managing RKN. A study by Sun et al. (2021) highlighted the active substances found in bacteria were useful for managing *Meloidogyne incognita*. The nematicidal activity of *Pseudomonas simiae* strain MB751 against *M. incognita* was recorded at around 80% mortality. A cyclic dipeptide Cyclo (L-Pro-L-Leu) was the main nematicidal component in *P. simiae* strain MB751, which effectively killed the RKN. This BCA was found effective in suppressing RKN and promoting plant growth. The culture filtrates of a gram-negative, rod-shaped bacterium, *Pseudomonas fluorescens*, along with eight species of nematode-trapping fungi, namely *Arthrobotrysconoides*, *A. cookedichison*, *A. eudermata*, *A. microscaphoides*, *A. oligospora*, *A. thaumasia*, *Clonostachysrosea* and *Trichodermaharzianum* were employed as BCAs against RKNs (Lafta and Kasim 2019). The treatment of *P. fluorescens* filtrates alone was observed to cause maximum juvenile mortality, and in combination with *T. harzianum*filtrates, it showed minimum hatching of eggs.

Besides this, soil bacteria secrete lipopeptide molecules, which are known to serve as natural antibiotics. A lot of research work is being done in this direction where the potential of lipopeptide molecules is examined for their role in increasing resistance to plant pests. Nadeem et al. (2021b) studied isolated lipopeptide molecules and tested them at several concentrations against J2. They found significant J2 mortality of *Meloidogyne incognita*. Both the bacterial strains, *Bacillus subtilis* (MTCC-441) and *Pseudomonas putida* (MTCC-102), were used in J2-inoculated tomato plants and found to have significant growth improvement. Further application of a gram-negative bacterium, *Pseudomonas putida*, was observed to control the root-knot disease in rice cv PS-5 significantly. The antagonistic potential of *P. putida* against *M. graminicola* was observed to be more than both *Purpureocillium lilacinum* and *P. fluorescens*. The BCA resulted in an increase of tillering as well as the yield of rice. A low-cost powder formulation of bacterial BCA was developed to control RKNs of chickpeas (*Cicer arietinum* L.) and pigeon peas (*Cajanuscajan* L.), and it also lowered the RKN population in soils (Khan et al. 2011). The commercial BCA formulation is prepared using sawdust fly-ash soil and a mixture of molasses. *Trichoderma harzianum*, *Pochonia chlamydosporia* and *Pseudomonas fluorescens* were used at 5 g/kg seeds to control fusarium wilt and *Meloidogyne* sp. (RKN) in pulses, namely chickpea and pigeon pea.

The RKNs were controlled by cucumber seed treatment with diverse valuable microbes in cucumber grown under sheltered cultivation practices inside the polyhouse (Mohammed and Khan 2021). The cucumber seeds treated with *Bacillus subtilis* RP24 and *Pseudomonas fluorescens* ITCC-B0034 along with *Trichoderma harzianum*, *T. longibrachiatum*, *T. viride* and carbofuran (nematicide) have been reported to reduce the RKN egg mass production. These treatments also increased the growth of the cucumber plants. Bacteria as BCAs were employed to manage soilborne diseases in pulses (Kunda et al. 2020).

Abd-Elgawad et al. (2018), in their in vitro study, found that the biocides BM-Root Pan and Velum Prime, a chemical pesticide, were helpful in controlling RKN. The biocide Bm-Root Pan was observed to be most effective against RKN at 1×10^6 CFU. Another study recommended that fungal and bacterial BCA can be used as nematicides in integrated nematode management strategies (Abd-Elgawad et al. 2018). The synergistic and additive effect of BCAs help in integrated pest management programs, and harnessing such BCAs will uplift sustainable agriculture and help with understanding their biology, mode of action and ecology interactions (Figure 12.3). Furthermore, exhaustive information on active dose, active nematicide ingredient, chemical structure, name, origin and producer, type and constitution of formulation, targeted pests, crop and more will provide knowledge about employing the bacterial BCAs in the management of RKNs.

12.4.2.1 Antagonistic Mechanism of Bacteria

A diverse range of bacterial species has been defined to trigger plant defence responses to various biotic stresses. However, the underlying mechanism of their action at molecular levels remains

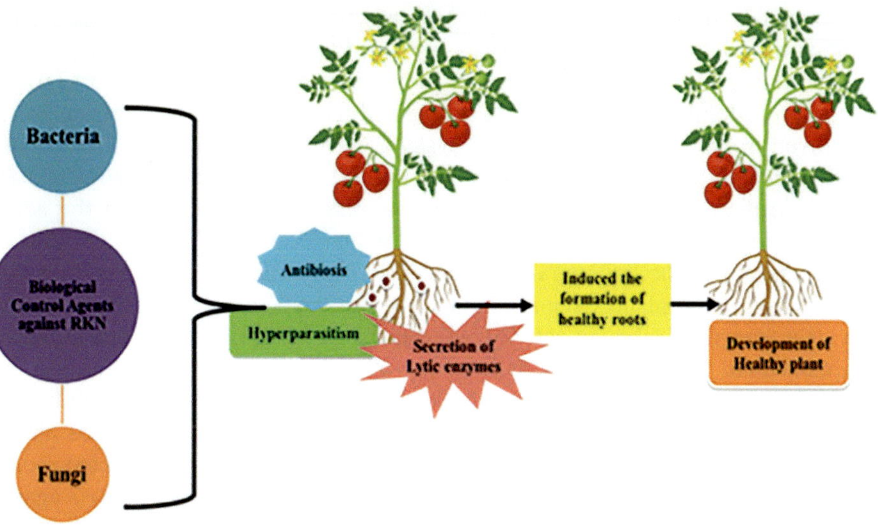

FIGURE 12.3 Root-knot nematode illustration of various nematophagous fungus strains with an inhibitory mode of action against nematode.

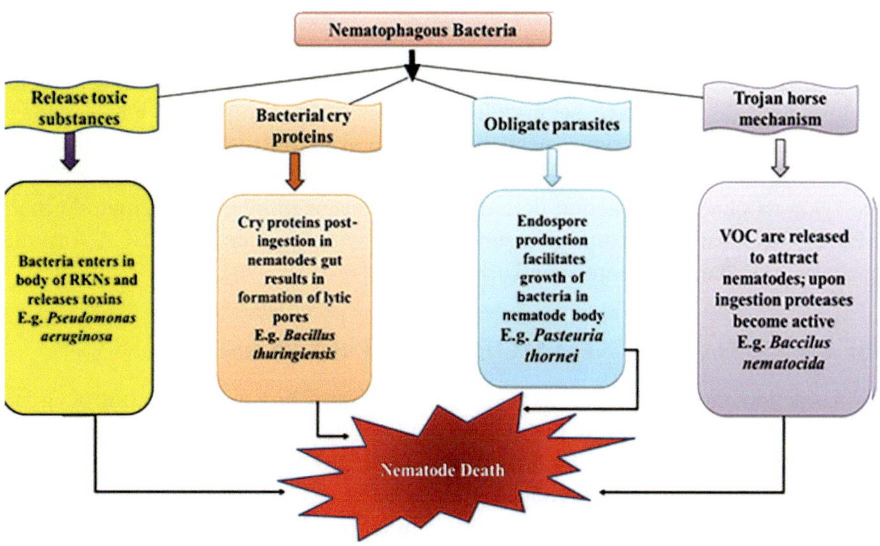

FIGURE 12.4 Diagram representation of mechanism involved in control of root-knot nematodes using bacteria.

uncertain. Molinari and Paola (2019) emphasized that BCA manages RKN by activating plant immune responses in tomato plants. The molecular studies revealed the regulation of expression of six key salicylic acid (SA)–dependent pathogenesis-related defence gene (*PR*-gene), namely *PR-1b*, *PR-5*, *PR-1*, *PR-3*, *PR-5* and *ACO* in BCA pretreated tomato plants. The expression of the jasmonic acid/ethylene (JA/ET)–dependent gene *JERF3* (*Jasmonate Ethylene Response Factor 3*) was not upregulated, while expression of catalase gene (*CAT*) was overexpressed after nematode inoculation. In addition, the activities of glucanase and endochitinase enzymes were reported in the roots of pretreated inoculated plants. Therefore, BCA-mediated immunity in plants is dependent on SA, the stimulation of chitinase and glucanase enzyme activities and the deactivation of the enzymatic antioxidant defence system. The BCA-mediated immunity also relies on penetration and movements of invading RKN and the feeding site in the roots of the plants (Figure 12.4).

Another study by Molinari and Paola (2019) revealed the underlying mechanism of molecular signalling involved in BCA-mediated immune system activation against RKNs in tomato plants. During the preliminary stages of BCA-root interactions, *PR* genes like *PR1*, *PR3* and *PR5* were overexpressed, and the expression of these genes was directly associated with the activation of the enzyme 1-aminocyclopropane-1-carboxylic acid (ACC) oxidase (ACO). The enzyme ACO was involved in the catalysis of the last step of ethylene (ET) biosynthesis, and its level increased after nematode inoculation in the plants. Also, in plants, the enzymatic antioxidant defence system was repressed, evident from the decline in activity and catalase gene (*CAT*) expression in infected plant roots. Systemic-acquired resistance (SAR) is responsible for activating BCA-induced responses instead of induced-systemic resistance (ISR) at the initial stages of root–nematode interaction. Besides this, BCA-induced plant immunity is also associated with the improved ability of the infected plants to contrast the invasive juvenile's feeding site building in the infected roots. Various factors responsible for determining the variability of performance of different BCAs against RKNs in vegetables have been reported. The study revealed that high-performance variability is attributed to the stage of growth and development of treated plants and the method, as well as a dose of formulate treatments provided per plant. For an effective dose, the amount of formulates needs to be supplied at a specific range.

Recently, Boakye et al. (2022) highlighted the mechanism of antagonistic bioagents as an alternative to synthetic pesticides in managing *Meloidogyne* spp. The BCA are eco-friendly, less hazardous and important ecological enemies of RKN infestations. These bioagents effectively suppress the nematode population by facilitating the colonisation of antagonistic microbes and their activation in the rhizosphere area. The insights of biological antagonism as a safe and sustainable way to succeed against plant diseases have been well-documented. BCA-mediated RKN management is dependent on the activation of various mechanisms such as antibiosis, predation, cross-protection, induced resistance, competition and hyperparasitism, among others. The extensive knowledge of BCA's underlying mechanism of action determines its success in intensive nematode management strategies. Further exhaustive research is essential for developing innovative BCA formulations by identifying novel and potential microorganisms.

12.5 USE OF PREDATORY NEMATODES AGAINST RKNS

In the previous section, the discussion was focused on the promising use of biocontrol agents against nematode loss. In this section, we focus on predatory nematodes; they act as a vigorous agent to kill PPNs owing to their predation potential (Cobb 1917; Bilgrami et al. 2008). Researchers have now extensively explored predatory nematodes' action, including RKNs, to manage PPN infection (Esser 1987). The predator executes the symbiotic function to control the population of parasitic nematodes in the rhizospheric zone. They are differentiated into diverse types, namely mononchids, dorylaimids, aphelenchids and diplogasteridae, based on the behaviour of the feeding mechanism and mode of action. The high involvement of diplogasterids in future nematode management is due to their short life span, chemotaxis intelligence nature and their ability to withstand antagonistic conditions and prey specificity (Bilgrami 1997).

One of the recent studies shows that *Protogamasellus mica* (a mite) lessened nematodes in the habitat as *P. mica* grows rapidly on account of plentiful nematodes available as food and plays a vital role in regulating the populations of both predators and free-living nematodes. It is observed that these predatory nematodes generate an ecological approach together with plant parasitic nematodes as they feed on soil microflora and normalise the prey populations, which enhance the plant's nutrients (Stirling et al. 2017). The mode of action of these diverse groups also depends on characteristic soil and habitat environmental conditions. Predators inhabit diverse environments like mononchids in cold atmospheres and diplogasterids in organic-rich soil, whereas some vizdorylaims, aphelenchids and others grow in moist-clogging conditions (Kanwar et al. 2021). Hence, from the available few reports, it is concluded that predators play an important function against parasitic nematodes by circulating nutrients among plants. But the molecular mechanism to prove them a powerful biocontrol agent for the reduction of parasitic nematodes in crops needs to be explored in a deeper way

12.6 CONCLUSION AND FUTURE PERSPECTIVES

Farmers across the globe are under pressure to fulfil the ever-growing demand for food and coincidently maintain harmony with environment sustainability. For the successful management of the production of organic foods with disease resistance, we managed to highlight key agronomic practices and the use of potent natural pesticides in order to combat the challenges in managing RKNs in this chapter. Also, *in vitro* studies on the interaction between nematodes and their impact on cultural practices will serve as valuable information in the science of nematology. Inter kingdom microbe–microbe interaction studies will add to the knowledge about indigenous keystone microbial taxa and practices such as crop rotation, soil amendments and tillage. Altogether, understanding microbiome structure and its relationship with plants will provide more insights on mechanisms responsible for nematode suppression and managing soil biome for biocontrol of RKNs.

REFERENCES

Abd-Elgawad, Mahfouz M. M. "Plant-parasitic nematodes and their biocontrol agents: current status and future vistas." In *Management of phytonematodes: recent advances and future challenges*, pp. 171–203. Springer, Singapore, 2020.

Abd-Elgawad, Mahfouz M. M., and Tarique Hassan Askary. "Fungal and bacterial nematicides in integrated nematode management strategies." *Egyptian Journal of Biological Pest Control* 28, no. 1 (2018): 1–24.

Abdelrahman, Mostafa, and Sudisha Jogaiah. "Defense mechanism and diverse actions of fungal biological control agents against plant biotic stresses." In *Plant Biotechnology: progress in Genomic Era*, pp. 461–478. Springer, Singapore, 2019.

Ahmad, Gufran, Amir Khan, Abrar A. Khan, Asgar Ali, and Heba I. Mohhamad. "Biological control: a novel strategy for the control of the plant parasitic nematodes." *Antonie van Leeuwenhoek* 114, no. 7 (2021): 885–912.

Akhtar, Mohammad. "Current options in integrated management of plant-parasitic nematodes." *Integrated Pest Management Reviews* 2, no. 4 (1997): 187–197.

Akhtar, Mohammad, and Irshad Mahmood. "Potentiality of phytochemicals in nematode control: a review." *Bioresource Technology* 48, no. 3 (1994): 189–201.

Al-Ani, Laith Khalil Tawfeeq. "*Trichoderma*: beneficial role in sustainable agriculture by plant disease management pp. 105–126." In *Plant microbiome: stress response*, Springer, Singapore, 2018.

Albalkhi, M. F. F., A. Altoune, M. Salman, M. Attiah, and G. Zeeni. "Influence of olive cake forms against root-knot nematodes on tomato in greenhouses in Syrian coast." *Egyptian Journal of Agricultural Research* 82, no. 2 (2004): 25.

Al-Hazmi, A. S., F. A. Al-Yahya, O. A. AbdelRafaa, and H. A. Lafi. "Effects of humic acid, *Trichoderma harzianum*, and *Paecilomyces lilacinus* on *Meloidogyne javanica*." *International Journal of Agriculture environment and Bioresearch* 4 (2019): 61–74.

Allen, G. E., and J. E. Bath. "The conceptual and institutional aspects of integrated pest management." *Bioscience* 30, no. 10 (1980): 658–664.

Anver, Suhail. "Effect of different organic amendments with *Paecilomyces lilacinus* for the management of soil nematodes." *Archives of Phytopathology and Plant Protection* 36, no. 2 (2003): 103–109.

Arshad, Usman, Husnain Butt, Muhammad Amjad Ali, Muhammad Jabran, Muhammad Salman Zahid, and Sohaib Sarfraz. "Exploring the nematicidal activity of plant extracts for management of *Meloidogyne incognita* in local cultivars of eggplant (*Solanum melongena* L.) in Pakistan." *Archives of Phytopathology and Plant Protection* 54, no. 19–20 (2021): 2333–2344.

Ashraf, Mohd Shaikhul, and Tabreiz Ahmad Khan. *Efficacy of Gliocladium virens and Talaromyces flavus with and without organic amendments against Meloidogyne javanica infecting eggplant. Asian Journal of Plant Pathology* 1 (2007): 18–21.

Avelar, I. C. *Actinobacteria for the control of phytopathogenic and/or mycotoxinogenic fungi of cereals* (Doctoral dissertation, Université Montpellier), 2021.

Banage, W. B. "The ecological importance of free-living soil nematodes with special reference to those of moorland soil." *The Journal of Animal Ecology* (1963): 133–140.

Bao, Yong, Senyu Chen, Jeffery Vetsch, and Gyles Randall. "Soybean yield and *Heterodera glycines* responses to liquid swine manure in nematode suppressive soil and conducive soil." *Journal of Nematology* 45, no. 1 (2013): 21.

Baron, Noemi Carla, Andressa de Souza Pollo, and Everlon Cid Rigobelo. "*Purpureocillium lilacinum* and *Metarhizium marquandii* as plant growth-promoting fungi." *PeerJ* 8 (2020): e9005.

Basit, Abdul, Syed Tanveer Shah, Izhar Ullah, Inayat Ullah, and Heba I. Mohamed. "Microbial bioactive compounds produced by endophytes (bacteria and fungi) and their uses in plant health." In *Plant Growth-Promoting Microbes for Sustainable Biotic and Abiotic Stress Management*, pp. 285–318. Springer, Cham, 2021.

Bilgrami, A. áL. "Evaluation of the predation abilities of the mite *Hypoaspis calcuttaensis*, predaceous on plant and soil nematodes." *Fundamental and Applied Nematology* 20, no. 1 (1997): 96–98.

Bilgrami, Anwar L., Christopher Brey, and Randy Gaugler. "First field release of a predatory nematode, *Mononchoides gaugleri* (Nematoda: Diplogastrida), to control plant-parasitic nematodes." *Nematology* 10, no. 1 (2008): 143–146.

Boakye, Thomas Afriyie, Dominic Kwadwo Anning, Huixia Li, Richard Osei, Solomon Boamah, and Samuel Anim. "Mechanism of Antagonistic Bioagents in Controlling Root-Knot Nematodes (*Meloidogyne* sp.): A Review." *Asian Research Journal of Agriculture* 15, no. 2 (2022): 27–44.

Bonanomi, Gulliano, V. Antignani, C. Pane, and F. Scala. "Suppression of soilborne fungal diseases with organic amendments." *Journal of Plant Pathology* (2007): 311–324.

Borrego-Benjumea, Ana, M. José Basallote-Ureba, Pervaiz A. Abbasi, G. Lazarovits, and José M. Melero-Vara. "Effects of incubation temperature on the organic amendment-mediated control of Fusarium wilt of tomato." *Annals of Applied Biology* 164, no. 3 (2014): 453–463.

Branine, Margaret, Anna Bazzicalupo, and Sara Branco. "Biology and applications of endophytic insect-pathogenic fungi." *PLoS Pathogens* 15, no. 7 (2019): e1007831.

Cao, Xin, Qiang Xu, Xue-Mei Wan, Xue-Cai Yang, Xiao-Ye Jia, Xu-Jia Xue, Jiang-Long Du, and Kui-Zheng Cai. "Isolation, identification, and in vitro predatory activity of nematophagous fungus *Arthrobotrys musiformis* and its interaction with nematodes using scanning electron microscopy." *Biocontrol Science and Technology* 28, no. 1 (2018): 94–104.

Cassidy, G. H. "Some mononchs of Hawaii." *Hawaii Planters' Record* 35 (1931): 305–339.

Chaves, C., Fernando Pomares, Remedios Albiach, and Rodolfo Canet. "Rates of nitrogen mineralization of meat and bone meals in Mediterranean soils." *Communications in Soil Science and Plant Analysis* 45, no. 17 (2014): 2258–2267.

Cobb, N. A. "The mononchs (Mononchus Bastian, 1866), A genus of free-living nematodes." *Soil Scientist* 3 (1917): 431–486.

Conn, Kenneth L., Edward Topp, and George Lazarovits. "Factors influencing the concentration of volatile fatty acids, ammonia, and other nutrients in stored liquid pig manure." *Journal of Environmental Quality* 36, no. 2 (2007): 440–447.

Dackman, Carin, Hans-BÖrje Jansson, and Birgit Nordbring-Hertz. "Nematophagous fungi and their activities in soil." In *Soil biochemistry*, pp. 95–130. CRC Press, Boca Raton, 2021.

Dahlin, Paul, Reinhard Eder, Erika Consoli, Jürgen Krauss, and Sebastian Kiewnick. "Integrated control of *Meloidogyne incognita* in tomatoes using fluopyram and *Purpureocillium lilacinum* strain 251." *Crop Protection* 124 (2019): 104874.

Dann, Elizabeth, and Adéle McLeod. "Phosphonic acid: a long-standing and versatile crop protectant." *Pest Management Science* 77, no. 5 (2021): 2197–2208.

Deb, Anwesha, and Kingsuk Das. "*Trichoderma*: an antagonist with multifaceted mode of action." *TTPP* (2020): 409.

D'Errico, F. P., and F. Di Maio. "Effect of some organic materials on root-knot nematodes on tomatoes in field preliminary experiments." *Nematologia Mediterranea* 8 (1980): 107–110.

De Silva, Nimali I., Siraprapa Brooks, Saisamorn Lumyong, and Kevin D. Hyde. "Use of endophytes as biocontrol agents." *Fungal Biology Reviews* 33, no. 2 (2019): 133–148.

Dimkić, Ivica, Tamara Janakiev, Marija Petrović, Giuliano Degrassi, and Djordje Fira. "Plant-associated *Bacillus* and *Pseudomonas* antimicrobial activities in plant disease suppression via biological control mechanisms-a review." *Physiological and Molecular Plant Pathology* 117 (2022): 101754.

Duncan, Larry W. "Current options for nematode management." *Annual Review of Phytopathology* 29, no. 1 (1991): 469–490.

El-Ashry, Ramadan M., Abdelhadi AI Ali, and Salonaz E. Awad. "Enhancing application efficiency of *Pseudomonas* spp. and *Serratia marcescens* isolates against Meloidogyne incognita in tomato plants." *Egyptian Academic Journal of Biological Sciences, F. Toxicology & Pest Control* 12, no. 2 (2020): 127–145.

El-Nagdi, W. M. A., H. Abd-El-Khair, G. M. Soliman, H. H. Ameen, and Gh M. El-Sayed. "Application of protoplast fusants of *Bacillus licheniformis* and *Pseudomonas aeruginosa* on *Meloidogyne incognita* in tomato and eggplant." *Middle East Journal of Applied Sciences* 9 (2019): 622–629.

El-Sagheer, Atef M. "Plant responses to phytonematodes infestations." In *Plant health under biotic stress*, pp. 161–175. Springer, Singapore, 2019.

Engelbrecht, Gerhard. "Metabolite profiling of Bacillus species with nematicidal activity." PhD diss., North-West University (South Africa). Potchefstroom Campus, 2019.

Esser, R. P. *Biological Control of Nematodes by Nematodes: Dorylaims (Nematoda: Dorylaimina). I.* Fla. Department Agric. & Consumer Serv., Division of Plant Industry, 1987.

Fatima, Sabin, and Tehmina Anjum. "Identification of a potential ISR determinant from *Pseudomonas aeruginosa* PM12 against Fusarium wilt in tomato." *Frontiers in Plant Science* 8 (2017): 848.

Forghani, Fereidoun, and Hajihassani Abolfazl. "Recent advances in the development of environmentally benign treatments to control root-knot nematodes." *Frontiers in Plant Science* 11 (2020): 1125.

Ghahremani, Zahra, Nuria Escudero, Ester Saus, Toni Gabaldón, and F. Javier Sorribas. "*Pochonia chlamydosporia* induces plant-dependent systemic resistance to *Meloidogyne incognita*." *Frontiers in Plant Science* 10 (2019): 945.

Gilardi, Giovanna, Stefano Demarchi, Maria Lodovica Gullino, and Angelo Garibaldi. "Evaluation of the short term effect of nursery treatments with phosphite-based products, acibenzolar-S-methyl, pelleted *Brassica carinata* and biocontrol agents, against lettuce and cultivated rocket fusarium wilt under artificial inoculation and greenhouse conditions." *Crop Protection* 85 (2016): 23–32.

Giné, Ariadna, Marc Carrasquilla, Maira Martínez-Alonso, Núria Gaju, and Francisco J. Sorribas. "Characterization of soil suppressiveness to root-knot nematodes in organic horticulture in plastic greenhouse." *Frontiers in Plant Science* 7 (2016): 164.

Giri, Bhoopander, Renuka Rawat, Geeta Saxena, Preet Manchanda, Qiang-Sheng Wu, and Anuradha Sharma. "Effect of *Rhizoglomus fasciculatum* and *Paecilomyces lilacinus* in the biocontrol of root-knot nematode, *Meloidogyne incognita* in *Capsicum annuum* L." *Communicative & Integrative Biology* 15, no. 1 (2022): 75–87.

Goswami, B. K., Rajesh Kumar Pandey, Jaideep Goswami, and D. D. Tewari. "Management of disease complex caused by root knot nematode and root wilt fungus on pigeonpea through soil organically enriched with vesicular arbuscular Mycorrhiza, karanj (*Pongamia pinnata*) oilseed cake and farmyard manure." *Journal of Environmental Science and Health, Part B* 42, no. 8 (2007): 899–904.

Goswami, B. K., U. Rao, and M. Manton. "Efficacy of decomposed products of some oil-cakes and ABCD collected at different time intervals on hatching and mortality of *Meloidogyne incognita*." *Current Nematology* 4 (1993): 215–217.

Guerrero Abad, Juan Carlos, Ammer Padilla Domínguez, Elías Torres Flores, Carlos López Rodríguez, Roger Arbildo Guerrero Abad, Danny Coyne, Fritz Oehl, and Mike Anderson Corazón Guivín. "A pathogen complex between the root knot nematode Meloidogyne incognita and Fusarium verticillioides results in extreme mortality of the inka nut (Plukenetia volubilis)." *Journal of Applied Botany and Food Quality* 94 (2021): 162–168.

Hajieghrari, Behzad, Mousa Torabi-Giglou, Mohammad Reza Mohammadi, and Mahdi Davari. "Biological potential of some Iranian *Trichoderma* isolates in the control of soil borne plant pathogenic fungi." *African Journal of Biotechnology* 7, no. 8 (2008).

Hameed, S. F. "Relative efficacy of different oil cakes on the incidences of root-knot nematodes." *Madras Journal of Agricultural Science* 55 (1968): 168–171.

Herrera-Parra, E., J. Ramos-Zapata, C. Basto-Pool, and J. Cristóbal-Alejo. "Sweet pepper (*Capsicum annuum*) response to the inoculation of native arbuscular mycorrhizal fungi and the parasitism of root-knot *Meloidogyne incognita*." *Revista bio ciencias* 8 (2021).

Hildalgo-Diaz, L., and B. R. Kerry. "Integration of biological control with other methods of nematode management." In *Integrated management and biocontrol of vegetable and grain crops nematodes* (eds.) A. Ciancio, K. G. Mukerji, pp. 29–50, Springer, The Netherlands, 2008.

Iqbal, Mudassir, Mukesh Dubey, Kerstin McEwan, Uwe Menzel, Mikael Andersson Franko, Maria Viketoft, Dan Funck Jensen, and Magnus Karlsson. "Evaluation of Clonostachys rosea for control of plant-parasitic nematodes in soil and in roots of carrot and wheat." *Phytopathology* 108, no. 1 (2018): 52–59.

Islam, Waqar, Muhammad Adnan, Asad Shabbir, Hassan Naveed, Yakubu Saddeeq Abubakar, Muhammad Qasim, Muhammad Tayyab, Noman A, Nisar MS, Khan KA, Ali H. "Insect-fungal-interactions: a detailed review on entomopathogenic fungi pathogenicity to combat insect pests." *Microbial Pathogenesis* 159 (2021): 105122.

Ji, Zhao-Lin, Shuai Peng, Wei Zhu, Jing-Ping Dong, and Feng Zhu. "Induced resistance in nectarine fruit by *Bacillus licheniformis* W10 for the control of brown rot caused by *Monilinia fructicola*." *Food Microbiology* 92 (2020): 103558.

Jiang, Chunhao, Zhihang Fan, Zijie Li, Dongdong Niu, Yan Li, Mingzi Zheng, Qi Wang, Hailing Jin, and Jianhua Guo. "*Bacillus cereus* AR156 triggers induced systemic resistance against *Pseudomonas syringae* pv. tomato DC3000 by suppressing miR472 and activating CNLs-mediated basal immunity in Arabidopsis." *Molecular Plant Pathology* 21, no. 6 (2020): 854–870.

Juyal, Deepak, Shekhar Pal, Munesh Sharma, Vikrant Negi, Shamanth Adekhandi, and Manoj Tyagi. "Keratomycosis due to *Purpureocillium lilacinum*: a case report from sub-Himalayan region of Uttarakhand." *Indian Journal of Pathology and Microbiology* 61, no. 4 (2018): 607.

Kanwar, R. S., J. A. Patil, and Saroj Yadav. "Prospects of using predatory nematodes in biological control for plant parasitic nematodes–a review." *Biological Control* 160 (2021): 104668.

Kassam, Rami, Deeba Kamil, and Uma Rao. "Caenorhabditis elegans as bait for isolating promising biocontrol fungi against *Meloidogyne incognita* from soils across India." *Indian Phytopathology* 74, no. 3 (2021): 739–752.

Kenawy, Ahmed, Daniel Joe Dailin, Gaber Attia Abo-Zaid, Roslinda Abd Malek, Kugan Kumar Ambehabati, Khairun Hani Natasya Zakaria, R. Z. Sayyed, and Hesham Ali El Enshasy. "Biosynthesis of antibiotics by PGPR and their roles in biocontrol of plant diseases." In *Plant growth promoting rhizobacteria for sustainable stress management*, pp. 1–35. Springer, Singapore, 2019.

Khan, Mujeebur Rahman, Shahana Majid, Fayaz A. Mohidin, and Nabilah Khan. "A new bioprocess to produce low cost powder formulations of biocontrol bacteria and fungi to control fusarial wilt and root-knot nematode of pulses." *Biological control* 59, no. 2 (2011): 130–140.

Kiriga, Agnes W., Solveig Haukeland, George M. Kariuki, Danny L. Coyne, and Nikolai V. Beek. "Effect of *Trichoderma* spp. and *Purpureocillium lilacinum* on *Meloidogyne javanica* in commercial pineapple production in Kenya." *Biological Control* 119 (2018): 27–32.

Kunda, Pranamita, Abhishek Mukherjee, and Paltu Kumar Dhal. "Bacterial biological control agents for soilborne diseases management in pulses: present status and future prospects." *Microbial Mitigation of Stress Response of Food Legumes* (2020): 231–243.

Lafta, Anfal A., and Ali A. Kasim. "Effect of nematode-trapping fungi, *Trichoderma harzianum* and *Pseudomonas fluorescens* in controlling meloidogyne SPP." *Plant Archives* 19, no. 1 (2019): 1163–1168.

Larkin, Robert P., and Timothy S. Griffin. "Control of soilborne potato diseases using *Brassica* green manures." *Crop Protection* 26, no. 7 (2007): 1067–1077.

Lastochkina, Oksana, Ludmila Pusenkova, Ruslan Yuldashev, Marat Babaev, Svetlana Garipova, Dar'ya Blagova, Ramil Khairullin, and Sasan Aliniaeifard. "Effects of Bacillus subtilis on some physiological and biochemical parameters of *Triticum aestivum* L.(wheat) under salinity." *Plant Physiology and Biochemistry* 121 (2017): 80–88.

Latif, Z. H., R. Ahmad, and M. I. Haq. "Effect of seed treatment with neem cake, neem oil and Latex of Aak on the germination of cowpea and its vulnerability, root-knot nematode *Melodogyne incognita*." *Pakistan Journal of Phytopathology* 11 (1999): 52–53.

Lazarovits, George. "Managing soilborne disease of potatoes using ecologically based approaches." *American Journal of Potato Research* 87, no. 5 (2010): 401–411.

Lazzeri, L., R. Tacconi, and S. Palmieri. "In vitro activity of some glucosinolates and their reaction products toward a population of the nematode *Heterodera schachtii*." *Journal of Agricultural and Food Chemistry* 41, no. 5 (1993): 825–829.

Lee Marzano, Shin-Yi, Michelle M. Wander, Maria B. Villamil, Carmen M. Ugarte, Edmond R. Zaborski, and Darin M. Eastburn. "Organic amendment and transitional cropping system effects on crop diseases." *Agronomy Journal* 106, no. 2 (2014): 519–527.

Liang, Lian-Ming, Cheng-Gang Zou, Jianping Xu, and Ke-Qin Zhang. "Signal pathways involved in microbe–nematode interactions provide new insights into the biocontrol of plant-parasitic nematodes." *Philosophical Transactions of the Royal Society B* 374, no. 1767 (2019): 20180317.

Litwin, Anna, Monika Nowak, and Sylwia Różalska. "Entomopathogenic fungi: unconventional applications." *Reviews in Environmental Science and Bio/Technology* 19, no. 1 (2020): 23–42.

Los, Agata, Dana Ziuzina, Daniela Boehm, and Paula Bourke. "Effects of cold plasma on wheat grain microbiome and antimicrobial efficacy against challenge pathogens and their resistance." *International Journal of Food Microbiology* 335 (2020): 108889.

Mazzola, Mark, Jack Brown, Antonio D. Izzo, and Michael F. Cohen. "Mechanism of action and efficacy of seed meal-induced pathogen suppression differ in a Brassicaceae species and time-dependent manner." *Phytopathology* 97, no. 4 (2007): 454–460.

Mazzuchelli, Rita de Cassia Lima, Eduardo Henrique Lima Mazzuchelli, and Fabio Fernando de Araujo. "Efficiency of *Bacillus subtilis* for root-knot and lesion nematodes management in sugarcane." *Biological Control* 143 (2020): 104185.

Mian, I. H., G. Godoy, R. A. Shelby, R. Rodriguez-Kabana, and G. Morgan-Jones. "Chitin amendments for control of *Meloidogyne arenaria* in infested soil." *Nematropica* (1982): 71–84.

Migunova, Varvara D., and Nicola Sasanelli. "Bacteria as biocontrol tool against phytoparasitic nematodes." *Plants* 10, no. 2 (2021): 389.

Mitiku, Misgana. "Plant-parasitic nematodes and their management: a review." *Agricultural Research & Technology* 8 (2018): 30–38.

Mohammed, Rafeeq Kasim Abdu, and Mujeebur Rahman Khan. "Management of root-knot nematode in cucumber through seed treatment with multifarious beneficial microbes in polyhouse under protected cultivation." *Indian Phytopathology* 74, no. 4 (2021): 1035–1043.

Molinari, Sergio, and Paola Leonetti. "Bio-control agents activate plant immune response and prime susceptible tomato against root-knot nematodes." *PloS one* 14, no. 12 (2019): e0213230.

Mostafa, Fatma A. M., Ashraf E. Khalil, Ahmed H. Nour El-Deen, and Dina S. Ibrahim. "The role of *Bacillus megaterium* and other bio-agents in controlling root-knot nematodes infecting sugar beet under field conditions." *Egyptian Journal of Biological Pest Control* 28, no. 1 (2018): 1–6.

Mukhtar, Tariq, Muhammad Tariq-Khan, and Muhammad Naveed Aslam. "Bioefficacy of *Trichoderma* species against javanese root-knot nematode, *Meloidogyne javanica*, in green gram." *Gesunde Pflanzen* 73, no. 3 (2021): 265–272.

Nadeem, Hera, Pieter Malan, Amir Khan, Mohd Asif, Mansoor Ahmad Siddiqui, Simon Tuhafeni Angombe, and Faheem Ahmad. "New insights on the utilization of ultrasonicated mustard seed cake: chemical composition and antagonistic potential for root-knot nematode, *Meloidogyne javanica*." *Journal of Zhejiang University-Science B* 22, no. 7 (2021a): 563–574.

Nadeem, H., P. Niazi, M. Asif, G. Kaskavalci, and F. Ahmad. "Bacterial strains integrated with surfactin molecules of *Bacillus subtilis* MTCC441 enrich nematocidal activity against *Meloidogyne incognita*." *Plant Biology* 23, no. 6 (2021b): 1027–1036.

Najafi, Faezeh, Saied Reza Naddaf, Sasan Rezaie, Eshrat Beigom Kia, and Gholamreza Mowlavi. "In vitro ovicidal activity of nematophagous fungus *Paecilomyces lilacinus* on the eggs of parasitic helminths." *Journal of Medical Microbiology and Infectious Diseases* 4, no. 3 (2016): 52–56.

Narasimhamurthy, H. B., H. Ravindra, M. Sehgal, S. D. Ekabote, and G. Ganapathi. "Bio-management of rice root-knot nematode (*Meloidogyne graminicola*)." *Journal of Entomology and Zoology Studies* 5, no. 4 (2017): 1433–1439.

Ochiai, N., M. L. Powelson, R. P. Dick, and F. J. Crowe. "Effects of green manure type and amendment rate on Verticillium wilt severity and yield of russet Burbank potato." *Plant Disease* 91, no. 4 (2007): 400–406.

Odelade, Kehinde Abraham, and Olubukola Oluranti Babalola. "Bacteria, fungi and archaea domains in rhizospheric soil and their effects in enhancing agricultural productivity." *International Journal of Environmental Research and Public Health* 16, no. 20 (2019): 3873.

Palyzová, Andrea, Kateřina Svobodová, Lucie Sokolová, Jiří Novák, and Čeněk Novotný. "Metabolic profiling of *Fusarium oxysporum* f. sp. conglutinans race 2 in dual cultures with biocontrol agents *Bacillus amyloliquefaciens*, *Pseudomonas aeruginosa*, and *Trichoderma harzianum*." *Folia microbiologica* 64, no. 6 (2019): 779–787.

Parihar, Kavita, Bushra Rehman, Mohamad Ashraf Ganai, Mohd Asif, and Mansoor A. Siddiqui. "Role of oil cakes and *Pochonia chlamydosporia* for the management of *Meloidogyne javanica* attacking *Solanum melongena* L." *International Journal of Plant Pathology* 6 (2015): 12–20.

Peiris, P. Upamali S., Yujuan Li, Philip Brown, and Chengyuan Xu. "Fungal biocontrol against *Meloidogyne* spp. in agricultural crops: a systematic review and meta-analysis." *Biological Control* 144 (2020): 104235.

Philbrick, Ashley N., Tika B. Adhikari, Frank J. Louws, and Adrienne M. Gorny. "*Meloidogyne enterolobii*, a major threat to tomato production: current status and future prospects for its management." *Frontiers in Plant Science* 11 (2020): 606395.

Postnikova, Olga A., Maria Hult, Jonathan Shao, Andrea Skantar, and Lev G. Nemchinov. "Transcriptome analysis of resistant and susceptible alfalfa cultivars infected with root-knot nematode *Meloidogyne incognita*." *PLoS One* 10, no. 2 (2015): e0118269.

Poveda, Jorge, Patricia Abril-Urias, and Carolina Escobar. "Biological control of plant-parasitic nematodes by filamentous fungi inducers of resistance: trichoderma, mycorrhizal and endophytic fungi." *Frontiers in Microbiology* 11 (2020): 992.

Pugliese, Massimo, Giovanna Gilardi, Angelo Garibaldi, and Maria Lodovica Gullino. "Organic amendments and soil suppressiveness: results with vegetable and ornamental crops." In *Organic amendments and soil suppressiveness in plant disease management*, pp. 495–509. Springer, Cham, 2015.

Qtu, J., J. Hallmann, N. Kokalis-Burelle, D. B. Weaver, R. Rodríguez-Kábana, and S. Tuzun. "Activity and differential induction of chitinase isozymes in soybean cultivars resistant or susceptible to root-knot nematodes." *Journal of Nematology* 29, no. 4 (1997): 523.

Raghavan, Ramya, Gomathi Chithra, Sanal Fernandez, Bettadpura Shamana Suryanarayana, and Rakesh Singh. "An unusual case of hyalohyphomycosis due to *Purpureocillium lilacinum* in a patient with myasthenia gravis." *Current Medical Mycology* 4, no. 2 (2018): 36.

Reddy, P. Parvatha. "Crop residue management and organic amendments." In *Agro-ecological Approaches to Pest Management for Sustainable Agriculture*, pp. 29–41. Springer, Singapore, 2017.

Renčo, M. "Organic amendments of soil as useful tools of plant parasitic nematodes control." *Helminthologia* 50, no. 1 (2013): 3–14.

Rich, J. R., and G. S. Rahi. "Suppression of *Meloidogyne javanica* and *M. incognita* on tomato with ground seed of castor, crotalaria, hairy indigo, and wheat." *Nematropica* (1995): 159–164.

Rizvi, Rose, Irshad Mahmood, Sartaj A. Tiyagi, and Zehra Khan. "Conjoint effect of oil-seed cakes and *Pseudomonas fluorescens* on the growth of chickpea in relation to the management of plant-parasitic nematodes." *Brazilian Archives of Biology and Technology* 55 (2012): 801–808.

Rosskopf, Erin, Francesco Di Gioia, Jason C. Hong, Cristina Pisani, and Nancy Kokalis-Burelle. "Organic amendments for pathogen and nematode control." *Annual Review of Phytopathology* 58 (2020): 277–311.

Seid, Awol, Chemeda Fininsa, Tesfamariam Mekete, Wilfrida Decraemer, and Wim M. L. Wesemael. "Tomato (*Solanum lycopersicum*) and root-knot nematodes (*Meloidogyne* spp.)–a century-old battle." *Nematology* 17, no. 9 (2015): 995–1009.

Shalaby, Hatem, Heba Ashry, S. A. A. D. Moataza, and Tarek Farag. "In vitro effects of *Streptomyces tyrosinase* on the egg and adult worm of *Toxocara vitulorum*." *Iranian Journal of Parasitology* 15, no. 1 (2020): 67–75.

Sharma, Meenakshi, Ishan Saini, Prashant Kaushik, Mona Mohammed Aldawsari, Thamer Al Balawi, and Pravej Alam. "Mycorrhizal fungi and *Pseudomonas fluorescens* application reduces root-knot nematode (*Meloidogyne javanica*) infestation in eggplant." *Saudi Journal of Biological Sciences* 28, no. 7 (2021): 3685–3691.

Sharma, Satyawati, Monica Verma, and Abhishek Sharma. "Utilization of non edible oil seed cakes as substrate for growth of *Paecilomyces lilacinus* and as biopesticide against termites." *Waste and Biomass Valorization* 4, no. 2 (2013): 325–330.

Sikora, Richard A., and Emilio Fernandez. "Nematode parasites of vegetables." In *Plant Parasitic Nematodes in Subtropical and Tropical Agriculture*, pp. 319–392. CABI Publishing, Wallingford, 2005.

Silva, Silas D., Regina M. D. G. Carneiro, Marcos Faria, Daniela A. Souza, Rose G. Monnerat, and Rogerio B. Lopes. "Evaluation of *Pochonia chlamydosporia* and *Purpureocillium lilacinum* for suppression of Meloidogyne enterolobii on tomato and banana." *Journal of Nematology* 49, no. 1 (2017): 77–85.

Singh, Rajendra, Manoj Kumar, Anshumali Mittal, and Praveen Kumar Mehta. "Microbial metabolites in nutrition, healthcare and agriculture." *3 Biotech* 7, no. 1 (2017): 1–14.

Singh, Simranjeet, Vijay Kumar, Daljeet Singh Dhanjal, and Joginder Singh. "Biological control agents: diversity, ecological significances, and biotechnological applications." In *Natural bioactive products in sustainable agriculture*, pp. 31–44. Springer, Singapore, 2020.

Sitaramaiah, K. "Effect of temperature on decomposition of margosa cake and on survival of *Tylenchorhynchus elegans*." *Indian Journal of Nematology* 7, no. 2 (1979): 167–169.

Soliman, M. S., M. M. El-Deriny, D. S. S. Ibrahim, H. Zakaria, and Y. Ahmed. "Suppression of root-knot nematode, *Meloidogyne incognita* on tomato plants using the nematode trapping fungus *Arthrobotrys oligospora* Fresenius." *Journal of Applied Microbiology* 131, no. 5 (2021): 2402–2415.

Stirling, G. R., E. Wong, and S. Bhuiyan. "Pasteuria, a bacterial parasite of plant-parasitic nematodes: its occurrence in Australian sugarcane soils and its role as a biological control agent in naturally-infested soil." *Australasian Plant Pathology* 46, no. 6 (2017): 563–569.

Strauss, S. L., and D. A. Kluepfel. "Anaerobic soil disinfestation: a chemical-independent approach to pre-plant control of plant pathogens." *Journal of Integrative Agriculture* 14, no. 11 (2015): 2309–2318.

Subedi, Pratima, Kaitlin Gattoni, Wenshan Liu, Kathy S. Lawrence, and Sang-Wook Park. "Current utility of plant growth-promoting rhizobacteria as biological control agents towards plant-parasitic nematodes." *Plants* 9, no. 9 (2020): 1167.

Sun, Xiaowen, Run Zhang, Meijuan Ding, Yongxuan Liu, and Lin Li. "Biocontrol of the root-knot nematode *Meloidogyne incognita* by a nematicidal bacterium *Pseudomonas simiae* MB751 with cyclic dipeptide." *Pest Management Science* 77, no. 10 (2021): 4365–4374.

Tawidian, P., V. L. Rhodes, and K. Michel. "Mosquito-fungus interactions and antifungal immunity." *Insect Biochemistry and Molecular Biology* 111 (2019): 103182.

Tazi, Hafssa, Mohamed Ait Hamza, Ayoub Hallouti, Hinde Benjlil, Amine Idhmida, James Nicholas Furze, Timothy C. Paulitz, El Hassan Mayad, Hassan Boubaker, and Abdelhamid El Mousadik. "Biocontrol potential of nematophagous fungi against *Meloidogyne* spp. infecting tomato." *Organic Agriculture* 11, no. 1 (2021): 63–71.

Tiyagi, Sartaj A., Irshad Mahmood, Zehra Khan, and Hilal Ahmad. "Biological control of soil-pathogenic nematodes infecting mungbean using *Pseudomonas fluorescens*." *Archives of Phytopathology and Plant Protection* 44, no. 18 (2011): 1770–1778.

Vallejos-Torres, Geomar, Elvis Espinoza, Jesús Marín-Díaz, Reynaldo Solis, and Luis A. Arévalo. "The role of arbuscular mycorrhizal fungi against root-knot nematode infections in coffee plants." *Journal of Soil Science and Plant Nutrition* 21, no. 1 (2021): 364–373.

Verma, Maya, Jitendra Mishra, and Naveen Kumar Arora. "Plant growth-promoting rhizobacteria: diversity and applications." In *Environmental biotechnology: for sustainable future*, pp. 129–173. Springer, Singapore, 2019.

Xia, Yanfei, Shen Li, Xueting Liu, Chong Zhang, Jianqiang Xu, and Yingwu Chen. "*Bacillus halotolerans* strain LYSX1-induced systemic resistance against the root-knot nematode *Meloidogyne javanica* in tomato." *Annals of Microbiology* 69, no. 12 (2019): 1227–1233.

Xu, Wen-Feng, Jia-Lin Yang, Xiang-Kun Meng, Zhi-Guang Gu, Qi-Lin Zhang, and Lian-Bing Lin. "Understanding the transcriptional changes during infection of *Meloidogyne incognita* eggs by the egg-parasitic fungus *Purpureocillium lilacinum*." *Frontiers in Microbiology* 12 (2021): 617710.

Yadav, Ramesh S., D. Mandal, and Amritesh C. Shukla. "Fungal diseases of medicinal and aromatic plants and their biological management." *Sustainable Horticulture, Volume 2: Food, Health, and Nutrition* (2018): 291.

Yassin, M. Y., and A. E. Ismail. "Effect of Zinnia elegans as a mix-crop along with tomato against *Meloidogyne incognita* and *Rotylenchulus reniformis*." *Anzeiger für Schädlingskunde, Pflanzenschutz, Umweltschutz* 67 (1994): 41–43.

Yogev, A., M. Raviv, Y. Hadar, R. Cohen, and J. Katan. "Plant waste-based composts suppressive to diseases caused by pathogenic *Fusarium oxysporum*." *European Journal of Plant Pathology* 116, no. 4 (2006): 267–278.

Zasada, I. A., S. L. F. Meyer, and M. J. Morra. "Brassicaceous seed meals as soil amendments to suppress the plant-parasitic nematodes *Pratylenchus penetrans* and *Meloidogyne incognita*." *Journal of Nematology* 41, no. 3 (2009): 221.

Zehra, Andleeb, Namita Anant Raytekar, Mukesh Meena, and Prashant Swapnil. "Efficiency of microbial bio-agents as elicitors in plant defense mechanism under biotic stress: a review." *Current Research in Microbial Sciences* 2 (2021): 100054.

Zhao, Jinjie, Shuai Wang, Xiaofeng Zhu, Yuanyuan Wang, Xiaoyu Liu, Yuxi Duan, Haiyan Fan, and Lijie Chen. "Isolation and characterization of nodules endophytic bacteria *Pseudomonas protegens* Sneb1997 and *Serratia plymuthica* Sneb2001 for the biological control of root-knot nematode." *Applied Soil Ecology* 164 (2021): 103924.

13 Oilseed Cake

Bioactive Compounds and Their Detrimental Effect on Root-Knot Nematodes

Saba Fatima, Faryad Khan, Mohammad Shariq, Arshad Khan and Mansoor Ahmad Siddiqui

CONTENTS

13.1 INTRODUCTION

Organic soil amendment has importance in the management of plant diseases. It is also helpful in the control of plant parasitic nematodes by enhancing and improving soil fertility as well as structure, the release of nemato-toxic compounds, alteration of the level of plant resistance and instigating the activities of microorganisms that are plant parasitic nematodes (McSorley and Gallaher 1995; Oka 2010). Various organic additives derived from the plants, such as oil cakes, plant extracts and chopped plant leaves, have been used to control nematodes (Tiyagi et al. 1988; Akhtar and Alam 1993). The present chapter is a step to update the available research on the use of oilseed cakes for the management of root-knot nematodes (RKNs) in agriculturally important crops.

Oilseed cakes are obtained in the form of solid residues as a by-product after the extraction of oil from the plant seeds through expelling or solvent extraction. The oilseed cake residues are a rich source of protein, and most are beneficial feeds for livestock. The oil cakes contain many biochemical constituents, initially present in the oil-yielding seeds (Karmelreetha and Muthukumar 2020). The oilseed cakes play an important role in reducing disease austerity and enhancing the soil fertility, antagonists and yield of the crop to a significant level. The oilseed cakes do not pollute the

DOI: 10.1201/9781003319252-13

environment but improve the field soil's sustainability (Karmelreetha and Muthukumar 2020). Thus, oilseed cakes can be used as a management tool to improve soil structure, cation exchange capacity and water holding capacity; stimulate plants' growth; and suppress soil-borne phytopathogens (Figure 13.1).

13.1.1 Agroecology of Oilseed Cake

Deguine et al. (2008) suggested that agroecology is mainly based on optimizing regulations in the agroecosystem, biological interactions and its application for crop protection. Agroecosystem diversity at various scales is one of the two pillars of the agroecological approach, along with soil quality (Ferron and Deguine 2005; Deguine et al. 2008). Managing the agroecosystem along with diversified vegetation exploits the principles mechanisms as follows:

a. allelopathic effects of plants,
b. disruption of life cycles of pests,
c. physiological resistance due to improved crop nutrition and
d. stimulation of specific belowground antagonists of pests or induction of general soil suppressiveness.

The utilization of oilseed cakes as manure at the rate of 2,500 kg/ha proved a highly effective control of RKNs increasing the crop yield (okra and tomato) and enhancing the properties of the soil (Singh and Sitaramaiah 1973). Soil amendment with mustard cake reduced the number of galls on tomato roots infected by *Meloidogyne arenaria*. The quantity of edible oil cakes also has the potential to limit the nematode population when used in combination with other biofertilizers (Tarini et al. 2020).

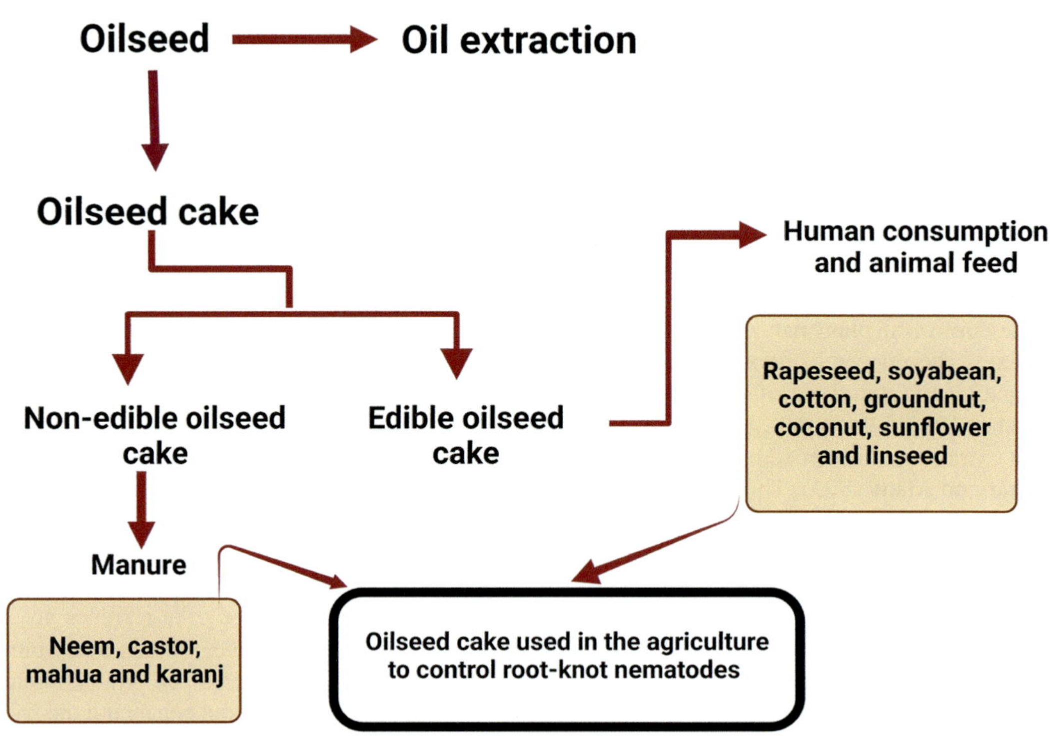

FIGURE 13.1 Oilseed cakes, types, their sources and their importance.

13.1.2 Bioactive Compounds from Oilseed Cakes

Secondary metabolites such as flavonoids and phenolic compounds are found in oil cakes (Karmelreetha and Muthukumar 2020). Oilseed cakes become a good source for the production of bioactive compounds, including phenolic and flavonoid (Ramachandran et al. 2007; Teh and Birch 2014). Oilseed cakes are rich sources of lignin, cellulose and hemicellulose. Besides, a high content of nitrogen, phosphorus and potassium is present in oilseed cakes, and hence, oil cakes are utilized as a great source of organic fertilizers. Neem is a rich source of limonoids (Kavathekar 2003) such as azadirachtin, and mahua is a main source of fatty acids as well as other phytoactive compounds like saponins (Kumar et al. 2011). Oilseed cakes of mahua and neem are being administered into the soil to manage phytonematodes (Ashraf and Khan 2010; Rizvi et al. 2012). VasudhaUdupa et al. (2021) identified many bioactive compounds in neem, mahua and simarouba oilseed cakes. The bioactive compound includes 5-hydroxymethyl furfural, 6-hydroxy-4-methyl-3-phenylcoumarin, α-monoacetin, α-D-glucopyranoside, methyl, methyl gallate, angelic acid, vanillic acid, vinyl sulphide, indole and cyclic octa-atomic sulphur. Simarouba oil cake contains eight bioactive compounds, namely nanonal, 2,4-Di-tert-bytlphenol, canthin-5-one, trans-2-decenal, mevalolactone, cis-13-eicosenoic acid, xilitol and diacetone. These bioactive compounds have antioxidant (Al-Douri and Shakya 2019), antimicrobial (Gazoni et al. 2018), anticancerous (https://pubchem.ncbi.nim.nih.gov), insecticidal and nematicidal (Caboni et al. 2012) bioactive properties. Oilseed cakes consist of a variety of bioactive compounds with biological significance. The isolation and utilization of nematicidal compounds from oilseed cakes could be a promising strategy in agriculture.

Gas chromatography-mass spectrometry (GCMS) analysis of neem cake displayed 101 volatile organic bioactive compounds that belong to various chemical groups such as aldehydes and ketones, alcoholic groups, nitrogenous compounds, sulphur-containing compounds, some acids, esters, hydrocarbons, oxides and others (Shivashankar et al. 2012). Acidic compounds are found in higher proportions in cakes (18.54%) followed by aldehyde and ketones (17.53%), while fatty acid like palmitic acid (14.97%) present abundantly in neem cake. Many compounds viz., 4-methyl-1-penten-3-ol, 2-cyclohexen-1-ol, α-methyl-1,4-benzenedimethanol and Nerolidol are alcohols whereas, the compounds like 2-propenenitrile,3-(2-quinolinyl1)-,(E), (9Z)-octadec-9-enenitrile, and 3,4-Dihydro-2H-pyran are nitrogenous (Shivashankar et al. 2012). Some compounds, such as furfural, 3-furaldehyde, octenal, isolongifolen-5-one, benzophenone and 1,4-cis-1,7-cis-acorenone are alhydes and ketones. The chemical group, that is, esters, is found in the form of phenethyl acetate, benzyl benzoate, ethyl palmitate, methyl hexadecanoate, methyl linolenate and ethyl linolenate. The types of hydrocarbons also identified in neem cake are as follows: zingibrene, acoradiene, α-longipinene, toluene, 1,3,7-octatrien-5-yne, naphthalene, ylangene, copaene, valencene and β-eudesmene. A few sulfur-containing bioactive compounds of neem cake are methyl-thiirane, methyl-3-thiophene, thiophene,2,5-dimethyl and carbon disulphide.

Phenolic compounds play a crucial role in reducing various plant diseases (Singh et al. 2005). Seed cake of *Brassica* species secretes biocidal compounds, which are toxic to soil-borne pests and diseases (Rahman and Somers 2005; De Corato et al. 2015). Corrales et al. (2008) reported that the polyphenols of oilseed seed cakes are found to be enclosed in hard and insoluble structures like lipoprotein bilayers, vacuoles, lignin, cell walls and the hull.

13.2 NEMATICIDAL EFFECT OF OILSEED CAKES ON RKNS

The antagonistic impact of oilseed cakes on nematodes is mainly associated with the emergence of various forms of nitrogen and the stimulation of microbial activity in the soil (Sumbul et al. 2015). The nematicidal effect of seed cakes on RKNs can be investigated *via in vitro* testing in terms of egg hatching inhibition and juvenile mortality of RKNs (Figure 13.2) and in the soil under pot experiments as an organic amendment, as shown in Figure 13.3 (Kumar et al. 2006; Tarini et al. 2020).

13.2.1 Impact of Oilseed Cake on Survival of RKNs under In Vitro Study

Much research has been conducted based on evaluating oilseed cake's nematicidal potential against RKNs. Oilseed cakes from *Brassica* species, namely *B. juncea* (brown mustard), *Sinapis alba* (white mustard) and *B. napus* (rapeseed), had a deleterious impact on the RKN (Mazzola et al. 2001; Reardon et al. 2013). Khan et al. (2011) reported that the isolated water-soluble fractions of oilseed cakes extracted from mahua, groundnut, neem and castor were toxic to the juvenile of *M. incognita*. The

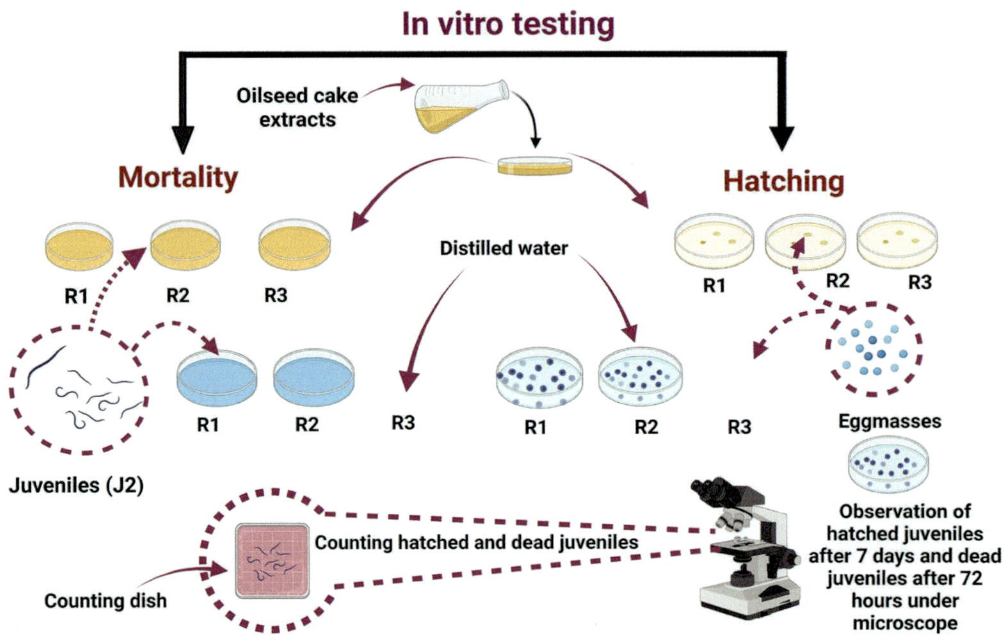

FIGURE 13.2 In vitro testing for nematicidal action of oilseed cakes on root-knot nematodes.

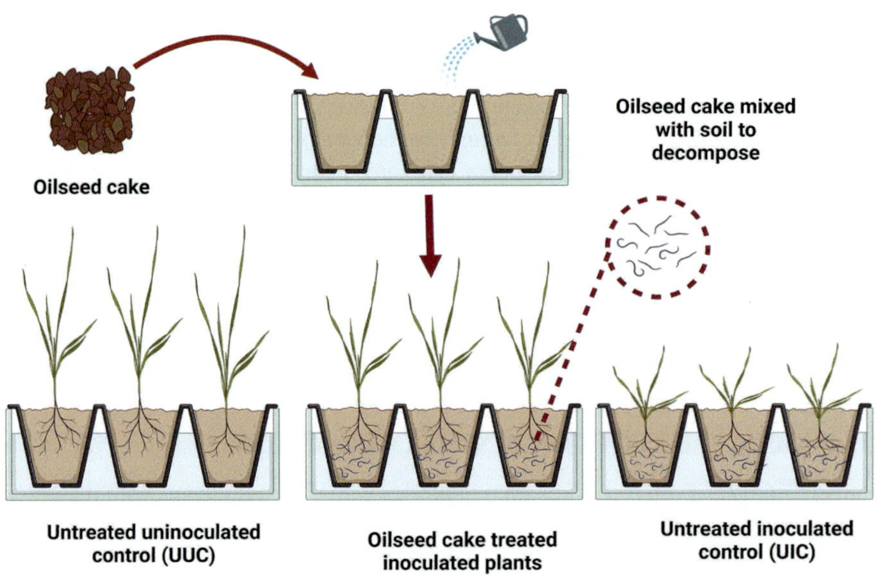

FIGURE 13.3 Pot application of oilseed cakes as an organic amendment against root-knot nematodes.

higher concentration of oil cake extract displayed excellent results due to the presence of metabolites in higher concentrations (Sumbul et al. 2015). *In vitro* testing of some oilseed cakes including neem, jatropha and castor revealed a significant reduction in egg hatching, and an increased percentage of mortality of juveniles of *M. incognita* was achieved (Kalaiarasan et al. 2007). The inhibition of egg hatching of *M. incognita* by the *in vitro* application of neem and karanj oil cakes was also reported (Rao et al. 1986). Oilseed cakes successfully killed nematodes by releasing some compounds that have nematicidal action.

13.2.2 RKN Suppression via Oilseed Cake Application as Organic Soil Amendments

Previous research has shown that using different seed oil cakes has a lot of success in inhibiting nematodes in the soil. The nematicidal efficacy of *Brassica* species is based on their decomposition in soil. During the decomposition process, a compound known as glucosinolates is secreted and reacts with the enzyme myrosinase, forming bioactive metabolites including epinitriles, nitriles (Mayton et al. 1996) and isothiocyanates (Mazzola et al. 2001; Zasada and Ferris 2004; Wu et al. 2011). The decomposed products of mahua, mustard, groundnut, neem and karanj oilseed cakes affected the second-stage juvenile (J2) mortality of *M. incognita* (Goswami et al. 1993). It has also been well understood that the mortality of RKN *M. incognita* was found to increase with the increasing period of decomposition of oilseed cake (Khan and Goswami 1995). Tiyagi et al. (2001) applied different oilseed cakes, such as neem, castor, groundnut, mustard, linseed and duan, against plant parasitic nematodes on lentils and mung beans against nematodes and explored the nematicidal effects. They found oilseed cakes effectively reduced root galling and nematode population. Goswami et al. (2006) studied the significant improvement in plant growth by applying numerous oilseed cakes such as mustard and neem. In a pot study, the investigation of various oil cakes (mustard, neem, linseed, castor cakes) displayed nematoxic biopotency against RKN *M. incognita*. The suppression of the RKN population was due to the presence of several bioactive compounds that played a crucial role in impeding the nematode (Rehman et al. 2014). Amending soil with various oil cakes such as mustard, mahua, karanj, castor, cotton and neem has shown remarkable potential to manage RKNs (Devi and Das 2016; Sahu et al. 2018; Baheti et al. 2019).

13.3 OILSEED CAKES FOR THE MANAGEMENT OF RKNS

13.3.1 Neem Cake

Azadirachta indica (neem) is a member of the family Meliaceae. Neem has a variety of biologically active components (Akhtar 2000). A by-product left after oil extraction from neem seeds is called neem cake. The availability of neem in simple formulations, for example, neem seed kernel powder, neem seed powder, neem seed cake powder, dry leaf powder and aqueous extracts. An experiment was conducted to examine the protective and curative effect of neem formulations (neem cake and dry neem leaf powder) on the development of RKN, *M. javanica* (Javed et al. 2007). In their treatment, the fecundity of the females and the number of egg masses were affected due to the presence of active ingredients of neem cake and leaves. The participation of phenolic compounds absorbed systemically by the roots when exposed to neem formulations may have induced tolerance against RKNs and prevented the multiplication of nematodes (Alam et al. 1980). Neem cakes are more effective in inhibiting J2 development (Javed et al. 2007). Gowda and Gowda (1999) examined the efficiency of neem cake extract against RKN *M. incognita*, and it was found to be greatly effective in killing J2. When neem oilcake combined with urea was inserted into the soil, the bioactive compounds like triterpenes retarded the multiplication and growth of the nitrifying bacteria that results in the delayed transformation of ammoniacal nitrogen into nitrate (Akhtar and Alam 1993); see Table 13.1.

13.3.2 Castor, Peanut, Sesame, Flax and Cotton Oilseed Cakes

Oilseed cake, namely castor (*Ricinus communis*) member of Euphorbiaceae, peanut (*Arachis hypogaea*) belonging to the family Fabaceae, sesame (*Sesame indicum*) family Pedaliaceae, flax (*Linum usitatissimum*) family Linaceae and cotton (*Gossypium hirsutum*) genus Malvaceae are being used as organic fertilizers. Akhtar and Mahmood (1996) examined castor oilseed cake's effect on controlling nematodes and obtained promising results from their investigation. Khan et al. (2011) demonstrated that the castor and mustard oilseed cake have enough efficiency to control RKNs, including *M. javanica* and *M. incognita*. Ashraf and Khan (2010) conducted a greenhouse experiment to examine the efficacy of peanut, castor, linseed, neem and mahua against *M. javanica*. They found that oilseed cakes remarkably reduced RKN multiplication and curtailed root galling in the infected plant when combined and treated with a biocontrol agent. The analysis of cotton oilseed cake showed that cotton oilseed cake contains a high amount of ash and greater mineral composition (Aziz et al. 2018). Lopes et al. 2009 investigated the role of VOCs obtained from castor cake in controlling root-knot nematode *M. javanica* (Table 13.1). Seed cake is a rich source of lignocellulosic components, amino acids, proteins and lipids (Melo et al. 2008).

13.3.3 Mustard Cake

A choline ester of sinapic acid called sinapine (phenolic choline ester) is found in high concentrations in Brassicaceae seeds (Zhou et al. 2005). Mustard cake has high nutritive value and is rich in glucosinolates (Sonowal et al. 2018) and their derivatives such as isothiocyanates (Nadeem et al. 2021). Mustard cake is considered a good protein source for livestock because it consists of

TABLE 13.1
Evaluation of Nemato-Toxic Efficacy of Different Oilseed Cakes as a Soil Amendment on Root-Knot Nematodes

Oilseed Cakes	Tested Crop	Root-Knot Nematode	Reference
Neem	*Solanum lycopersicon*	*Meloidogyne incognita*, *M. javanica*	Singh and Sitaramaiah (1971); Jacob and Illaque (1998)
Mustard, mahua, linseed, neem and castor	*Glycine max*	*M. incognita*	Mishra and Gupta (1997)
Groundnut, duan, sesame and mustard	*Capsicum annum*	*M. incognita*	Trivedi et al. (1978)
Pasteria pentrans + castor, mustard and neem	*Capsicum annum*	*M. incognita*	Chaudhary and Kaul (2013)
Fennel, anise and sesame	*Solanum melongena*	*M. incognita*	El-Sherif et al. (2006)
Paecilomyces lilacinus + neem, groundnut, castor and linseed	*Solanum melongena*	*M. javanica*	Ashraf and Khan (2010)
VAM + neem	*Solanum melongena*	*M. incognita*	Borah and Phukan 2004
AMF + mustard and neem	*Solanum melongena*	*M. incognita*	Bhardwaj and Sharma (2006)
Neem, castor and *Simarouba glauca*	*Oryza sativa*	*M. graminicola*	Prasad et al. (2005)
Neem	*Arachis hypogaea*	*M. arenaria*	Vaishnava et al. (1993)

comparable metabolisable energy substances to peanut cake with great availability of nitrogen and lysine and a better gross protein value (Sonowal et al. 2018). Mustard cake is easily available and often utilised to control nematodes. Nadeem et al. (2021) experimented to explore the nemato-toxic efficacy of ultrasonicated-assisted mustard cake against juveniles of *M. javanica*. A pot experiment revealed that mustard cake significantly inhibited penetration of infective stage J2 juveniles of root-knot nematode *M. incognita* in infected tomato roots (Goswami and Meshram 1991). The decomposed form of mustard cake in the soil efficiently increases the growth of the plant, providing it with essential nutrients, whereas the survival of juveniles decreases in terms of the increased mortality of *M. incognita* (Goswami et al. 1993; Chaudhary and Kaul 2013).

13.3.4 Mahua Cake

Madhuca latifolia (mahua) belongs to the family Sapotaceae and is economically significant due to the widespread uses of each part (Gupta et al. 2012). The biodiesel production from mahua seed oil has recently attracted researchers' interest. Mahua cake is a rich source of proteins, nitrogen and sugars, but the presence of one toxic secondary metabolite known as saponins limits its extensive usage (Gupta et al. 2012). Saponins in the crude form were found to be a mixture of three compounds, and among them, Mi-saponin A and Mi-saponin B were identified (Gupta 2013). The cake has been used as a bio-pesticide and low-grade fertilizer. Gupta (2013) investigated that saponins, purified prosapogenin, and sapogenin were observed to be effective against root-knot nematode *M. incognita*. The root galling was reduced by treating mahua cake in the soil (Pandey 1994); Table 13.1. Saponins and eight types of volatile organic compounds (VOCs), mainly 4-methyl phenol, exhibit effective nematicidal properties (Yang et al. 2015). The nematicidal VOCs from different chemical groups identified in caster bean cake such as phenols, ketones (2-nonanone, 2-undecanone), alcohols (2-octanol, terpeneol), sulphur compounds (dimethyl disulphide) and aldehydes (benzeneacetaldehyde, decanal) efficiently inhibit nematodes (Gu et al. 2007; Huang et al. 2010).

13.4 MECHANISM OF ACTION

A study reported that three major biological events are engaged in soil-amended (oilseed cake) mechanism of action against nematodes (Bridge 1996; Oka 2010). (1) They enhance the soil's capability for water and nutrient holding, which ultimately promotes plant health and tolerance to the nematode. (2) They secrete a specific chemical compound that may be nematicidal. (3) They induce microbial activities in the soil to nematode antagonists. Also, they indirectly stimulate parasites and predators of the nematode, such as nematophagous fungi and micro-arthropods. Oilseed cakes' mechanism of action is depicted in Figure 13.4.

Many possible mechanisms are recorded to be involved in the inhibition of nematode development by neem cake including Water-soluble nematicidal compounds in neem (Egunjobi and Afolami 1976). Alam et al. (1980) demonstrated that the presence of phenolic bioactive compounds absorbed systemically by the roots of the tomato when exposed to neem formulations. It might have induced tolerance towards nematodes, and such plants can halt nematode multiplication. Azadirachtins, the nemato-toxic compounds of the neem plant, are released by exudation, volatilization, leaching and decomposition (Akhtar 2000). Javed et al. (2007) studied that neem cake has curative and protective applications. They found that azadirachta was effective in blocking the invasion of J2 and their further development in the root. Neem cake was recorded as the most effective against nematodes, showing an induced defence mechanism inside the plants, which detained the development of the nematode (Javed 2007). Khan et al. (1974) suggested that the nematicidal action of neem formulations is not only based on the presence of compounds such as thionimone and nimbidin but also because of the by-products, namely formaldehyde, ammonia, fatty acids and phenols, released during the decomposition of neem formulations (Huber 1991).

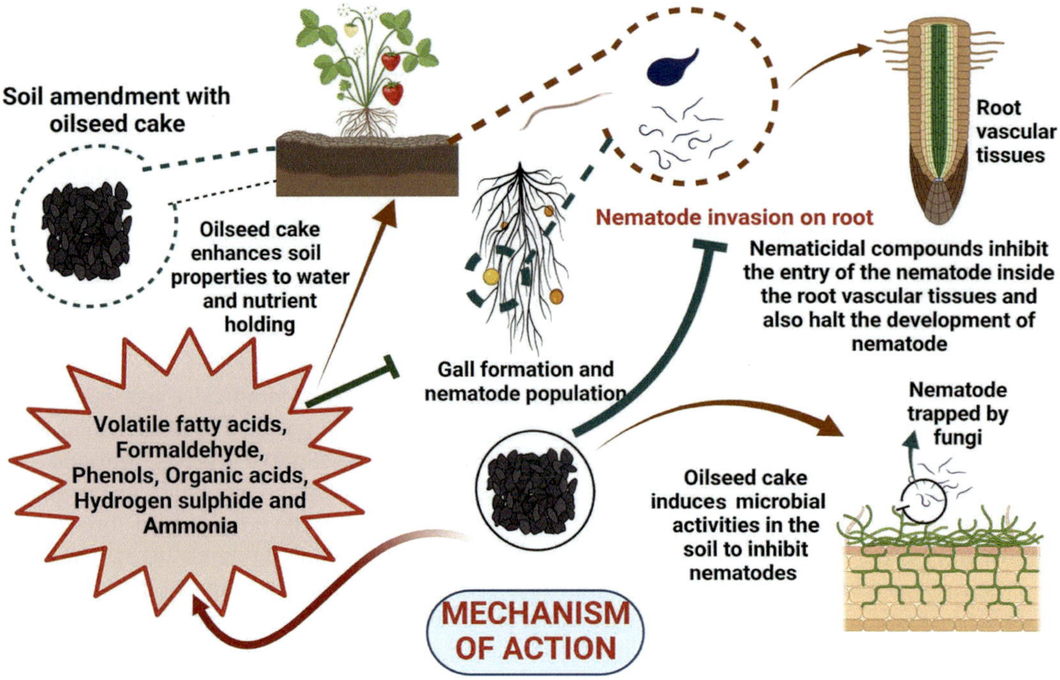

FIGURE 13.4 Mechanism of action of different oilseed cakes on root-knot nematode through secretion of various bioactive nematicidal compounds by stimulating bioagents' activity in the soil and enhancing soil properties.

13.5 CONCLUSION

Obviously, with the rising awareness of the adverse impact of synthetic pesticides, most people are looking for organically produced agricultural products. Oilseed cakes are the most beneficial and readily available substrates for inducing some biocontrol agents' growth and indirectly promoting plant health in various ways. Oilseed cakes consist of many diverse groups of different bioactive compounds. Out of them, some proved nematicidal in nature. Among oilseed cakes, neem has occupied the first rank because it is rich in nematicidal compounds like azadiractin and nimbidin, followed by castor. Research on oilseed cakes has explored microorganisms assisting decomposition rapidly and excellently to curtail nematode populations. Ultimately, we can conclude that for the sustainable management of RKNs that damage valuable crops, it is necessary to integrate various biological strategies with the combination of oilseed cake that plays a vital role in preventing nematodes in agricultural fields.

REFERENCES

Akhtar, M. 2000. Nematicidal potential of the neem tree *Azadirachta indica* (A. Juss). *Integrated Pest Management Reviews* 5:57–66.

Akhtar, M., and M.M. Alam. 1993. Utilization of waste materials in nematode control: A review. *Bioresource Technology* 45:1–7.

Akhtar, M., and I. Mahmood. 1996. Control of plant-parasitic nematodes with organic and inorganic amendments in agricultural soil. *Applied Soil Ecology* 4:243–247.

Alam, M.M., M. Ahmad, and A. M. Khan. 1980. Effect of organic amendment on the growth and chemical composition of tomato, eggplants and chilli and their susceptibility to attack by *Meloidogyne incognita*. *Plant and Soil* 57:231–236.

Al-Douri, N.A., and A.K. Shakya. 2019. Fatty acids analysis and antioxidant activity of a lipid extract obtained from *Mercurialis annua* L. grown wildly in Jordan. *Acta Poloniae Pharmaceutica-Drug Research* 76:275–281.

Ashraf, M.S., and T.A. Khan. 2010. Integrated approach for the management of *Meloidogyne javanica* on eggplant using oil cakes and biocontrol agents. *Archives of Phytopathology Plant Protection* 43:609–614.

Aziz, S., M.A.B. Siddique, and M.H.A. Begum. 2018. Cotton seed oil cake as a valuable source of plant nutrients for sustainable agriculture. *The Pharmaceutical and Chemistry Journal* 5:39–45.

Baheti, B.L., S.S. Bhati, and H. Singh. 2019. Efficacy of different oil-cakes as soil amendment for the management of root-knot nematode, *Meloidogyne incognita* infecting okra (*Abelmoschus esculentus* L.). *International Journal of Current Microbiology and Applied Science* 8:799–808.

Bharadwaj, A., and S. Sharma. 2006. Biocontrol of *Meloidogyne incognita* in *Lycopersicon esculentum* with AM fungi and oil cakes. *Plant Pathology Journal* 5:166–172.

Borah, A., and P.N. Phukan. 2004. Comparative efficacy of *Glomus fasciculatum* with neem cake and carbofuran for the management of *Meloidogyne incognita* on brinjal. *Indian Journal of Nematology* 34:129–132.

Bridge, J. 1996. Nematode management in sustainable and subsistence agriculture. *Annual Review of Phytopathology* 34:201–225.

Caboni, P., N.G. Ntalli, N. Aissani, I. Cavoski, and A. Angioni. 2012. Nematicidal activity of (E, E)-2, 4-decadienal and (E)-2-decenal from *Ailanthus altissima* against *Meloidogyne javanica*. *Journal of Agricultural and food Chemistry* 60:1146–1151.

Chaudhary, K.K., and R.K. Kaul. 2013. Efficacy of *Pasteuria penetrans* and various oil seed cakes in management of *Meloidogyne incognita* in Chilli pepper (*Capsicum annuum* L.). *Journal of Agricultural Science and Technology* 15:617–626.

Corrales, M., S. Toepfl, P. Butz, D. Knorr, and B. Tauscher. 2008. Extraction of antho-cyanins from grape by-products assisted by ultrasonics, high hydrostatic pressure or pulsed electric fields: A comparison. *Innovative Food Science and Emerging Technologies* 9:85–91.

De Corato, U., C. Pane, G.L. Bruno, F.A. Cancellara, and M. Zaccardelli. 2015. Co-products from a biofuel production chain in crop disease management: A review. *Crop Protection* 68:12–26.

Deguine, J.P., D. Russell, and P. Ferron. 2008. *Crop Protection: From Agrochemistry to Agroecology*, Science Publishers, Enfield, NH, 192.

Devi, T.H., and D. Das. 2016. Effect of organic amendments on root-knot nematode (*Meloidogyne incognita*) in cucumber. *Pest Management in Horticultural Ecosystems* 22:176–181.

Egunjobi, O.A., and S.O. Afolami. 1976. Effects of neem (*Azadirachta indica*) leaf extracts on populations of *Pratylenchus Brachyurus* and on the growth and yield of maize. *Nematologica* 22:125–132.

El-Sherif, A., A.R. Refaei, M.E. El-Nagar, and H.A. El-Ghnam. 2006. Impact of certain organic amendments, *Bacillus thuringiensis* and oxamyl on *Tylenchulus semipenetrans* infecting lemon plant. *Journal Plant Protection Pathology* 31:6761–6770.

Ferron, P., and J.P. Deguine. 2005. Towards an agroecological conception of crop protection. In: *Phytosanitary issues for agriculture and the environment*, ed. C. Regnault-Roger, 347–366. Lavoisier, Paris.

Gazoni, V.F., S.O. Balogun, K. Arunachalam, D.M. Oliveira, V. Cechinel Filho, S.R. Lima, and D.T. de Oliveira Martins. 2018. Assessment of toxicity and differential antimicrobial activity of methanol extract of rhizome of *Simaba ferruginea* A. St.-Hil. and its isolate canthin-6-one. *Journal of Ethnopharmacology* 223:122–134.

Goswami, B.K., and N.J. Meshram. 1991. Studies on comparative efficacy of mustard and karanj oil seed cakes with a nematicide, carbofuran against root-knot nematode, *Meloidogyne incognita* in tomato. *Indian Journal of Nematology* 21:66–70.

Goswami, B.K., R.K. Pandey, K.S. Rathour, C. Bhattacharya, and L. Singh. 2006. Integrated application of some compatible biocontrol agents along with mustard oil seed cake and furadan on *Meloidogyne incognita* infecting tomato plants. *Journal Zhejiang University Science B* 7:873–875.

Goswami, B.K., U. Rao, and M. Manton. 1993. Efficacy of decomposed products of some oil-cakes and ABCD collected at different time intervals on hatching and mortality of *Meloidogyne incognita*. *Current Nematology* 4:215–217.

Gowda, P.D., and D.N. Gowda. 1999. In-vitro studies of oil-cake extracts on the survival of larvae of *Meloidogyne incognita* (Kofoid and White) Chitwood. *Plant Protection Bulletin* 49:10–11.

Gu, Y.Q., M.J. Mo, J.P. Zhou, C.S. Zou, and K.Q. Zhang. 2007. Evaluation and identification of potential organic nematicidal volatiles from soil bacteria. *Soil Biology and Biochemistry* 39:2567–2575.

Gupta, A. 2013. Integrated approach for mahua seed cake utilization (Doctoral dissertation, IIT Delhi).

Gupta, A., R. Chaudhary, and S. Sharma. 2012. Potential applications of mahua (*Madhuca indica*) biomass. *Waste and Biomass Valorization* 3:175–189.

Huang, Y., C. Xu, L. Ma, K. Zhang, C. Duan, and M. Mo. 2010. Characterization of volatiles produced from *Bacillus megaterium* YFM325 and their nematicidal activity against *Meloidogyne incognita. European Journal of Plant Pathology* 126:417–422.

Huber, D.M. 1991. The use of fertilizers and organic amendments in the control of plant disease. CRC handbook of pest management in agriculture. ed. D. Pimentel, vol. 1, 405–494. CRC Press, Boca Raton.

Jacob, A.J., and M.M. Illaque. 1998. Effect of neem products on the suppression of root-knot nematode *Meloidogyne incognita* in tomato. In III Intl. Afro Asian Soc. Nematologists, April 16–19, 1998, Coimbatore, India, p. 85.

Javed, N., S.R. Gowen, M. Inam-ul-Haq, and S.A. Anwar. 2007. Protective and curative effect of neem (*Azadirachta indica*) formulations on the development of root-knot nematode *Meloidogyne javanica* in roots of tomato plants. *Crop Protection* 26:530–534.

Kalaiarasan, P., M. Senthamarai, D. Ramesh, and M.J. Sudheer, 2007. Jatropha: An efficient organic amendment for the management of root-knot nematode, *Meloidogyne incognita* in tomato. *Indian Journal of Nematology* 37:115–118.

Karmelreetha, A., and A. Muthukumar. 2020. Role of oil cakes against soil born pathogens an eco-friendly approach. In: *Organic Farming for Sustainable Agriculture*, JPS Scientific Publishers, India, 120–127.

Kavathekar, K.Y. 2003. *Neem in India*, NISCAIR, New Delhi, 21–23.

Khan, A., S.S. Shaukat, and M. Sayed. 2011. Control of nematodes associated with almond using oil-cakes in Balochistan. *Pakistan Journal of Nematology* 29:52–58.

Khan, A.M., M.M. Alam, and R. Ahmad. 1974. Mechanism of the control of plant parasitic nematodes as a result of the application of oil cakes to the soil. *Indian Journal of Nematology* 4:93–96.

Khan, M.R., and B.K. Goswami. 1995. Nematicidal efficacy of decomposed neem cake and fungal filtrates of *Paecilomyces lilacinus* and *Aspergillus flavus* against *Meloidogyne incognita.* In Proceedings of National Symposium on Sustainable Agriculture in Sub-humid Zone, March, pp. 3–5.

Kumar, K.P., G. Vidyasagar, D. Ramakrishna, I.M. Reddy, V.A. Gupta, and C. Raidu. 2011. Screening of *Madhuca indica* for antidiabetic activity in streptozotocin and streptozotocin–nicotinamide induced diabetic rats. *International Journal of PharmTech Research* 3:1073–1077.

Kumar, P., and H.M. Poehling. 2006. Persistence of soil and foliar azadirachtin treatments to control sweet potato whitefly *Bemisia tabaci* Gennadius (Homoptera: Aleyrodidae) on tomatoes under controlled (laboratory) and field (netted greenhouse) conditions in the humid tropics. *Journal of Pest Science* 79:189–199.

Lopes, E.A., S. Ferraz, O.D. Dhingra, P.A. Ferreira, and L.G. Freitas. 2009. Soil amendment with castor bean oilcake and jack bean seed powder to control *Meloidogyne javanica* on tomato roots. *Nematologia Brasileira* 33:106–109.

Mayton, H.S., C. Olivier, S.F. Vaughn, and R. Loria. 1996. Correlation of fungicidal activity of *Brassica* species with allyl isothiocyanate production in macerated leaf tissue. *Phytopathology* 86:267–271.

Mazzola, M., D.M. Granatstein, D.C. Elfving, and K. Mullinix. 2001. Suppression of specific apple root pathogens by *Brassica napus* seed meal amendment regardless of glucosinolate content. *Phytopathology* 91:673–679.

McSorley, R., and R.N. Gallaher. 1995. Cultural practices improve crop tolerance to nematodes. *Nematropica* 25:53–60.

Melo, W.C., D.B.D. Silva, N. Pereira Jr., L.M.M.S. Anna, and A.S.D. Santos. 2008. Ethanol production from castor bean cake (*Ricinus communis* L.) and evaluation of the lethality of the hydrolyzed cake for mice. *New Chemistry* 31:1104–1106.

Mishra, S.M., and P. Gupta. 1997. Effect of some oil cakes and saw dust application on root-knot (*Meloidogyne incognita*) infection and root nodulation of soybean. *Indian Phytopathology* 50:131–132.

Nadeem, H., P. Malan, A. Khan, M. Asif, M. Ahmad Siddiqui, S. Tuhafeni Angombe, and F. Ahmad. 2021. New insights on the utilization of ultrasonicated mustard seed cake: Chemical composition and antagonistic potential for root-knot nematode, *Meloidogyne javanica*. *Journal of Zhejiang University-Science B* 22:563–574.

Oka, Y., 2010. Mechanisms of nematode suppression by organic soil amendments-a review. *Applied Soil Ecology* 44:101–115.

Pandey, R. 1994. Comparative performance of oil seed cakes and pesticides in management of root-knot nematode disease of davana. *Nematologia Mediterranea* 22:374–377.

Prasad, J.S., K.S. Varaprasad, Y.R. Rao, E.S. Rao, and M. Sankar. 2005. Comparative efficacy of some oil seed cakes and extracts against root-knot nematode (*Meloidogyne graminicola*) infection in rice. *Nematologia Mediterranea* 33:191–194.

Rahman, L., and T. Somers. 2005. Suppression of root knot nematode (*Meloidogyne javanica*) after incorporation of Indian mustard cv. Nemfix as green manure and seed meal in vineyards. *Australian Plant Pathology* 34:77–83.

Ramachandran, S., S.K. Singh, C. Larroche, C.R. Soccol, and A. Pandey. 2007. Oil cakes and their biotechnological applications-a review. *Bioresource Technology* 98:2000–2009.

Rao, C.V., A. Mani, and P.K. Rao. 1986. Effect of plant products on egg hatch and larval mortality of *Meloidogyne incognita. Proceedings: Animal Sciences* 95:397–401.

Reardon, C.L., S.L. Strauss, and M. Mazzola. 2013. Changes in available nitrogen and nematode abundance in response to *Brassica* seed meal amendment of orchard soil. *Soil Biology and Biochemistry* 57:22–29.

Rehman, B., M.A. Ganai, K. Parihar, M. Asif, and M.A. Siddiqui. 2014. Biopotency of oilcakes against *Meloidogyne incognita* affecting *Vigna mungo. European Journal of Applied Science* 6:57–63.

Rizvi, R., I. Mahmood, S.A. Tiyagi, and Z. Khan 2012. Conjoint effect of oilseed cakes and *Pseudomonas fluorescens* on the growth of chickpea in relation to the management of plant-parasitic nematodes. *Brazillian Archives of Biology Technology* 55:801–808.

Sahu, S., M.K. Patra, and B. Dash. 2018. Management of root knot nematode (*Meloidogyne incognita*) in tomato (cv. Pusa ruby) using different oil cakes. *International Journal of Current Microbiology Applied Science* 7:2527–2532.

Shivashankar, S., T. Roy, and P. Krishna. 2012. Headspace solid phase micro extraction and GC/MS analysis of the volatile components in seed and cake of *Azaridachta indica* A. juss. *Chemical Bulletin of Politehnica University of Timisoara* 57:1–6.

Singh, R.S., and K. Sitaramaiah. 1971. Control of root knot through organic and inorganic amendments of soil: Effect of sawdust and inorganic fertilizers. *Indian Journal of Nematology* 1:80.

Singh, R.S., and K. Sitaramaiah. 1973. Effect of organic amendments, green manuring and inorganic fertilizers on root knot of vegetable crop. Research Bulletin, Experiment Station and College of Agriculture, GBPUAT. *Pant Nagar* 6:289.

Singh, U.P., S. Maurya, and D.P. Singh. 2005. Phenolic acids in neem (*Azadirachta indica*) a major preexisting secondary metabolites. *Journal of Herbal Pharmacotherapy* 5:35–43.

Sonowal, M., H.F. Ahmed, and P. Borah. 2018. Effect of different dietary levels of untreated and copper sulphate treated mustard oil cake (*Brassica campestris*) on growth performance and nutrient utilization of broiler chickens. *International Journal of Livestock Research* 8:226–230.

Sumbul, A., R. Rizvi, I. Mahmood, and R.A. Ansari. 2015. Oil-cake amendments: Useful tools for the management of phytonematodes. *Asian Journal of Plant Pathology* 9:91–111.

Tarini, G., A.S. Melo, L.F. Fontana, E. da Silva Júnior, B.C. Bolanho, B.P. Moreno, and C.R. Dias-Arieira. 2020. Aqueous extracts of *Crambe abyssinica* seed cake: Chemical composition and potential for nematode control. *Industrial Crops and Products* 156:112860.

Teh, S.S., and E.J. Birch. 2014. Effect of ultrasonic treatment on the polyphenol content and antioxidant capacity of extract from defatted hemp, flax and canola seed cakes. *Ultrasonics Sonochemistry* 21:346–353.

Tiyagi, S.A., M. Bano, and M.M. Alam. 1988. Evaluation of nematicidal potential in some plant species belonging to family Compositae. *Indian Journal of Nematology* 18:228–231.

Tiyagi, S.A., A.V. Khan, and M.M. Alam. 2001. Role of oilseed cakes for the management of plant-parasitic nematodes and soil-inhabiting fungi on lentil and mung bean. *Archives of Phytopathology and Plant Protection* 33:453–472.

Trivedi, P.C. 1978. Control of root-knot nematode on *Capsicum annum* by application of oil-cakes. *Pascal and Francis Bibliographic Databases* 31:75–76.

Vaishnava, M.U., I.U. Dhruj, and R.P. Lakhataria. 1993. Efficacy of cakes in the management of root-knot nematode (*M. arenaria*) in groundnut. Proceeding of World Neem Conference, February 24–28, Bangalore, India.

VasudhaUdupa, A., B. Gowda, B.E. Kumarswammy, and M.B. Shivanna. 2021. The antimicrobial and antioxidant property, GC–MS analysis of non-edible oil-seed cakes of neem, madhuca, and simarouba. *Bulletin of the National Research Centre* 45:1–14.

Wu, Q., Y.C. Song, H. Xu, Y. Guo, J. Li, and R.X. Tan. 2011. Medium optimization for enhanced co-production of two bioactive metabolites in the same fermentation by a statistical approach. *Journal of Asian Natural Products Res*earch 13:1110–1121.

Yang, X., X. Wang, K. Wang, L. Su, H. Li, R. Li, and Q. Shen. 2015. The nematicidal effect of *Camellia* seed cake on root-knot nematode *Meloidogyne javanica* of banana. *PloS One*, 10(4):e0119700.

Zasada, I.A., and H. Ferris. 2004. Nematode suppression with brassicaceous amendments: Application based upon glucosinolate profiles. *Soil Biology Biochemistry* 36:1017–1024.

Zhou, H., Y. Huang, T. Hoshi, Y. Kashiwagi, J.I. Anzai, and G. Li. 2005. Electrochemistry of sinapine and its detection in medicinal plants. *Analytical and Bioanalytical Chemistry* 382:1196–1201.

14 Sunflower and Mahagoni Oilseed Cake for the Management of Plant-Parasitic Nematodes

R. Sharmila, M. Shanmuga Priya, S. Prabhu and Mohd Ikram

CONTENTS

14.1 INTRODUCTION

Plant-parasitic nematodes caused crop losses on important crops around the world, ranging from 8% to 30%. The nematodes cannot be eliminated after they have been introduced to a field. Chemical nematicides have been used for a very long time, but they have some significant disadvantages, including direct toxicity to humans, aquatic organisms and beneficial insects; nematicide-induced resistance; and environmental hazard. Currently considered biodegradable, systemic, environmentally friendly and non-toxic to animals, some botanical antimicrobial substances derived from plants are considered harmless. Plant-based antimicrobials, such as pyrethroids, essential oils and neem products, are already widely utilized to control nematode populations in soil. Some plant leaf, stem, seed, root, bulb, rhizome, unripe fruit and flower head parts are used to extract botanical nematicides. Research findings have revealed that several plants have effective nematicidal properties. Farmers can grow botanicals, namely *Zantho xylumrhetsa*, *Ipomoea maxima*, *Polygonum hydropiper*, *Melia sempervirens*, *Eucalyptus globules*, *Datura stramonium*, *Cynodon dactylon*, *Melia sempervirens*, *Barringtonia*, *Pongamia pinnata* and *Aphanamixis polystachya*. These plant compounds can be used in place of chemical pesticides. Chemical insecticides are being replaced with these botanical ingredients.

The application of oil cakes into the soil causes the enhancement of antagonistic activity, improving the soil nutrient efficiency (soil texture, soil structure, water-holding capacity, cation-exchange capacity) and increasing crop yield. According to several reports, secondary metabolites of oilseed cakes, such as terpenoids and flavonoids, are released (Siddarth et al. 2014). The oilseed cake's phytochemicals (total phenol, flavonoids) suppress the soil-borne pathogens (Terpinc et al. 2012; Teh and Birch 2014).

The control of *Meloidogyne graminicola* is treated with oil cakes that contain a nemato-toxic substance. This technology is developed for integrated pest management strategies in vegetable

DOI: 10.1201/9781003319252-14

nurseries. When oil cakes were applied to the soil to increase the population of soil microorganisms, these organisms created a defence mechanism against nematodes. Somehow, these bacteria released the antimicrobial substance that causes agricultural plants to develop resistance, as shown by Bostock (2005). The discoveries to improve soil structure and porosity to lessen nematode movement and promote plant growth were published by Queneherve and Chotte (1994) and Cleyet Marcel et al. (2001). In order to reduce the nematode population, oilseed cakes were applied to the soil. This boosted predacious nematode activity (Southey 1986; Tiyagi et al. 2002). The nematode problem in vegetables belonging to Solanaceous and Cruciferous families can be effectively overcome with oilseed cakes (Goswami and Swarup 1971). One of the major threats to crops worldwide is nematodes that infest crop plants. It has been discovered that nematodes can damage every component of a plant, including roots and seeds, either on their own or in association with other soil microbes. The oil cakes used for managing plant-parasitic nematodes are an alternative to chemical control.

14.2 SUNFLOWER OILSEED CAKE

Helianthus annus L., also known as the sunflower, is an important oilseed crop. It comprises about 10% of the oil that might be extracted from this plant. Sunflower oilseeds rank third among all agricultural produce in the world. Sunflower oil cakes provide the highest levels of protein, antioxidants, vitamins, fibre and minerals (Table 14.1). Sunflower seed cake, which has a high polyphenol activity, is utilized to regulate several aspects of agriculture productivity in regions where vegetables are grown. Plant-parasitic nematode populations in soil are significantly reduced in tomatoes, and the nematode population is also reduced considerably due to phenolic compounds from oilseed cakes (Alam et al. 1980). According to the findings, high phenol content in the roots of tomatoes has a repellent effect and reduces the population of *M. javanica* (Sitaramaiah and Singh 1978). Vasudha and Sarla (2021) reported that sunflower oil cakes contain tyrosine, leucine, isoleucine, valine, lysine, arginine, phenylalanine, threonine, methionine, histidine, cysteine and tryptophan in addition to other active amino acid components. Cakes obtained from sunflower seeds are high in nitrogen, phosphorus and potash. Oil cakes applied to moist soil will take 1–2 more weeks to decompose than on dry soils. Oil cakes are used in agriculture production, including nitrogen (N), phosphorus (P) and potassium (K) (NPK) and improving N uptake to combat soil nematodes (Tiyagi et al. 2002). Additionally, this provides antagonistic activity against nematodes, gradual and steady nutrition and crop protection (Akhtar and Malik 2000). In combination, oil cakes and biocontrol agents also effectively control nematodes and induce resistance mechanisms in crop plants.

14.2.1 Sunflower Oilseed Cake and Its Role in Nematode Management

Sunflower oilseed cakes enhanced microbial-enzyme activity and physiological changes in plant roots that inhibit nematodes. Nematotoxic chemicals, such as ammonia, phenols and aldehyde, were produced during the breakdown of oil cakes (Khan et al. 1974). According to many scientists, these substances are reportedly particularly efficient against nematodes (Abrar et al. 1974; Alam et al. 1978, 1979). The effectiveness of sunflower (*Helianthus annuus*) oilseed cakes with *Pseudomonas fluorescens* increased plant height, weight, percentage of pollen fertility, number of pods, nodule formation in root and chlorophyll content in chickpea and decreased nematode populations under field conditions (Table 14.2).

Application of oilseed cakes into the soil significantly reduced *Meloidogyne incognita*, *R. reniformis* and other pests on chickpeas. Rizvi et al. (2012) also used sunflower, castor and groundnut cake to control the nematodes in the okra plant. Ram et al. (2009) and Ganaie et al. (2011) found that oilseed cakes decreased the *M. incognita* population and increased plant growth parameters and parasitic nematodes controlled with biocontrol agents and oil cakes. Oilseed cakes of the following plants, namely *Azadirachta indica* (neem), *Linum usitatissimum* (linseed), *Arachis hypogaea*

TABLE 14.1
An Average Nutrient Content in Sunflower Oilseed Cake

Sunflower Oilseed Cake	Nutrient Content (%)		
	N	P_2O_5	K_2O
Oilseed Cake (un-decorticated)	4.9	1.4	1.2
Oilseed Cake (decorticated)	7.9	2.2	1.9

TABLE 14.2
The Application of Sunflower Oilseed Cake for the Management of Plant-Parasitic Nematodes in Various Agricultural Plants

Host Plant	Nematode Species	Pot/Field/Lab	Reference
Okra	*Meloidogyne incognita*	Pots	(Bhosle et al.2006; Ganaie et al. 2011)
Fenugreek and mung bean	*Meloidogyne incognita*, *Rotylenchulus reniformis*, *Tylenchorhynchus brassicae*	Field	(Tiyagi et al. 2010, 2011)
Tomato	*Meloidogyne incognita*	Lab and pots	(Silva et al. 2018)
Lentil	*Meloidogyne incognita*	Pots	(Shankar et al. 2016)
Rice	*Meloidogyne graminicola*	Pots	(Kumar et al. 2018)
Chickpea	Plant-parasitic nematodes	Pots	(Rizvi et al. 2012)
Cowpea	*Rotylenchulus reniformis*	Pots	(Yassin and Ismail 1994)
Eggplant	*Meloidogyne incognita*, *Meloidogyne javanica*	Pots	(Singh and Veena 1988)
Chickpea	*Helicotylenchus indicus*, *Rotylenchulus reniformis*, *Tylenchorhynchus brassicae*, *Tylenchus filiformis Pratylenchus coffeae*, *Dorylaimus* spp., *Trichodorus mirzai*, *Longidorus elongatus*, *Xiphinema basiri*, *Hoplolaimus indicus*, *Meloidogyne incognita*, *Hemicriconemoides mangiferae*	Field	(Tiyagi and Ajaz 2004)

(groundnut), *Ricinus communis* (castor), *Glycine max* (soybean) and *Helianthus annuus* (sunflower), have been found to be very successful at reducing the growth of soil nematodes. The nematode damage was reduced when *Paecilomyces lilacinus* and oilseed cakes were introduced. As a result, the growth of the plant increased noticeably. Nematode development was lower in the presence of *P. lilacinus*. Other cakes and *P. lilacinus* were also successful combinations in field conditions for management of nematodes (Tiyagi and Ajaz 2004).

The plant growth–promoting characteristics of sunflower oilseed cakes have been studied (Rizvi et al. 2012). The decomposition of sunflower seed cake releases toxic components harmful to nematodes and alters the soil's physical, chemical and biological characteristics (Abbas et al. 2015). Healthy crop plants result from combining organic matter and oil cakes to reduce nematode population density and enhance the microbial population in the soil. Oilseed cakes increase soil fertility, the soil's organic matter content and antagonistic activity. More biomass is produced after the breakdown process because organic manures supply micronutrients (nitrate reductase) and increase photosynthetic activity by producing more carbohydrates and metabolites (Mahmood et al. 2007). Oilseed cake effectively controls nematodes and promotes plant growth significantly. These oilseed cakes efficiently reduce the nematode population and are environmentally safe (Sumbul et al. 2015).

14.3 MAHAGONI OILSEED CAKE

The plant *Swietenia mahagoni* (L.) Jacq. belongs to the Meliaceae family. The seeds of this plant are used to control insect pests. Mahogoni seed cake contains phenols that are useful for controlling soil pests (Gojo Cruz et al. 2018). According to Guevara et al. (1996), the crude extract from mahagoni seeds has biological properties, including antagonistic properties (Sahgal et al. 2009). The *S. mahogoni* oilseed cake contains some fatty acid components.

14.3.1 Physiochemical Properties

Phytochemical substances such as phenols, flavonoids, alkaloids, phospholipids, terpenoids, glycosides, saponins and essential oils, triterpenes, steroids, coumarines, tannins, anthrones, anthraquinones and fatty acids are found in mahogoni (Kadota et al. 1990). Mahagoni seed oil, which has a harsh flavour, a moderate drying oil content and a high level of unsaturated fatty acids, is thought to be a valuable resource. *S. mahagoni* also contains cyclomahogenol, a tetracyclic triterpene (Chakraborty et al. 1971). Figure 14.1 shows the important phytochemical contents of oilseed cake of *S. mahagoni* L.

14.3.2 Mahagoni Oilseed Cake and Its Role in Nematode Management

Mahagoni plant cakes had the highest (79.89%) mortality rate against nematode in the soil which had been treated with Bishkatali (*Polygonum hydropiper*; 74.82%) followed by neem cake (71.57%). Mahogani cake revealed a remarkable level of nematode toxicity (Mamun et al. 2014). According

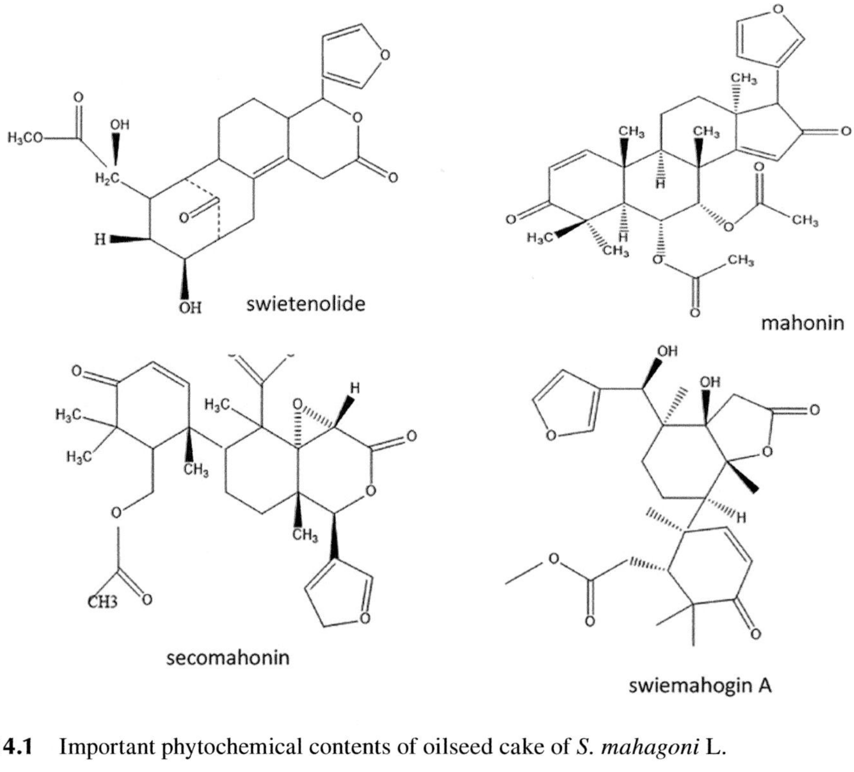

FIGURE 14.1 Important phytochemical contents of oilseed cake of *S. mahagoni* L.

TABLE 14.3
The Application of *S. mahagoni* Oilseed Cake for the Management of Plant-Parasitic Nematodes in Various Agricultural Plants

Host Plant	Nematode Species	Pot/Field/Lab	Reference
Chickpea	*Meloidogyne incognita*	Pots	(Khoraniya and Baheti 2020)
Tea	*Meloidogyne incognita*, *M. hapla* and *M. javanica*	Pots	(Hoque et al. 2017)
Tea	Plant parasitic nematodes	Field	(Mamun and Ahmed 2011)
Tea	Root lesion nematode, *Pratylenchus loosi*, Spiral nematode, *Meloidogyne* spp. *Helicotylenchus* spp.	Pots	(Mamun et al. 2014)

to the results, it may be suggested to utilize Rynaxapyr and mahogani cake as soil treatments successfully control the nematode population. Crude plant cakes of *S. mahagoni* have also been used for nematode management as an alternative to chemical nematodes in conventional and organic tea plantations. (Table 14.3). Thus, mahagoni oilseed cakes can be effectively utilized as an alternative to chemical nematicides. Its use in integrated pest management (Mamun et al. 2014) would also benefit and be safe for our environment.

14.4 CONCLUSION

Nematode management in organic agriculture requires economic support for the farmers and logistical challenges. Alternative to chemical control, sunflower and mahogoni oilseed cake is an efficient management solution for keeping the nematode population below the threshold. These oilseed cakes also increase soil fertility and promote plant growth and development. The secondary metabolites of mahogoni oilseed cake, such as phenolic compounds, triterpenoids, tetranortriterpenoids, saponins and alkaloids, remarkably manage the nematode population when used as soil amendments. The uniqueness and usefulness of the oilseed cake of mahogoni still are not well explored. Therefore, in-depth scientific studies are required to know the potential utilization in agriculture, especially in pest and disease management.

REFERENCES

Abbas, H., N. Javed, S.A. Khan, and S. Ahmad. 2015. Exploitation of the nematicidal potential of bio- and synthetic chemicals against *Meloidogyne incognita* and their impact on phytotoxicity and nematode reproduction. *Pakistan Journal of Zoology* 47:1587–1600.

Abrar, M.K., M.M. Alam, and R. Ahmad. 1974. Mechanism of the control of plant parasitic nematodes as a result of the application of oil-cakes to the soil. *Indian Journal Nematology* 93–96.

Akhtar, M., and A. Malik. 2000. Roles of organic amendments and soil organisms in the biological control of plant-parasitic nematodes: A review. *Bioresource Technology* 74:35–47.

Alam, M.M., M. Ahmed, and A.M. Khan. 1980. Effect of organic amendments on the growth and chemical composition of tomato, eggplant and chilli and their susceptibility to attack by *Meloidogyne incognita. Plant and Soil* 57:231–236.

Alam, M.M., A.M. Khan, and S.K. Saxena. 1978. Mechanism of control of plant parasitic nematodes as a result of the application of organic amendments to the soil. IV. Role of formaldehyde and acetone. *Indian Journal of Nematology* 8:172–174.

Alam, M.M., A.M. Khan, and S.K. Saxena. 1979. Mechanism of control of plant parasitic nematodes as a result of the application of organic amendments to the soil. V. Role of phenolic compounds. *Indian Journal of Nematology* 2:136–142.

Bhosle, B.B., S. Mukesh, S.N. Puri, H.R. Sardana, and D.K. Singh. 2006. Efficacy of organic amendment in management of root-knot nematode *Meloidogyne incognita* on okra. *Indian Journal of Nematology* 36:37–40.

Bostock, R.M. 2005. Signal crosstalk and induced resistance: Straddling the line between cost and benefit. *Annual Review of Phytopathology* 43:545–580.

Chakraborty, D.P., S.P. Basak, B.C. Das, and R. Beugelmans. 1971. Cyclomahogenol, a new tetracyclic triterpene from *Swietenia mahagoni*. *Phytochemistry* 10:1367–1372.

Cleyet Marcel, J.C., M. Larcher, H. Bertrand, S. Rapior, and X. Pinochet. 2001. Plant growth enhancement by rhizobacteria. In: Morot-Gaudry JF, editor. *Nitrogen Assimilation by Plants: Physiological, Biochemical and Molecular Aspects*. Science Publishers Inc., Plymouth, 185–197.

Ganaie, M.A., A.A. Rather, and M.A. Siddiqui. 2011. Pathogenicity of root knot nematode *Meloidogyne incognita* on okra and its management through botanicals. *Archives of Phytopathology Plant Protection* 44:1683–1688.

Gojo Cruz, P.H., R.L. Nunez, C.A. Anulacion, N.A. Gregorio, K. Niwan, E.J. Nolasco, R.M. Pallarca, and K.G.D. Waing. 2018. Insecticidal property and phytochemical screening of mahogany (*Swietenia macrophylla* King) leaves, barks and seeds as an alternative insecticide against fungi growing termites (*Macrotermes gilvus* (hagen, 1858)). *International Journal of Biology, Pharmacy and Allied Science* 7:1521–1528.

Goswami, B.K., and G. Swarup. 1971. Effect of oil-cake amended soil on the growth of tomato and root-knot nematode population. *Indian Phytopathology* 24:491–494.

Guevara, A.P., A. Apilado, H. Sakurai, M. Kozuka, and H. Takuda. 1996. Antiinflamatory, antimutagenicity, and antitumor promoting activities of mahogany seeds, *Swietenia macrophylla* (Meliaceae). *Philippine Journal Science* 125:271–277.

Hoque, A.K.M.R., A.F.M. Aslam, A.J. Howlader, and M.S.A. Mamun. 2017. Performance of some plant cakes against plant parasitic nematodes in tea. *Journal of Entomology and Zoology Studies* 5:1428–1432.

Kadota, S., L. Marpaaung, T. Kikuch, and H. Ekimoto. 1990. Constituents of the seeds of *Swietenia mahagoni* JACQ. II. Structures of swietemahonia A,B,C,D,E,F and G and swietemahonilide. *Chemical and Pharmaceutical Bulletin* 38:894–901.

Khan, A.M., M.M. Alam, and S.K. Saxena. 1974. Mechanism of the control of plant parasitic nematodes as a result of the application of oil-cakes to the soil. *Indian Journal of Nematology* 4:93–96.

Khoraniya, P., and B.L. Baheti. 2020. Bioefficacy of botanical-based seed treatment against root-knot nematode, *Meloidogyne incognita* infesting chickpea (*Cicer arietinum* L.). *Journal of Entomology and Zoology Studies* 8:9–12.

Kumar, A., K. Khilari, S.K. Jain, A.K. Yadav, and A.K. Mishra. 2018. Efficacy of different oil cakes on management of root knot nematode (*Meloidogyne graminicola*) in rice crop in Western Uttar Pradesh, India. *International Journal of Current Microbiology and Applied Science* 7(12):3524–3530.

Mahmood, I., S.A. Tiyagi, and M.F. Azam. 2007. Efficacy of latex bearing plants for the management of plant-parasitic nematodes and soil-inhabiting fungi on chickpea and mungbean. *Environmental Biology and Conservation* 12:23–27.

Mamun, M.S.A., and M. Ahmed. 2011. Prospect of indigenous plant extracts in tea pest management. *International Journal of Agricultural Research, Innovation and Technology* 1:16–23.

Mamun, M.S.A., M. Ahmed, I. Ahmad, M.S. Uddin, and S.K. Paul. 2014. Comparative studies on the performances of some plant cakes and synthetic chemicals against nematodes in tea in Bangladesh. *World Journal of Agriculture Research* 2(1):1–4.

Queneherve, P., and J.L. Chotte. 1994. Distribution of nematodes in vertisol aggregates under a permanent pasture in Martinique. *Applied Soil Ecology* 4:193–200.

Ram, L., A.U. Siddiqui, and A. Parihar. 2009. Management of root-knot nematode, *Meloidogyne incognita* infecting okra using oil cakes. *Indian Journal of Nematology* 39:125–127.

Rizvi, R., I. Mahmood, S.A. Tiyagi, and K. Zehra. 2012. Conjoint effect of oil-seed cakes and *Pseudomonas fluorescens* on the growth of chickpea in relation to the management of plant-parasitic nematodes. *Brazilian Archives of Biology and Technology* 55:801–808.

Sahgal, G., S. Ramanathan, S. Sasidharan, M.N. Mordi, S. Ismail, and S.M. Mansor. 2009. Phytochemical and antimicrobial activity of *Swietenia mahagoni* crude methanolic seed extract. *Tropical Biomedicine* 26:274–279.

Shankar, P., P. Naresh, and S.K. Biswas. 2016. Integrated effect of oil cakes, bioagent and nematicide on root knot namatode and root nodulation in lentil. *Journal of Pure and Applied Microbiology* 10:2949–2955.

Siddarth, N.R., V.P. Singh, R.K. Singh, S. Kumar, and K. Khilari. 2014. Effect of different organic amendments on the radial growth of *Rhizoctonia solani* Kuhn causing black scurf of potato. *Trends in Biosciences* 7:3874–3876.

Silva, J.C.P., V.P. Campos, A.F. Barros, M.P. Pedroso, W.C. Terra, L.E. Lopez, and J.T. de Souza. 2018. Plant volatiles reduce the viability of the root-knot nematode *Meloidogyne incognita* either directly or when retained in water. *Plant Disease* 102:2170–2179.

Singh, S.P., and S. Veena. 1988. Evaluation of oilcakes and nematicides for the control of *Meloidogyne incognita* infecting egg-plants. *Indian Journal of Nematology* 18:366–368.

Sitaramaiah, K., and R.S. Singh. 1978. Effect of organic amendment on phenolic content of soil and plant response of *Meloidogyne javanica* and its host to related compounds. *Plant Soil* 50:671–679.

Southey, J.F. 1986. *Laboratory methods for work with plant and soil nematodes.* London: Ministry of Agriculture, Fisheries and Food, HMSO.

Sumbul, A., R. Rizvi, I. Mahmood, and R.A. Ansari. 2015. Oil-cake amendments: Useful tools for the management of phytonematodes, Asian. *Journal Plant Pathology* 9:91–111.

Teh, S.S., and E.J. Birch. 2014. Effect of ultrasonic treatment on the polyphenol content and antioxidant capacity of extract from defatted hemp, flax and canola seed cakes. *Ultrason Sonochemistry* 21:346 353.

Terpinc, P., B. Ceh, N.P. Ulrih, and H. Abramovic. 2012. Studies of the correlation between antioxidant properties and the total phenolic content of different oil cake extracts. *Indian Crop Production* 39:210–217.

Tiyagi, S.A., and S. Ajaz. 2004. Biological control of plant parasitic nematodes associated with chickpea using oil cakes and *Paecilomyces lilacinus. Indian Journal of Nematology* 34:44–48.

Tiyagi, S.A., A.V. Khan, and M.M. Alam. 2002. Biodegradable effect of oil seed cakes on plant-parasitic nematodes and soil inhabiting fungi infesting *Trigonella foenumgreacum* and *Phaseousaureus. Indian Journal of Nematology* 32:47–57.

Tiyagi, S.A., I. Mahmood, Z. Khan, and H. Ahmad. 2011. Biological control of soil-pathogenic nematodes infecting mungbean using *Pseudomonas fluorescens. Archives of Phytopathology and Plant Protection* 44:1770–1778.

Tiyagi, S.A., I. Mahmood, and R. Rizvi. 2010. Application of organic additives for the management of plant-parasitic nematodes and soil-pathogenic fungi on fenugreek (*Trigonella foenum-graecum*) and mungbean (*Phaseolus aureus*). *Journal of Medicinal and Aromatic Plant Sciences* 32:420–431.

Vasudha, C., and L. Sarla. 2021. Nutritional quality analysis of sunflower seed cake. *The Pharma Innovation Journal* 10:720–728.

Yassin, M.Y., and A.E. Ismail. 1994. Effect of some oilseed cakes as soil amendments and Aldicarb on cowpea plants infected with *Rotylenchulus reniformis* Lin. & Ol. in relation to soil type. *Anzeiger für Schädlingskunde, Pflanzenschutz, Umweltschutz* 67:176–178.

15 Importance of Oilseed Cakes in Agriculture

Niloufar Mahmoudi and Yousef Naserzadeh

CONTENTS

15.1 INTRODUCTION

Oilseed cakes left after the oil extraction for different purposes are chiefly used as cattle feed, compost amendment, or plant conditioner. In this context, using oilseed meals as nitrogen (N) organic fertilizers may offer a possibility to decrease the environmental impact associated with the manufacture and application of chemical fertilizers (high fossil fuel use, substantial greenhouse gas emissions, N leaching, nitrous oxide volatilization, etc.), as well as to promote plant growth by increasing the supply of organic carbon (C) and boosting soil biological activity (Zaccardelli et al. 2013). Even though oilseed meals are typically used to feed animals, multiple studies have shown that seed meals from cotton and various Brassicaceae, such as Ethiopian mustard, are effective fertilizers (Gale et al. 2006; Marchetti et al. 2008). *Camelina sativa* (CS), known as false flax, has increased interest as a substitute oilseed crop over the past ten years (Rizvi et al. 2012). The high concentration of omega-3 fatty acids and essential fatty acids in CS seed oil distinguishes it from others. The oil is appropriate for human and industrial uses for food and non-food products (Chaudhary and Kaul 2013). Human ingestion of CS seed began in the Neolithic period. According to archaeological digs, the seed was a significant component of the human diet from 500 BCE to 300 CE. The seed was added to bread and used in porridge. *Camelina* was occasionally grown during the Middle Ages.

DOI: 10.1201/9781003319252-15

15.1.1 What Is Oilseed Cake, and What Ingredients Is It Made of?

Oilseed cake, a coarse by-product of oil extraction from various oilseeds, is beneficial as chicken and other animal feed because it is high in protein and minerals. It could be sold in pieces or processed into an oil meal. Castor beans, soybeans, peanuts, flaxseed (linseed), rapeseed, cottonseed, coconuts (copra), oil palm, and sunflower seeds are among the oilseeds that are used to make oil cakes used as animal feed (Gupta et al. 2018; Sharma 2013). The woody hulls and shells of cottonseed and peanuts are often removed before processing. Before using as feed for non-ruminant livestock like pigs and chickens, the pressed cake from the cottonseed oil manufacturing must also be treated to remove poisonous substances (Kushwaha 2010). Numerous unfavourable situations have surfaced during the past few years, including resource overuse and misuse, unsustainable consumption patterns, environmental deterioration, ecosystem balance, and climate change. If we want to reduce these problems, we must focus on innovative tactics and methods to optimally use natural resources and eradicate the concept of waste in the supply chain. Recycling, or the idea that waste can be used by turning it into value-added items, is a recent innovation (Bárta et al. 2021).

The edible oil industry processed the raw seeds to obtain the oil and the residual cakes of solid by-products known as oilseeds. India is the world's greatest producer of oilseeds, producing more than 25 million tons of oilseed cake yearly (Nevara et al. 2022). The largest agricultural by-product in the world by volume is soybean oilseed cake, followed by canola cake, and this problem is getting worse every day (Naserzadeh et al. 2022).

15.2 APPLICATION AND USES OF OILSEED CAKES

Oilseed cakes were once employed as soil compost and animal feed due to their high quality, low cost, and benefits. According to a study by Hessle et al. (2009), oil cake enhances the weight of young calves. Because hemp cakes are high in protein and amino acids, they have several extra benefits as animal feed (Hessle 2009). Oilseed cakes also have substantial scientific value because they are crucial to human nutrition. However, they are underutilized due to a lack of knowledge and preliminary study. Oilseed cakes offer much potential for producing many products with added value and high nutritional value. Oilseed cake is an excellent alternative to satisfy the growing need for food high in protein, minerals, and bioactive components. Evidence for using oilseed cakes as food sources for human consumption has been obtained. Meyer (1971) reported protein extraction, processing, and the functional and nutritional properties of soybean and other oilseed cakes. There is a growing demand to supply food that is rich in protein. The recommended dietary allowance (Kumar et al. 2021) for protein is 0.8 grams per kilogram of body weight, representing the minimum protein requirement for growth and development. It can be said that vegetable proteins are versatile and non-toxic. Additionally, vegetable protein sources may not provide all essential amino acids in adequate amounts, which can limit their utilization as a sole source of protein (Singh et al. 2022). Therefore, it is recommended to combine different vegetable protein sources to achieve a balanced amino acid profile and maximize their nutritional value. However, the taste, colour and texture of foods are affected, because vegetable protein does not contain sulfur amino acids.

The protein composition of oilseed cakes has not been thoroughly studied. In order to supplement the existing protein sources for human consumption, this section critically reviews the literature on the quality and functional properties of proteins extracted from oilseed cakes (Greiling et al. 2018). It also discusses issues with oilseed cake disposal and suggests ways to increase the value of oilseeds. The Protein Digestibility Modified Amino Acid Score (PDCAAS) technique, used by the World Health Organization/Food and Agriculture Organization (WHO/FAO) to assess the nutritional value of protein, is preferred by crop growers and processors. The basic idea behind this approach is to contrast the first necessary amino acid's concentration with the reference amino acid concentration pattern (Bárta et al. 2021). In a nutshell, it is used to assess the protein quality of food and is typically used to determine the per cent daily value listed on the label in research and development foods. This scoring system is based on the small age range of preschoolers' need for critical amino acids. The actual faecal digestibility of the test protein is adjusted to the chemical score that was acquired by this item (Kotecka-Majchrzak et al. 2021). PDCAAS scores higher than 100% should be trimmed to 100% because they are unacceptable. Since there has not been much study on amino acid needs since the PDCAAS technique was adopted, the scoring model, which excludes conditionally essential amino acids, needs to be validated (Ely and Fike 2022).

15.3 USE OF OILSEED CAKES AND PLANT MATERIALS AGAINST PLANT PESTS AND DISEASES

Insect pests, nematodes, and plant diseases of vegetables and other agricultural products are generally controlled by the frequent use of chemical pesticides to increase crop productivity and earn more profit in conventional agriculture. In global markets, products with less pesticide use are preferred (Gahukar 2012). For this purpose, crude plant-derived products or formulated pesticides can be environmentally friendly, effective, and cost-effective for the average producer. One of the critical uses of oilseed cakes is their use as organic pesticides, which is a great help to farmers. For example, neem seed oil cake is widely used because it is an egg-laying inhibitor, disinfectant, anti-nutritional, repellent, and pesticide against insects and a spore inhibitor. They affect the growth and development of pathogens and hinder in root penetration of nematodes. Further research should be focused on the synthesis and commercial formulation of oilseed insecticides for pest and pathogen control.

15.4 OILSEEDS AS BY-PRODUCTS IN PEST CONTROL AND USE AS A SOURCE OF PLANT NUTRIENTS

Oilseed cakes of plant products have received more attention for preventing several species of plant-parasitic nematodes. Numerous experiments have shown the potential nematicidal value of plant parts and their by-products when incorporated into the soil or when seed cake is incorporated into crops. Various products (oils, cakes, extracts, etc.) are prepared from the leaves and seeds of the neem plant (*Azadirachta indica* A. Juss; Meliaceae family) as adequate protection against nematode pests when used as roots and seeds. It has been reported that nematode toxic compounds, especially azadirachtins, are released through volatilization, secretion, leaching, and decomposition (Nadeem et al. 2021). The modes of action of these compounds are complex, and several mechanisms related to nematode management are still not fully understood. This chapter critically evaluates the potential of these products in managing nematodes in tropical agriculture (Anirudh et al. 2022). Figure 15.1 shows oilseed cake as a biofertilizer.

FIGURE 15.1 Oilseed cake as a biofertilizer.

The use of oilseed cake as a source of plant nutrients and nematode control is common among crop and fruit growers. Using neem, castor, mustard, and rocket salad oil cakes suppressed the root-knot nematode and reniform nematode in okra. It improved the plant growth and water absorption capacity of its roots. Mainpueira, a by-product of cassava flour production, was effective against *M. incognita* under greenhouse conditions and in the field. Citrus cannery waste in the form of oilseed cake as a soil conditioner had reasonable control over tomato nematodes. Using cassava root peeler and locust bean shell as soil conditioners will reduce the sugarcane root-knot and increase its growth.

15.4.1 Nematicidal Properties of Oilseed Cake

A large number of oilseed cakes have a direct effect on plant nematodes because they have nematicidal properties. Oilseed cake contains numerous secondary metabolites which have antagonistic activity for plant-parasitic nematodes. Nadeem et al. (2021) studied the oilseed cake application against J2 (second-stage juveniles) of root-knot nematodes to test the nematicidal potential. They found that oilseed cake showing a higher number of metabolites causes significant immobilization in the J2 population. The killing efficiency of the oil cake was also higher when incubating the J2 for a more extended period. However, the higher concentration strength was also an essential factor. The J2 were killed at a maximum in a higher concentrations of oilseed cake extract. Thus, they reported that concentration strength and incubation period were directly proportional to the killing of J2.

15.5 THE ROLE OF OLIVE OILSEED CAKE IN LIVESTOCK NUTRITION

If we take into account that each year, oil extraction facilities in 42 different nations, including Iran, Italy, France, Turkey, and others, produce roughly 20,000 tonnes of waste (olive skin and kernel) during the extraction of oil from 65,000 tonnes of olive seeds (Hazreen-Nita et al. 2022). Thus, it is possible to forecast the role of olive waste or other agricultural waste in producing roughly 5,000 tons of red meat and 1,000 tons of eggs by studying this waste (amount of protein, fat, dry matter, etc.; Obeidat 2021).

One of animal production's primary and fundamental issues is the lack of animal feed in emerging and even developed nations. Since the most significant importance of the ruminant animal is the consumption and digestion of forages rich in cellulose and fibrous carbohydrates that are suitable for animals, it is necessary to provide animal feed that makes the best use of non-conventional feed sources that are available or extracted and do not compete with human foods (Hazreen-Nita et al. 2022). However, a significant amount of agricultural waste is produced each year that can be utilized as animal feed, such as olive pomace (skin and core).

15.5.1 Stray Olive Waste

Olive pomace, a by-product of the oil extraction process, is composed of olive kernels, skins and pomace, and some fat. Waste (olives) is one of the challenges of modern life in industrialized countries because, according to data from all over the world, a significant part of agricultural production–between 5% and 10%–is thrown away as waste and can instead be used for feeding (de Lima et al. 2021). A large population is estimated to be 2 million to 900,000 tons of olive pomace produced worldwide.

The conversion of the by-products into materials with higher nutritional value has effectively used available resources and solved environmental issues. However, aside from olive oil, two types of by-products of olive pomace (used in its processing and used in animal feeding) and oilseed cake have been the focus of animal breeders and experts in nutritional sciences for years (Kranthi et al. 2012). Olive pomace contains more fibre and less crude protein than a meal made from other oilseeds. Pomace is one of the principal by-products produced during the oil extraction process from the olive fruit. Processing and animal feed manufacturers now buy this by-product in limited quantities (Anver 2006). Due to the simultaneous production of olive pomace (about 6–8 weeks per year), it is important to maintain its quality, risk-free and healthy, which is protected from damage, and during the storage period, its materials and components remain intact. Remain oilseed cake can be processed to be stored and then used as animal food while simultaneously reducing its environmental impact. Because raw olive oilseed cake has a high crude fibre content, rarely used to feed dairy cows (Prasad et al. 2005).

15.6 EFFECT OF OLIVE OILSEED CAKE ON DIFFERENT ANIMAL TRAITS

Experts believe that using 25% olive seed cake in the diet of fattening calves will reduce the cost of the diet without having any adverse effects on the animal's performance. So far, much research has been conducted in the direction of using industrial and agricultural wastes and by-products, and various experiences have been obtained (Goswami et al. 2007). Also, studies show the use of agricultural wastes and residues such as pistachio skins, pomegranate seeds, date seeds, corn and wheat waste from pasta production factories, and olive waste. These are enriched in energy and protein needed in animal feed. It is useful, but in any case, we must first analyse the olive pomace and finally the olive oilseed cake (dry matter, fat, protein, etc.; Ramachandran et al. 2007), but it should also be checked in the areas of livestock production and consumption and the market to be operational. Among them, most of the available information is about the consumption and digestion of olive oil cake. Figure 15.2 shows turning olive pomace into oilseed cake suitable for animal feed.

Some researchers have shown that the use of oilseed cakes in the diet improves the digestibility of animal feed, improves growth and increases daily weight and body weight in all types of animals, especially lambs and goats, and improves the feed conversion ratio (Hazreen-Nita et al. 2022; Mahmoudi et al. 2020).

15.7 TURNING DIRTY GOLD INTO RICHES

Investigations show that it is possible to use the waste of pomegranate seeds, tomatoes, grapes, olives, and others (dirty gold) to produce all kinds of raw materials for hygiene, cosmetics, edible oil, oil for industrial purposes, animal feed, and even poultry feed in the form of oilseed cakes (Nadakavukaren Schefer 2018). Interestingly, dry olive pomace includes over 90% dry matter, certain vitamins and minerals, 7–10% oil, 6–9% crude protein, 44–48% cellulose fibre (absorbable), and lignin (ideal for animal rumination).

These oilseed cakes as by-products are included in the diets of dairy cows (3%), fattening cows (5%), and laying hens. However, the applications and advantages of oilseed cakes are still unclear

FIGURE 15.2 Turning olive pomace into oilseed cake suitable for animal feed.

because they have not been cultured, marketed, or encouraged for usage (Sousa et al. 2022). In the UK, a knowledge-based business transforms regional trash, including rotting fruit and vegetable remnants, agricultural waste, and food waste, into high-quality animal feed and oilseed cakes, which is also helpful for insect breeding (Dame-Korevaar et al. 2021). Taking into account hundreds of thousands of tons of occupation and urban waste every day, a step should be taken in this way to reach an acceptable level while doing less damage to the environment in the production of animal feed and without the need for imports (Salemdeeb et al. 2017).

15.8 EVALUATION OF OILSEED CAKE UTILIZATION IN FOOD SECTOR ITEMS

15.8.1 Two Categories of Components Can be Used to Create Novel Culinary Products

- First, it consists of brand-new, replacement, and unused food items.
- The second method is through valuing by-products.

You may successfully implement them by developing a marketing plan. Explicit transparency (improved communication, suitable labelling), new regulations, and client education are all part of this (Mirpoor et al. 2021). The function is the creation of unique food products that offer considerable advantages to consumers. Oil industry leftovers can be used as co-products for high-value commodities, food additives, or vitamins. In a review, oilseed cakes are sorted into three groups based on their physicochemical composition. The enriched sesame oil cake is predominantly made up of proteins and lipids. When they are predominant in oilcake, proteins and carbohydrates can be used as fillers in dairy and meat (as in the case of pumpkin and walnuts). They are mixed and crushed to create supplement-grade pills with beneficial properties (Zaky et al. 2021). Treatments were applied to oilseed cakes to enhance their flavour, appeal, and usefulness as dietary supplements for individuals. In this instance, coconut and sesame cakes are blended with water, and the mixture is then heated until the solid is separated from the liquid and dried. Next, it was determined how much food was in the cooked and raw residue (solid and extract; Moure et al. 2006). Nutrient retention is generally higher in boiling residues than in extracts. For instance, camellia and coconut solid cake were used to make four healthy snacks (Ni 2019). The first and third formulae contained 10% and 20% coconut cake. However, the other two contained coconut cake and camellia at 15% each. A sensory evaluation determined that all foods were suitable.

15.8.2 Packaging Composed of Biopolymers

Biodegradable or self-degrading polymeric materials were synthesized to improve the quality and shelf life of food, as well as to minimize packaging residues and use fewer petroleum derivatives. Films and coatings are biopolymeric packaging materials with the same chemical composition but different appearances: the former are other materials, while the latter is formed on the product's surface. The two processes to develop polymer films are casting (or wet process) and compression/extrusion (or dry process). The most-used one is the first one and involves the dispersion of the constituent macromolecules in a suitable medium followed by the evaporation of the solvent (Aung et al. 2018). Due to the high potential of oilseed cake production, these products can be used to produce biodegradable packaging.

The application of the film in packaging is influenced by its optical, thermal, mechanical microbiological, structural, sensory (must be tasteless, clean, colourless, completely transparent, and odourless), and barrier properties (against aromatic compounds, water, gases, and light penetration; Rhim and Kim 2014). These layers provide some advantages, such as biocompatibility, biodegradability, oxidation resistance, and resistance to microbial issues.

Lipids, proteins, and polysaccharides are used alone or in combination with other biopolymers to make biodegradable films and coatings. Meanwhile, remarkable films with good mechanical and gas barrier properties were produced using proteins (Kumar et al. 2022). A single natural polymer can be used to create films with a variety of high attributes but low quality. It is possible to improve polymers by combining two or more of them. These materials can be used to create biodegradable packaging due to the high potential of oilseed cake manufacturing.

15.8.3 Variability of Specific Components in CS Oilseed Cakes

False flax, also known as CS (Anirudh et al. 2022), has received increased interest as an oilseed substitute over the past ten years. The proportion of omega-3 fatty acids and necessary fatty acids in CS seed oil makes it unique (Alberghini et al. 2022). This oil can be used in industrial food and non-food applications and for human consumption. According to archaeological digs, seeds were a significant component of the human diet from 500 BCE to 300 CE. This seed was used in both porridge and bread. *C. annua* was infrequently grown during the Middle Ages. It was grown in Europe and Russia before World War II and shortly afterwards. Figure 15.3 shows CS plantation.

Compared to typical oilseeds like rapeseed (*Brassica napus*) and sunflower (*Helianthus annuus*), camelina has substantial technical agricultural advantages. It might be argued that this crop, which can be grown without pesticides or herbicides and with minimal input, is crucial for producing oilseed cakes (Embaye et al. 2018). In rational industrial exploitation, CS oilseed cake (CSOC) should be viewed as a significant economic product in the agricultural sector; for instance, how is CSOC operated? Animals' daily diets must include them as part of a balanced diet. Numerous studies have been conducted to support and validate the usage of CSOC as a feed (Brunharo et al. 2022). However, the use of CSOC in animal feed is prohibited in EU countries, with various regulations specifying undesirable substances in food because some of these substances may contain mycotoxins or other unknown toxins. Only a few classes of mycotoxins, including aflatoxins (AFs), fumonisins (FMs), ochratoxins (OTs), trichocenes (TRCs), and zearalenone (Andreadis et al. 2021), are thought to be clinically proven, despite being abundant in feed. Earlier research suggested that eating CS and CSOC seeds had negative consequences. Animals ingested less fodder, and modifications to the

FIGURE 15.3 *Camelina sativa* plantation.

texture and aroma of milk were observed. However, recent studies have shown that CSOC can be tolerated as a dietary component when used in tiny amounts. When used in pork feed, there were substantial effects on the growth rate, liver weight, carcass fat content, and meat flavour. Similar limitations are known to apply to the use of oilseed rape cakes in diets. Other cakes are incredibly valued, but it is important to remember that some might not include unhealthy substances (Obeng 2018). This guideline applies to all plants, including the CS. However, it might be a food or a great food substitute in the agricultural and food sectors.

15.9 CONCLUSION

The chapter covered the potential uses of enormous waste and by-products made from oilseeds in the form of oilseed cakes, primarily employed in pest and disease control, landfills, energy sources, and animal feed. Additionally, using these products to produce and consume agricultural goods might be seen as a green strategy. Oilseed cakes are leftovers from the process of extracting oil from seeds. Because it contains significant levels of bioactive compounds and can separate and utilize them, the synthesis of low-cost value-added goods like fuel, surfactants, enzymes, antibiotics, vitamins, natural colours, flavouring ingredients, and health-improving ingredients can be utilized efficiently. The proteins, phytochemicals, and flavonoids successfully implemented the zero waste and circular economy concepts and the effective utilization of garbage. The circular economy views the valuation of trash as a difficulty. However, it can also enable the reuse of resources in the supply chain since they increase food value, lower costs, foster economic growth, and lower environmental concerns associated with their disposal.

REFERENCES

Alberghini, B., F. Zanetti, M. Corso, S. Boutet, L. Lepiniec, A. Vecchi, and A. Monti. 2022. Camelina [*Camelina sativa* (L.) Crantz] seeds as a multi-purpose feedstock for bio-based applications. *Industrial Crops and Products* 182:114944.

Andreadis, S.S., N. Panteli, M. Mastoraki, E. Rizou, V. Stefanou, S. Tzentilasvili, E. Sarrou, S. Chatzifotis, N. Krigas, and E. Antonopoulou. 2021. Towards functional insect feeds: Agri-food by-products enriched with post-distillation residues of medicinal aromatic plants in tenebrio molitor (Coleoptera: Tenebrionidae) breeding. *Antioxidants* 11:68.

Anirudh, R., R. Archibald, M.S. Asif, M.M. Becker, S. Benkadda, P.T. Bremer, R.H. Budé, C.S. Chang, L. Chen, R.M. Churchill, and J. Citrin. 2022. Review of data-driven plasma science. *arXiv preprint arXiv* 2205:15832.

Anver, S. 2006. Control of plant parasitic nematodes associated with linseed by oil seed cakes and ploughing. *Indian Journal of Nematology* 36:173–175.

Aung, S., P. Sone, H.H.H. Shein, K.N. Aye, and N. Nwe. 2018. Environment-friendly biopolymers for food packaging: Starch, protein, and poly-lactic acid (PLA). In *Bio-based Materials for Food Packaging*, 173–195. Springer, Singapore.

Bárta, J., V. Bártová, M. Jarošová, J. Švajner, P. Smetana, J. Kadlec, V. Filip, J. Kyselka, M. Berčíková, and Z. Zdráhal. 2021. Oilseed cake flour composition, functional properties and antioxidant potential as effects of sieving and species differences. *Foods* 10:2766.

Brunharo, C., A.C.G. Roger Gast, V. Kumar, C.A. Mallory-Smith, B.D. Tidemann, and H.J. Beckie. 2022. Western United States and Canada perspective: Are herbicide-resistant crops the solution to herbicide-resistant weeds? *Weed Science* 1–15.

Chaudhary, K.K., and R.K. Kaul. 2013. Efficacy of *Pasteuria penetrans* and various oil seed cakes in management of *Meloidogyne incognita* in chilli pepper (*Capsicum annuum* L.). *Journal of Agricultural Science and Technology* 15:617–626.

Dame-Korevaar, A., I.J.M.M. Boumans, A.F.G. Antonis, E.V. Klink, and E. Mde. Olde. 2021. Microbial health hazards of recycling food waste as animal feed. *Future Foods* 4:100062.

de Lima, S., G. Cordeiro, S.G. Neto, L.R. Bezerra, A.N. de Medeiros, E.O. Cavalcante, and R.D. Signoretti. 2021. Palm oil cake in milk production and quality of dairy cows: Systematic review, meta-analysis and principal component analysis. *Livestock Science* 254:104760.

Ely, K., and J. Fike. 2022. Industrial hemp and hemp byproducts as sustainable feedstuffs in livestock diets. In *Cannabis/Hemp for Sustainable Agriculture and Materials*, 145–162, Springer, Singapore.

Embaye, W.T., J.S. Bergtold, D. Archer, C. Flora, G.C. Andrango, M. Odening, and J. Buysse. 2018. Examining farmers' willingness to grow and allocate land for oilseed crops for biofuel production. *Energy Economics* 71:311–320.

Gahukar, R.T. 2012. Evaluation of plant-derived products against pests and diseases of medicinal plants: A review. *Crop Protection* 42:202–209.

Gale, E.S., D.M. Sullivan, C.G. Cogger, A.I. Bary, D.D. Hemphill, and E.A. Myhre. 2006. Estimating plant available nitrogen release from manures, composts, and specialty products. *Journal of Environmental Quality* 35:2321–2332.

Goswami, B.K., R.K. Pandey, J. Goswami, and D.D. Tewari. 2007. Management of disease complex caused by root knot nematode and root wilt fungus on pigeonpea through soil organically enriched with vesicular arbuscular Mycorrhiza, karanj (Pongamia pinnata) oilseed cake and farmyard manure. *Journal of Environmental Science and Health, Part B* 42:899–904.

Greiling, A.M., R. Reiter, and M. Rodehutscord. 2018. Utilization of unprocessed and fibre-reduced oilseed cakes of rapeseed and sunflower seed in rainbow trout (*Oncorhynchus mykiss* W.) nutrition–Evaluation of apparent digestibility and growth performance. *Aquaculture Nutrition* 24:1133–1143.

Gupta, A., R. Sharma, S. Sharma, and B. Singh. 2018. Oilseed as potential functional food ingredient. In *Trends & Prospects in Food Technology, Processing and Preservation*, 1st ed., Prodyut Kumar, P., Mahawar, M.K., Abobatta, W., and Panja, P. eds., 25–58. Today and tomorrow printers and publishers.

Hazreen-Nita, M.K., Z.A. Kari, K. Mat, N.D. Rusli, S.A.M. Sukri, H.C. Harun, S.W. Lee, M.M. Rahman, N.H. Norazmi-Lokman, M. Nur-Nazifah, and M. Firdaus-Nawi. 2022. Olive oil by-products in aquafeeds: Opportunities and challenges. *Aquaculture Reports* 22:100998.

Hessle, A.K. 2009. Effects of social learning on foraging behaviour and live weight gain in first-season grazing calves. *Applied Animal Behaviour Science* 116:150–155.

Kotecka-Majchrzak, K., A. Sumara, E. Fornal, and M. Montowska. 2021. Proteomic analysis of oilseed cake: A comparative study of species-specific proteins and peptides extracted from ten seed species. *Journal of the Science of Food and Agriculture* 101:297–306.

Kranthi, V.S., D.M. Rao, and P. Jaganmohan. 2012. Production of protease by *Aspergillus flavus* through solid state fermentation using different oil seed cakes. *International Journal of Microbiological Research* 3:12–15.

Kumar, R., N.R. Nizampatnam, M. Alam, T.K. Thakur, and A. Kumar. 2021. Genome editing technologies for plant improvement: Advances, applications and challenges. *Omics Technologies for Sustainable Agriculture and Global Food Security* 1:213–240.

Kumar, S., I.B. Basumatary, A. Mukherjee, and J. Dutta. 2022. An overview of natural biopolymers in food packaging. In *Biopolymer-Based Food Packaging*, 1–28. Hoboken, NJ: John Wiley & Sons, Ltd.

Kushwaha, P. 2010. Need to discontinue the use of solvents for the extraction of edible oils for ensuring safety of public health. *International Journal of Pharmaceutical Research* 2:7–14.

Mahmoudi, N., M. Pridannikov, M. Zargar, Y. Naserzadeh, L. Limantceva, and E. Pakina. 2020. Molecular diagnostics of *Ditylenchus destructor* based on the ITSrDNA from Iran and Russia Federation. *Research on Crops* 21:151–155.

Meyer, E.W. 1971. Oilseed protein concentrates and isolates. *Journal of the American Oil Chemists' Society* 48:484–488.

Mirpoor, S. Fatemeh, C.V.L Giosafatto, and R. Porta. 2021. Biorefining of seed oil cakes as industrial co-streams for production of innovative bioplastics. A review. *Trends in Food Science and Technology* 109:259–270.

Moure, A., J. Sineiro, H. Domínguez, and J.C. Parajó. 2006. Functionality of oilseed protein products: A review. *Food Research International* 39:945–963.

Nadakavukaren Schefer, K. 2018. All that glitters: Conflict diamonds, dirty gold and the WTO legal framework on trade and human rights. In *European Yearbook of International Economic Law 2018*, 81–111. Springer Nature, Switzerland.

Nadeem, H., P. Malan, A. Khan, K.H.A.N. Amir, A.S.I.F. Mohd, M.A. Siddiqui, S.T. Angombe, and F. Ahmad. 2021. New insights on the utilization of ultrasonicated mustard seed cake: Chemical composition and antagonistic potential for root-knot nematode, *Meloidogyne javanica*. *Journal of Zhejiang University-SCIENCE B* 22:563–574.

Naserzadeh, Y., N. Mahmoudi, A.M. Nafchi, A.F. Hashim, and F.K. Ahmed. 2022. Antipathogenic effects of plant essential oil nanoemulsions. In *Bio-Based Nanoemulsions for Agri-Food Applications*, 245–257. Elsevier, UK.

Nevara, G.A., S.G. Ibrahim, S.K.S. Muhammad, N. Zawawi, N.A. Mustapha, and R. Karim. 2022. Oilseed meals into foods: An approach for the valorization of oilseed by-products. *Critical Reviews in Food Science and Nutrition* 2022:1–14.

Ni, Y. 2019. *Climate Change Mitigation Potential of Lignocellulosic Succinic Acid: Assessing Feedstock Supply and Integrated Land Use Options in a UK Wheat-Miscanthus Bio-Succinic Acid-Based Bioplastics Production System*. PhD Thesis, Imperial College London, UK.

Obeidat, B.S. 2021. Olive Cake in Livestock Nutrition. *Jordan Journal of Agricultural Sciences* 17:187–197.

Obeng, E. 2018. *Assessing Camelina Sativa as a Fallow Replacement Crop in Wheat Production Systems*. K-State Electronic Theses, Dissertations, and Reports, Kansas State University, USA.

Prasad, J.S., K.S. Varaprasad, Y.R. Rao, E.S. Rao, and M. Sankar. 2005. Comparative efficacy of some oil seed cakes and extracts against root-knot nematode (*Meloidogyne graminicola*) infection in rice. *Nematologia Mediterranea* 33:191–194.

Ramachandran, S., S.K. Singh, C. Larroche, C.R. Soccol, and A. Pandey. 2007. Oil cakes and their biotechnological applications–A review. *Bioresource Technology* 98:2000–2009.

Rhim, J.W., and Y.T. Kim. 2014. Biopolymer-based composite packaging materials with nanoparticles. In *Innovations in Food Packaging*, 413–442. Elsevier, Singapore.

Rizvi, R., I. Mahmood, S.A. Tiyagi, and Z. Khan. 2012. Conjoint effect of oilseed cakes and *Pseudomonas fluorescens* on the growth of chickpea in relation to the management of plant-parasitic nematodes. *Brazilian Archives of Biology and Technology* 55:801–808.

Salemdeeb, R., E.K.H.J.Z. Ermgassen, M.H. Kim, A. Balmford, and A. Al-Tabbaa. 2017. Environmental and health impacts of using food waste as animal feed: A comparative analysis of food waste management options. *Journal of Cleaner Production* 140:871–880.

Sharma, N.K. 2013. Chemical composition of oilseed cakes and deoiled cakes in Nepal. *Online Journal of Animal Feed Resource* 3:74–76.

Singh, R., S. Langyan, S. Sangwan, B. Rohtagi, A. Khandelwal, and M. Shrivastava. 2022. Protein for human consumption from oilseed cakes: A review. *Frontiers in Sustainable Food Systems* 6:856401.

Sousa, D., J.M. Salgado, M. Cambra-López, A.C.P. Dias, and I. Belo. 2022. Degradation of lignocellulosic matrix of oilseed cakes by solid-state fermentation: Fungi screening for enzymes production and antioxidants release. *Journal of the Science of Food and Agriculture* 102:1550–1560.

Zaccardelli, M., D. Villecco, G. Celano, and R. Scotti. 2013. Soil amendment with seed meals: Short term effects on soil respiration and biochemical properties. *Applied Soil Ecology* 72:225–231.

Zaky, A., A. J. Shim, and A.M. Abd El-Aty. 2021. A review on extraction, characterization, and applications of bioactive peptides from pressed black cumin seed cake. *Frontiers in Nutrition* 8:743909.

Index

Page numbers in italic indicate a figure and page numbers in bold indicate a table on the corresponding page

D

E

F

G

H

I

J

K

L

M